Transforming Energy
Solving Climate Change with Technology Policy

Climate change will be an ecological and humanitarian catastrophe unless we move quickly to eliminate greenhouse gas emissions. Policy experts advise us that we need to make major changes to our lifestyles, and our governments need to agree to globally binding treaties and implement market instruments like carbon taxes. This advice is a mistake: it treats technological innovation as being at the periphery of the climate policy challenge. Instead, technological innovation needs to be at its core; we will phase out emissions when and only when the technologies to replace fossil fuels are good enough, and policies need to support these new technologies, quickly and directly. Anyone with an interest in climate change and energy policy will find this book forward-thinking and invaluable. Professional policy makers, climate and energy policy researchers, and students of energy and public policy, economics, political science, environmental studies, and geography will find this book especially stimulating.

Anthony Patt is Professor of Climate Policy at ETH Zurich, the Swiss Federal Institute of Technology. He had been a practicing environmental lawyer in the United States when his concern about climate change led him to Harvard University's Kennedy School of Government, where he earned a PhD for research on the relationships among science, engineering, and climate change governance. Since moving to Europe in 2006, he has turned his attention to the policy challenges associated with scaling up low-carbon technologies. In 2012, he received the prestigious European Research Council award to support his team's research on the environmental, social, and institutional challenges of solar energy development. From 2011 to 2014, he participated in numerous capacities in the preparation of the Intergovernmental Panel on Climate Change (IPCC) Fifth Assessment Report, including as Lead Author on the role of risk and uncertainty in climate policy and as a member of the writing team of the Summary for Policymakers for the IPCC's *Mitigation of Climate Change* report. He lives with his family in a highland farming community near Zurich and spends his free time in the mountains, gardening, and making things with wood.

Transforming Energy

Solving Climate Change with Technology Policy

ANTHONY PATT
ETH Zurich

CAMBRIDGE
UNIVERSITY PRESS

32 Avenue of the Americas, New York, NY 10013-2473, USA

Cambridge University Press is part of the University of Cambridge.

It furthers the University's mission by disseminating knowledge in the pursuit of education, learning, and research at the highest international levels of excellence.

www.cambridge.org
Information on this title: www.cambridge.org/9781107614970

First published 2015

Printed in Great Britain by Clays Ltd, St Ives plc

A catalog record for this publication is available from the British Library.

Library of Congress Cataloging in Publication Data
Patt, Anthony G.
Transforming energy : solving climate change with technology policy / Anthony Patt, ETH Zurich.
pages cm
Includes bibliographical references and index.
ISBN 978-1-107-02406-9 (alk. paper)
1. Renewable energy sources. 2. Energy conservation. 3. Climate change mitigation. 4. Energy policy. I. Title.
TJ808.P38 2015
333.79–dc23 2015014128

ISBN 978-1-107-02406-9 Hardback
ISBN 978-1-107-61497-0 Paperback

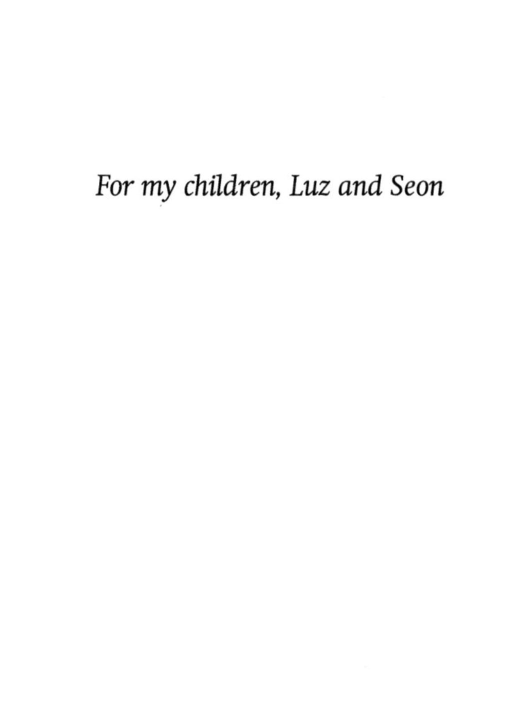

For my children, Luz and Seon

Contents

Foreword

Barely a month goes by without some more bad news about global climate change. The bad news about the impacts of torqueing the climate system is easy enough to understand. Humans are adding increasingly larger amounts of warming gases to the atmosphere, and the climate is now responding. Ice sheets are melting – some irrevocably, it seems – the seas are rising, and weather patterns are changing. Given the size and rate of the human thumb on the climate, the impacts are, for the most part, harmful. The climate is a complex system whose interactions are not understood perfectly. By pushing around a complex system, humans are setting themselves up for unpredictable and possibly horrible outcomes. And since that system is important – nothing less than the planet's life support system – we shouldn't do that lightly.

In recent years, though, a string of bad news has also appeared on the political front. Despite more than two decades of diplomacy and national policy discussions about climate change, there's almost no evidence that emissions – the root cause of the problem – have responded. Global emissions from the energy sector are higher than ever before and not set to reverse any time soon. Sure, a few jurisdictions – notably in Europe – have made cuts, but those have come often at huge cost and concern only a small fraction of the global total. Growth elsewhere, especially in the emerging economies, has been overwhelming.

It is easy to despair. The science around global climate change seems to suggest that the problem is getting worse quickly – with harms that are, on balance, worse than previously thought. And the experts on political systems are finally realizing that solving this problem will be a lot harder than anyone thought.

So what should be done? This book suggests some answers.

The goal of climate policy must be deep cuts in emissions. And to get deep cuts, there must be radical systemic change, foremost in the energy sector. On this front, Anthony Patt sings with the chorus. Politically and practically, there is almost zero chance that humanity will solve the climate crisis by tinkering at the margins with our energy systems or simply asking people to use less energy.

But as this book unfolds, a message appears that is much more challenging to conventional wisdom. The policy instruments that most experts think will boost innovation and cut emissions don't work. Market-based systems - such as cap-and-trade or emissions taxes - are lovely in theory, but, in practice, they don't seem to have much impact on behavior. The root of this failure isn't just that the prices for emissions are too low. Instead, as Patt shows, the problems are structural. Market-based systems are good at sending signals to individual actors over short periods of time, but they aren't very good at coordinating and creating confidence around massive system-wide changes in whole infrastructures. Yet that's what is needed.

Market policies are important. But even more important is a strategy that requires government to take a much more active role. That is the essence of this book. At a time when the policy experts are lambasting policies that involve picking winners, Patt offers us something different: the case for well-designed, careful interventions to support key technologies and ideas that could be the building blocks for a low-emission future. Real change comes from creating niches in which profoundly new technologies and business practices can thrive; and from those niches, the best approaches can spread. That's what will drive a fundamental transition - not efforts to tinker at the margins with small changes in carbon pricing or national emissions targets that represent only a small departure from business as usual.

So far, we haven't seen that kind of profound change in the world's energy systems, and the schemes most experts suggest are unlikely to deliver what is needed. At the same time, innovation in some key technologies has exceeded expectations of even a few years ago, to an extent that may be fundamentally altering the playing field. So, how should government intervene more decisively, building on this latter fact? What is the right role for firms - including state-owned firms - that must manage investment risks? How should policy makers design their interventions so they learn from mistakes and adjust? What is the right blend of carbon pricing and direct regulation? How can we lock in low-emission trajectories so that early innovations build upon themselves into larger markets while locking out dirtier pathways? These are the questions that Anthony Patt forces us to think about, with elegant prose and informative vignettes from his own career. They are the questions we must all grapple with as we search for new models to guide climate policy.

David G. Victor

Professor in the School of International Relations and Pacific Studies and Director of the Laboratory on International Law and Regulation, University of California, San Diego, La Jolla, California

Acknowledgments

Day in, day out, the person who has supported me the most in writing this book is my wife and professional colleague, Dagmar Schröter. She has helped me at so many levels. She has pushed me on the ideas as I have been developing them, pointing out new connections between them and places where my reasoning is incomplete or ignorant of important facts. She has continued to read the manuscript as it has developed, suggesting numerous ways to improve the writing to make it clearer and more succinct. She has encouraged me to prioritize writing it, telling this story, over other commitments. And she has supported my confidence in the value of telling the story I have. It is not a perfect story, but it doesn't need to be.

I am indebted to a whole lot of other people for what I have learned from them, ideas that have coalesced into this book. The first of these is Dan Schrag, one of my PhD advisors and a friend ever since. About ten or twelve years ago, at a time when I was convinced that the *sine qua non* of effective climate policy was a set of market instruments embedded in a strong international framework, Dan challenged me on that. The strength of his challenge caused me to begin to question and, ultimately, turn my thinking around.

Several colleagues at my former institute, the International Institute for Applied Systems Analysis (IIASA), helped me in major ways. Most of all, my supervisor and mentor there, JoAnne Bayer, encouraged me to extend my research agenda outside of climate adaptation and to apply the tools for thinking about social attitudes toward risk to the study of climate mitigation. IIASA was a perfect place to do so, simply because it was a meeting point for many of the leading minds in energy policy. The people who made that happen, and who supported me in starting to think about energy issues, were Arnulf Grübler, Keywan Riahi, Markus Amman, Michael Obersteiner, and Nabojsa Nakicenovic. Finally, at IIASA, Iain Stewart helped me immensely through the process of writing the book proposal and finding a publisher.

Where I work now, at ETH Zurich, the Swiss Federal Institute of Technology, a number of people have helped me in huge ways on this manuscript, as well as on the research agenda that supports it. This includes several members of my immediate working group, notably Anna Geddes, Anna Scolobig, Carmenza Robledo, Johan Lilliestam, Kerstin Damerau, Leonhard Späth, Merce Labordena, Nadya Komendantova, Oscar van Vliet, Stefan Pfenninger, and Susanne Hanger. All of them have, at one point or another, read and commented on pieces of this manuscript, helped track down the answers to particular questions that have arisen, or simply inspired me with the results of their research. I am indebted to Sandro Bösch for creating the figures and to Rasha Ahmed for keeping us all working together.

Beyond IIASA and ETH, many people with whom I have collaborated on projects or in workshops have provided me with ideas that are embedded in this book. Some of their contributions have been slight and unknowing, others huge, including commenting on drafts of this manuscript. Moving left to right across the map, they include David Victor, Morgan Bazilian, Howard Kunreuther, Elke Weber, Steve Rayner, Gus Schellekens, Gwyn Prins, Mike Hulme, Frans Berkhout, Terry Barker, Martin Grosjean, Bernd Siebenhüner, Benjamin Pfluger, Valentina Bossetti, Nico Bauer, Ottmar Edenhofer, and Antonella Battaglini.

Most of the scientific research that led to this book took place within one or more externally funded research projects. These include the ADAM, RESPONSES, and INSPIRE-Grid projects funded by the European Commission; the DESERTECTION project funded by the European Research Council; the ALICE project funded by the German Ministry for Education and Research; and the ANORAK project funded by the European Climate Foundation. I am indebted to the support staff at IIASA and ETH for keeping these projects going, most importantly Jun Watabe, Monica Manchanda, and Olympia Stefani. Perhaps the single biggest source of ideas and insights for this book came from being able to serve as an author and editor to the Intergovernmental Panel on Climate Change, in both Working Groups II and III. I am indebted to IIASA and to the climate research funding agencies of Austria and Switzerland for bearing the costs of my participation.

Finally, underlying everything, I want to thank my parents: Donald, who is no longer with us, and Gail. They inspired me to do good science, but, more importantly, they counted on me to split wood.

[illegible] for [illegible] of any [illegible]

[illegible] my [illegible]

[illegible] with [illegible] They inspired [illegible] do

[illegible] counted on me to [illegible]

PART 1

Setting the stage

This is a book about the strategies that society can take to solve climate change. Some strategies, or policies, are better than others. The insight I hope you will agree with and remember is that those strategies that have dominated discussions of climate change both in the media and in academic circles – carbon taxes and fees, global emissions targets, and steps to reduce our energy use – actually belong to the list of less effective ones. I describe the reasons that they haven't worked and why climate change continues to grow worse. At the same time, there is a new set of strategies that has started to work and can continue to work. All of these focus the power of the people on innovating and implementing a new set of technologies that will give us the energy we want without the carbon dioxide emissions and with no major effects on the economy. The difference is one of sticks and carrots. In Part 2, I explain why the sticks don't work, and, in Part 3, I explain why the carrots do.

Before jumping into theories of strategy and policy, however, I need to briefly set the stage, clarifying the problem for which we need a good discussion of strategy – of carrots and sticks – in the first place. That is what these first three chapters are about. Chapter 1 describes the state of hopelessness that many people feel toward solving climate, leading a good number of them to deny that it is a problem at all, and how this contradicts what others might see as an easy problem to fix. Chapter 2 digs into the science of climate change. What is important here is to distinguish what we know pretty well from what we can only make educated guesses at. The fact that the latter category is so large, and unlikely to become much smaller anytime soon, does in fact define some of the goals we ought to set. Chapter 3 blazes through two approaches toward coping with the symptoms of climate change once they are upon us without really solving the underlying problem. These need to be gotten out of the way so that they don't distract us later on.

1

From optimism to pessimism and back again

It was the middle of an oil crisis and the belief that the world was rapidly running out of everything. My parents, college professors in Boston, started growing an organic vegetable garden and bought a diesel Volkswagen that accelerated to highway speed in about five minutes. They put in a wood stove that entailed their teenage son getting to work splitting and stacking logs, and they installed a solar water heater on the roof of the house. As that teenage son, I was more concerned with swim practice than with world peace and the environment, but I nevertheless learned and remembered from my parents the name of the person guiding these choices. Amory Lovins. He was their guiding light.

Not long ago, I met and talked with him for the first time. Amazingly, he looked exactly the age that I had imagined him thirty-five years ago. Lovins had made his big first splash in 1973, with his book *World Energy Strategies*, when he was only twenty-six years old.[1] By the time he was thirty-five, he and his wife Hunter had founded the Rocky Mountain Institute, in Snowmass, Colorado, as a base from which to do research and demonstrate practice on sensible energy technologies, pathways, and policies. Lovins's two themes at the time were *energy efficiency*, the idea that we could get more value out of using less energy by simply using it more intelligently, and *renewable energy*, which could supply us with what we needed from sunlight, wind, wood, and water. The two themes came together under the name of *soft energy paths*.[2] Lovins's books of the 1970s and 1980s showed how these made more sense than the hard energy paths we were on. And it was in the 1980s that just about everyone in my school, even those of us on the swim team, were terrified of one particular hard energy source because it had the word "nuclear" there, right before the word "energy."

In the early years of the Reagan presidency, before Mikhail Gorbachev ended up as the Soviet leader, the teenagers of America whom I knew were convinced we were going to die during a nuclear attack. The Australian activist and physician Helen Caldicott came and spoke at our school; she suggested that even if war were not to kill us, we still had better not eat the chocolate from the town of Hershey, Pennsylvania, because it contained milk infused with radioactive fall-out from the Three Mile Island nuclear power station.

But all that is history – or at least it used to be. It became history because what Amory Lovins said actually worked, and the United States and other countries reduced their consumption of oil to the point where prices fell. Energy became cheap again, people realized that it had never been all that scarce in the first place, and everyone stopped caring. Renewable energy faded from the picture, and when my family's solar hot water heater broke, my parents didn't fix it. The threat of nuclear war went away as well, a product of the same forces. As the geopolitical expert Daniel Yergin describes, the Soviet Union in its last years had grown completely dependent on revenues from oil exports to support its inefficient and bloated economy.[3] The collapse of oil prices in the mid-1980s was the final blow to a fragile empire. Food shortages became common and severe. Discontent rose. And then, on Christmas Day, 1991, Gorbachev went on national television to announce that he would no longer be president of the USSR because the USSR would cease to exist.

But the fact that the energy crisis became history is now itself history. The relevance of soft energy paths, or the same thing by any other name, has returned, although for different reasons. In 2011, the Tohoku earthquake off the coast of Japan killed more than 15,000 people as a tsunami that many of us watched on live Internet television swept across coastal farmland and obliterated the towns and cities in its path. It also flooded the backup diesel generator of the Fukushima Daiichi nuclear power station. With no power to keep pumping water into the containment vessels of the separate reactors, a series of seven meltdowns occurred. There were no immediate deaths, and may never be any deaths, but the accident spooked the world about the safety of nuclear power, which many analysts had suggested was on the verge of a dramatic upswing. Chancellor Angela Merkel of Germany, in the wake of the crisis, announced that she would accelerate the shutdown of Germany's fleet of aging nuclear power plants, replacing their capacity with an *Energiewende*, or energy transition, of exactly the kind that Lovins had advocated way back when.

And then there is climate change, which is my own motivation for wanting to change the energy system. Unlike a generation ago, Chancellor Merkel could not announce a move from nuclear energy to increased reliance on Germany's own domestic coal reserves, even if to some extent that is what has actually occurred. This is because it had become accepted in German political circles, as in scientific circles worldwide, that carbon dioxide (CO_2) from the burning of coal and gas is in fact one of the biggest threats humanity faces, far worse than Hershey's chocolate or even high energy prices. The centerpiece of Merkel's strategy had to be an effort to promote renewable energy.

And so Amory Lovins, who was never really gone, is back in business. *Time* magazine in 2009 listed him as one of the 100 most influential people on the planet,[4] and by now he has received ten honorary doctorate degrees and national and international prizes beyond counting, including the MacArthur "genius" prize, and has briefed nineteen heads of state.

I met Lovins in 2012, when he came to my institute to promote the ideas in his latest book, *Reinventing Fire*.[5] That book, the result of a group effort among his scientists at the Rocky Mountain Institute, people from government, and, perhaps most importantly, business executives from the private sector, showed once again – indeed more clearly than ever – that the United States could eliminate the burning of coal and oil and nearly eliminate the burning of natural gas while saving money, increasing economic growth, and creating new net employment, and, paradoxically once again, enjoy falling energy prices. The trick, as he explained at my institute, is to combine a million possible improvements in energy efficiency – which save money – with a switch to temporarily more expensive renewable sources of energy. Lovins documented how, over the past decades, sometimes behind the scenes and sometimes up front and center, new technologies have been developed that can do the work of fossil fuels, ultimately for less money. Businesses want to invest in these and rely on them to benefit themselves while helping the planet. Indeed, they are doing so, but often cannot do so quickly enough because of a patchwork of barriers, from legal to institutional to simply having more important things to worry about. To make it happen faster would require no major legal changes from Washington, but rather the continued development of new policies and programs at the state and local level, policies and programs designed to stimulate businesses to think a bit more ahead, work together a bit better, and ultimately make a bit more money.

The tone of his book and of the talk he gave was full of optimism and hope. Why, after all, wouldn't people do the obvious? Near the end of his talk, Lovins said that he was confident that in fifty years we would look back at climate change as one of those problems that had been solved relatively painlessly and easily.

Widespread pessimism

Amory Lovins may be too optimistic on some points. But there is also a great deal of evidence to support his core qualitative proposition, that we could make the transition to a carbon-free energy system at no noticeable net economic cost. So, then, why are so many people so pessimistic about humanity's collective ability to solve climate change?

Pessimism is what you see and hear from just about every expert in the business. In 2009, the world was gearing up for major international negotiations in the Danish capital of Copenhagen on a new climate treaty. The British newspaper *The Guardian* conducted a poll of climate experts to find out from them whether they thought the problem would be solved.[6] By then, "solving the problem" had been taken to mean constraining the global rise in temperature in comparison to preindustrial times, to no more than 2°C. As of 2009, the world had already warmed by about 0.8°C, and another 0.5°C warming was seen as inevitable, even if carbon emissions were to cease entirely. That left 0.7°C of wiggle room. Two hundred scientists and sixty-one other experts, from a total of twenty-six countries, responded to the survey. Despite the wiggle room, 39 percent of the respondents said that achieving the 2°C target was simply impossible, that there was no way governments could get people to stop burning fossil fuels fast enough. Another 47 percent believed that it was technically possible, but didn't think that the 2°C target would be achieved. Only 14 percent thought the 2°C target would be reached.

The majority of respondents said that temperatures would rise by more than 3°C, about 15 percent said by a catastrophic 4°C or more, and a handful said by more than 6°C. And those few souls who thought that the 2°C target was realistic? "This optimism is not primarily due to scientific facts, but to hope," said one. "As a mother of young children I choose to believe this, and work hard towards it," said another. Others who thought the target was possible did not see it happening because of reduced emissions, but rather because of efforts to "geoengineer" the planet to keep the temperature under control, things like shooting reflective particles into the stratosphere or seeding the oceans with iron filings to boost photosynthesis in the water and thus sequester carbon.

That was seven months before negotiations in Copenhagen on a new global climate treaty, which started with a large contingent of the countries walking out and ended even worse. And that was before another five years of climate negotiations that have so far produced roughly nothing.[7] Even the global recession, which chilled energy demand in 2009 and caused emissions to sink, made barely a dent. As *The Guardian* reported in May 2011, by the end of 2010, emissions were up to their highest levels yet, close to being back onto a business as usual track, as if the recession had never happened.[8] The article reported the Chief Economist at the International Energy Agency, Fatih Birol, describing the achievement of the 2°C target as "a nice Utopia." Lord Nicolas Stern, an economist who is one of the most respected climate experts in the United Kingdom, suggested that continuing in the observed path would likely result in temperature rising by more than 4°C by 2100. "Such warming would disrupt the lives and livelihoods of hundred of millions of people across the planet," he said, "leading to widespread mass migration and conflict."[9]

In June 2012, Matthew Wald of the *New York Times* posted to his blog an article entitled "On not reaching carbon goals."[10] There, he reported how emissions were higher than ever and that the International Energy Agency (IEA) suggested that achieving the 2°C target was still technically possible but that the "trends in energy use are running in the wrong direction." Coal consumption was rising, while the United States and other countries were also "feverishly building" new natural gas installations. And while the evidence for climate change was only growing stronger, by 2012, the issue had, if anything, " fallen further down the political agenda."[11] Chris Berg of the Australian Broadcasting Company wrote on his blog that "we can't stop climate change."[12] Accepting climate change and adapting to it, he suggested, are now the tasks for humanity.

Denying the problem

In the face of something that can't be solved, one useful strategy is to deny that a problem exists in the first place. Indeed, that is what has happened: climate change has not only moved out of the political spotlight, but the number of people who say that they believe in it has declined.

The seeds of the latest upswing of disbelief in climate change began to take root even as climate change was at its most visible politically. The Intergovernmental Panel on Climate Change (known as the IPCC), a collection of roughly 2,000 scientists appointed by governments and working voluntarily to synthesize the latest scientific knowledge, won the Nobel Peace Prize for work completed in 2007. The IPCC shared the prize with Al Gore, following the success of the film documenting his speaking tour, *An Inconvenient Truth*. But, at the same time, British documentary filmmaker Martin Durkin had put together a film of his own, *The Great Global Warming Swindle*. The documentary cast doubt on the basic science of climate change. It also suggested that the motivation for climate scientists spreading the nasty rumors was financial and that the policies aimed to stop it would curse millions of people – the whole continent of Africa and more – to poverty.

It premiered on the United Kingdom's Channel 4 on March 8, 2007, and then later, in other countries. I first heard about it while vacationing on a farm in the German Alps. The farmers had seen it on German television, and, knowing that I worked on climate change, they wanted to share with me what they had learned. "It's all got something to do with sunspots," they told me. A few months later, I joined a group of climate scientists to watch it and discuss the arguments. The climatologists in our viewing audience pointed out that the scientific arguments in the film were based on work that had since been convincingly demonstrated to be inaccurate. The social scientists in our group, one by one, could demonstrate how the implications for society, committing us to poverty, were equally flawed. All of us introspected and asked ourselves whether what was motivating us to do our science and reach our conclusions was financial reward. We agreed this wasn't the case. But never mind. The film captured a large market share and won praise by climate skeptics. It touched a nerve that needed tickling.

That was before "Climategate." In November 2009, somebody hacked the email server at the Climate Research Unit at the University of East Anglia, home to a number of leading climatologists, copying thousands of emails to various Internet sites. Of these thousands, several had the appearance of covering up inconvenient holes in climate science. Or at least they seemed to, when taken out of context. Scientists describing "tricks" they had used to obtain a particular temperature record, where "tricks" in reality referred to valid and established statistical methods designed to enhance the accuracy of data. Once the emails were put back into context, they simply revealed a group of scientists communicating with each other about their legitimate findings. If you had the attention span to follow the emails this far, then you saw that the apparent cracks in the climate change armor were illusory.

Two months later, however, in January 2010, cracks looked like they might be appearing in the established science. Suddenly, the possibility seemed more real that climate change might indeed be a house of cards. The IPCC report, the one that had won the Nobel Peace Prize two years earlier, had reviewed the peer-reviewed scientific literature, summarizing the findings from across a vast body of literature and highlighting the most important. One of those findings was that a large number of glaciers in the Himalayas would vanish within thirty years. This sounded credible, except to many of the scientists who actually knew the glaciers. In fact, the scientific evidence for such a pace of glacier retreat was simply not there. The scientific paper on which the IPCC had relied for the claim was peer reviewed, but it was merely restating the result from another paper, which was not. When people discovered the error, the IPCC immediately issued a statement that the claim about the glacier retreat was an error and documented how it had occurred. But the damage was done in terms of credibility. If such an error had made it through the IPCC review process, then that process must have been flawed and then, certainly, other errors could be there as well. To those who wanted not to believe in climate change, then here was good ammunition to doubt the science.

The number of people who saw climate change as a hoax, or at least did not believe it to be a serious problem, started rising. In the United States, the Pew Research Center had been pooling Americans about their top environmental concerns. In 2007, when climate change was first included on the list, 38 percent listed it as a top concern. By 2010, that number had fallen to 28 percent.[13] A Stanford University poll in July 2012 showed the number to be down to 18 percent.[14] The Stanford group also asked people if they believed climate change were actually occurring. Eighty-four percent said yes in 2007; that declined to 80 percent in 2008, 74 percent in 2010, and then to 60 percent in 2012.[15] Taken together, the polling data showed that concern about climate change gradually rose from the 1990s until 2007, peaked, and then, by 2010, was back down below the 1990s values.[16]

The trend was not limited to the United States. In a series of polls taken in Hamburg, Germany, between 2008 and 2011, the proportion of people who felt that climate change was either not a problem or not a serious problem rose consistently from 37 percent to 54 percent, whereas the number who felt that it was a serious or very serious problem fell from 63 percent to 44 percent.[17] In the United Kingdom, polls at the time of Climategate showed a drop in the number of people believing climate change to be a reality from 83 percent to 75 percent, between November 2009 and February 2010. In New Zealand, the number of people who felt that climate change was not a problem more than doubled from 8 percent in 2007 to 17 percent in 2010; a similar trend could be seen in Australia.[18]

During the 2008 U.S. presidential campaign, candidate Barack Obama stated that he supported cutting greenhouse gas (GHG) emissions by 80 percent by 2050, while his Republican adversary, John McCain, agreed with him on the reality of the problem and suggested somewhat less ambitious cuts of 60 percent. Fast forward to the 2012 race. All of the Republican primary contenders either denied climate change as a problem or opposed government measures to deal with it.[19] Mitt Romney, the ultimate Republican challenger, changed his personal views in 2011, as the primary battle was heating up. In June 2011, he stated in a speech that he believed that climate change was real, that people were causing it, and that it was important to do something about it. He started backpedaling in August, and then, by October stated, "we don't know what's causing climate change on this planet."[20]

The state of North Carolina made news headlines in the spring of 2012 for a proposal that would "wish away" climate change. The official policy of the state is to recognize and respond to climate change, and, indeed, the website of the Department of Environment and Natural Resources describes the state's climate change initiative, including both measures to reduce GHG emissions and to adapt to sea level rise and other impacts.[21] But in late May 2012, Republicans in the state legislature proposed a new law that would require ignoring projections of climate change for coastal planning. Coastal planning is of high importance in North Carolina because the state includes the Outer Banks, a string of low-lying barrier islands that are popular tourism destinations and that are frequently battered by hurricanes making their way up the coast. Specifically, the law would require basing any projections of sea level and associated coastal flooding only on historical data from the first half of the twentieth century and in no cases on model projections of future sea levels.[22]

There are countless websites that provide the opinions and sounding boards for those who deny the climate change problem. A theme running through those websites and the discourse of denial more generally is to link the disbelief in climate change with a profound dislike of the policies that would be needed to stop it. The denialist website www.climategate.com states this linking of problem and feared solution especially well: "The goal of Climategate.com is to provide a daily dose of information regarding the world's greatest scam ... and other information and news to help you in your battle against the Religion of Settled Science to dispute their views on Anthropogenic Global Warming, and in addition, to battle the one-world socialist agenda, which is the movement's leaders' real goal."[23]

For some time, I have suspected that what really drives the denialist movement has very little to do with the thermometers on people's kitchen windows, but rather with the belief that solving climate change requires this "one-world socialist agenda." And then, one day I asked myself: is this sort of political agenda something that goes way beyond climate change and really has to do with restructuring society in some pretty radical ways, "the movement's leaders' real goal?" My first thought was that it could not be. And then I thought some more. And then, eventually, I came to the conclusion that actually it was – or at least it could certainly look that way.

How the experts tell us we will solve climate change

Since the early days of climate change existing as a policy problem, experts have been busy telling people how to solve it. All of them have said that the real need is to stop burning fossil fuels. That is obvious. The tricky part is the set of strategies that governments need to adopt to make that happen, and, for that, the experts dictated three changes: we need to change how the economy runs; we need to have a global agreement for CO_2, perhaps the strongest global agreement that has ever existed on anything; and, we need to change how people live their daily lives.

Let's start with the economy, which raises the issue of socialism. Almost every newspaper article or editorial touching on climate change policy has suggested that to stop burning fossil carbon, society needs to make it more expensive, taxing it at a rate equivalent to the harm it does to society – its social cost. Paul Krugman, the Nobel Prize-winning economist who is a columnist for the *New York Times*, put it quite clearly: use "market-based" environmental laws that make it more expensive to burn carbon and emit CO_2. These laws can either take the form of a direct tax on CO_2 emissions or can put a price on the emissions indirectly, by requiring government-issued permits to emit the CO_2, permits that people will buy and sell at some price.[24] This advice is not just Krugman's. Thomas Friedman, another well-read columnist, writes: "The only way to [deal with the problem] is if the developed countries, who can afford to do so, force their people to pay the full climate, economic and geopolitical costs of using gasoline and dirty coal. Those countries that have signed the Kyoto Protocol are starting to do that. But America is not."[25] And the *New York Times* editorial page itself? They have been clear and consistent that policy makers must, eventually, "put a price on greenhouse gas emissions in order to unlock private investment."[26] I could go through a list of other media outlets, but I won't, since the message would be essentially the same no matter what the mainstream source: if we want people to invest in alternatives to fossil fuel technologies, we must make the fossil fuel technologies themselves more expensive to operate. This isn't directly socialist, and certainly nobody is calling for the state taking over private industry. But it does mean regulating private markets not just to make them function more smoothly, but also so that they operate in a manner that benefits a wider social group.

If I were someone who saw socialism as a threat, then this kind of government action would appear to me like a dangerous step in socialism's direction.

Let's look at one-world government. Nobody is claiming that we need to get rid of countries and replace them with an all-powerful United Nations to handle all of our affairs, but most climate experts do suggest that there needs to be a global treaty limiting GHG emissions. The opinion and editorial pages of those same leading media outlets – such as the *New York Times*,[27] the *Economist*,[28] or the *Guardian*[29] – tell us that that the world needs an international treaty, likely under the UN umbrella but perhaps in other forums as well, to combat the problem. At first blush, that sounds fine, since, after all, there are a lot of international treaties out there, and they do not give us one global government. But an effective climate treaty would indeed be the mother of all treaties. It would have strict limits on how countries develop their own domestic energy systems, which happen to be the engines of their economic growth and development. Perhaps the only existing international regime with more teeth would be the World Trade Organization (WTO), which requires member countries to open their markets to imported goods of all kinds and punishes tariffs and subsidies that have the effect, directly or indirectly, of favoring domestic products over imported ones. But even the WTO is, most analysts claim, clearly in the economic interests (almost all of the time) of all the countries that are members. Not so an effective climate regime. At the required level of ambition, it would prevent countries, developed and developing, rich and poor, from scaling up their energy systems in the least-cost manner. All countries' energy systems would have to be designed to benefit not just the citizens of that country, but of the planet as a whole. That would represent the most significant forfeiture of national sovereignty yet in the history of the nation state.

If I were someone who saw an effective world government as a threat, then this scale of international treaty would appear to me like a giant step in the one-world direction.

Finally, let's look at how people live their lives. Everybody eats, breathes, and sleeps, but what separates the rich people of the planet from the poor is the amount of money spent buying things, the amount of time spent absorbing information from television and computer screens, and, perhaps most importantly, the amount of distance covered every day driving around in cars. At the heart of all this activity, especially the driving part, is modern energy, fossil fuels in particular. There is no shortage of experts suggesting that to stop burning fossil fuels, these habits have to change, especially the driving part. The editorial writers of the *New York Times*, for instance, have hailed efforts to reduce urban sprawl, "cutting back on the time people have to spend in their automobiles," as some of the most serious work on climate change.[30] They have similarly praised efforts to limit driving directly, such as by making it expensive to drive into downtown areas rather than by taking the subway.[31] Indeed, at the local level, the government policies that one notices most are those addressing our driving: bike lanes promoted as climate friendly, public transportation advertised as good for the environment, car pooling promoted as an act of climate stewardship. We all have learned that getting into a car alone to drive to work or to go shopping is bad for the climate and has to change. But where do you draw the line? When you get right down to it, just about everything a person does in modern society, from turning on the lights in the morning to going to work to relaxing the evening, uses energy. Do I need to be thinking about my CO_2 emissions, feeling guilty about my CO_2 emissions, every time I do anything that involves the use of energy?

If I were someone struggling to get along in life, which is to say just about anybody, then I would say that this climate change business has the potential to turn my daily existence into something really unpleasant. I may have all sorts of reasons to want to consume less, and less energy in particular. But doing so is often difficult in practice, and the choice to go in that direction needs to be a personal one, made because I see how greater simplicity adds value to my life. Being told that I have a moral obligation to reduce my energy consumption makes me want to scale back less, not more.

Resolving the paradox

There is a huge disconnect between the optimism that currently surrounds green growth and technology, the kind that Amory Lovins lives and breathes, and the pessimism on climate change that every opinion poll, every expert interview, suggests is growing. The key, which I explore in the pages that follow, lies first in the failure of the conventional wisdom on climate policy: the three pieces of wisdom that can sound an awful lot like socialism, one-world government, and the death of kids' getting their drivers license at the age of sixteen. On top of this, we have interpreted the failure to make fundamental changes to the economy, to our global governance institutions, and to our lives to mean that we have failed on climate. The fact that emissions have continued to rise simply confirms an obvious truth.

Pessimism as to whether any of the three established strategies will suddenly start working is well-founded and appropriate. There are important political, cultural, economic, and even scientific conditions that have so far prevented the three main policy prescriptions from taking root. These conditions have existed for good reasons, and they are not going away. If we expect for a turnaround in these three areas any time soon, we will be disappointed.

But recently, a "new" conventional wisdom has arisen. That wisdom says that we can find our way out of climate change through innovation, particularly with respect to energy technologies and energy systems. Andrew Revkin of the *New York Times* has endorsed this view. Popular books touch on it as well, such as Ramez Naam's *The Infinite Resource*[32] or Maggie Koerth-Baker's *Before the Lights Go Out*.[33] Even within the IPCC, on which I have spent a huge amount of time over the last three years, the view has started to emerge. I agree fully with this view, as I will argue in this book. But the curious thing is that even as some people have started to reframe the problem of climate change around the task of energy innovation, most of the time, they keep coming back to the same policy prescriptions. If you want to bring about innovation in the energy sector, they argue, the best way is to put a tax on carbon. Or promote more international cooperation. Or get people to change their lives so as to use less energy.

Here is where this book is different. I suggest here that there is no evidence that these three things work to enhance the kind of innovation and technological diffusion that will solve climate change. Meanwhile, there is growing evidence that other things do provide this boost.

The most important factor providing this boost is targeted public funding around a set of key new technologies. In some cases, this funding is for research and development (R&D), whereas in other cases it is to establish a critical mass in the market. The second most important factor is to identify and address the noneconomic factors that stand in the way of the widespread adoption of the new technology. These are often network barriers, like the fact that we have a wide network of gas stations but relatively few public charging or hydrogen refueling stations. Now, if the technologies that we need were several decades away from being market ready – things like cold-fusion – then this focus on supporting a select few of them would count as picking future winners, arguably a highly risky strategy. But the fact is that almost all the technologies we need already are winners. They are as poised, right now, to take over the energy system as the Internet was, back around 1990, to take over telecommunication. We just need public policy to reinforce and accelerate their harvesting of their advantages.

I will go into much more detail about the particular technologies and the policies to promote them in Part 3. Before getting there, I explain where the three policy prescriptions that have dominated so far – the wrong ones – came from, why they made sense at the time they were developed, why many people continue to advocate for them, and yet why, in fact, they are now quite wrong for climate change. We would do best to unlearn them, forget them, ignore them, as quickly as possible. Emissions can fall, and fall dramatically, even without the three policy prescriptions in place. Indeed, trying to hold on to these three policy prescriptions at the expense of other strategies can stand in the way of solving climate change.

Unlearning the three lessons means reconsidering how we view the core challenge that climate change presents us. It is not to reduce pollution by limiting GHG emissions, but rather to provide enough clean energy – the kind that does not create GHG emissions – for the world to function economically much the way it does today. When we consider the problem in this way, then we can see two things. First, we can see that the past twenty years of policy making around the world – in the United States, in Europe, even in developing countries like China, Brazil, and South Africa – have not been the disaster for the climate that we might believe but have actually put society in an excellent position from which to move forward today. Not the best possible position, but an excellent position nonetheless. Second, we can see that there actually is a set of fairly simple and politically feasible policies that can keep us moving forward at a pace that can stop climate change quite quickly. Not as quickly as many of us would like, but quite quickly nonetheless. Taking the conventional wisdom out of our view leaves room for optimism.

2

The natural and social science of climate change

The beginnings of humanity's modern understanding of the climate took shape in Switzerland in the 1830s. Three men played critical roles: a wild goat hunter who started to wonder about why the boulders near his home were shaped the way they were, a civil engineer who started to take note of sediment patterns in the river valley he was meant to manage, and a scientist who talked a lot with both of them and began to connect their dots with some of his own. Louis Agassiz was the scientist, and his idea was that the Swiss glaciers were part of something that had once been much bigger, a vast sheet of ice that had carved valleys and left behind lakes and ravines. He published his idea in 1840, that there had been an age of ice long before our current memory, and people started seeing other pieces of evidence all over the place. Evidence not just in Switzerland, but also in North America, where Agassiz later moved to become a professor at Harvard University. The moraines and kettle ponds he found in Massachusetts were linked to glaciers that had covered Europe and suggested that changes in climate had taken place at a global scale.

Until Agassiz, people had not seriously questioned whether the climate had been much different in the past than it is today. What Agassiz and then others showed was that it had. The last Ice Age ended about 12,000 years ago, and this has been only the latest part of a recurring cycle going back 2.7 million years and covering a period of time known as the Pleistocene Epoch, which has been marked most of the time by Ice Age conditions. Roughly every 100,000 years, there has been a respite from the ice of about 15,000–20,000 years, periods known as *interglacials*. We are living in an interglacial, one that ought to be ending in a few thousand years, to be replaced by another Ice Age for the next 80,000 years or so. Prior to the Pleistocene, the variations were less regular, but wider in magnitude, ranging from a world about

10°C warmer than our own and completely free of ice to a world that was completely frozen over.

Underlying the accumulating evidence of a nonstatic climate, and more recently a profoundly cyclical climate, is the question of what exactly has been causing the changes. The fact is, it is not completely understood, but two main sets of factors do appear to play the most important roles. The first is a set of changes to Earth's elliptical orbit around the Sun and the relationship between the shape of the ellipse and the tilt of its axis of rotation. This influences how much radiation reaches the Earth's surface and how much that changes over the course of a year. Indeed, it is these changes that appear to have provided the trigger for the cyclical advance and retreat of glaciers in the past. The second is the chemistry and physics of the atmosphere and of the Earth's surface. These influence how much of the radiation from the Sun gets retained. It appears to have dramatically magnified the effects of the cyclical changes in Earth's orbit, causing very minor changes in incoming radiation to have a very major effect on the climate. That chemistry is what is now undergoing long-term changes faster than at any time in the Earth's history and is of particular concern.

Greenhouse gases and radiative forcing

Understanding the atmospheric chemistry side of the climate story dates to 1859 and the work of a physicist named John Tyndall, who conducted experiments in his London basement. His work was about the process of *radiative forcing*. It is very well understood because people can study it in the laboratory in the context of controlled experiments.

The first aspect of radiative forcing is the fact that all things radiate heat, all the time. That radiation can act like waves. The radiation varies in intensity and in wavelength depending on the heat of the object that is radiating. Hotter objects radiate more strongly than colder objects, and hotter objects have shorter wavelengths on average. The second aspect is that all objects either absorb radiation (like a dark roof), reflect it (like a mirror), or let it pass through (like a glass window). Objects that reflect radiation or let it pass through keep their temperature unchanged, but if they absorb it, then they take on its energy and become hotter themselves. Whether an object absorbs, reflects, or lets radiation pass through depends both on the radiation's wavelength and on the molecular structure of the object itself.

Climate forcing results when we put these two processes together with what we know about the Sun, the atmosphere, and the Earth's surface. The Sun is hot, and so it emits a lot of short wavelength radiation. To reach the Earth's surface, it first has to pass through the atmosphere. The atmosphere is mainly nitrogen and oxygen, but it also has a lot of other gases in much smaller concentrations, such as water, carbon dioxide (CO_2), and ozone. Together, these absorb some wavelengths, reflect others, and let just a few pass through. Ozone, for example, absorbs radiation in the ultraviolet (UV) wavelength spectrum, which is why having a stratospheric ozone layer is a good thing. Most of the radiation that does pass through the atmosphere falls within a narrow wavelength spectrum, and this is what our eyes have evolved to see, perceiving slight differences in wavelength within this narrow band as differences in color.

The Earth's surface itself reflects or absorbs all of the visible light. What it reflects shoots upward with the same wavelength, and, since this wavelength made it in through the atmosphere, it can make it out again. Snow, ice, and light-colored sand reflect a lot – they have a high *albedo* – whereas water, dark soil, and plants reflect considerably less. What the surface of the Earth doesn't reflect, it absorbs. This makes it hotter.

Because the Earth is a lot cooler than the Sun, the radiation that it constantly emits falls within a spectrum of much longer wavelengths. But before it escapes into space, it encounters the atmosphere and, again, some is absorbed, some is reflected, and some passes all the way through. Neither nitrogen nor oxygen absorbs much longwave radiation, but other molecules present in smaller concentrations – making up less than one-tenth of 1 percent of volume in the atmosphere—do. That makes them hotter and then they radiate more strongly themselves. That radiation goes in all directions, both up toward space and down toward the Earth's surface again. That which makes it back down to the surface is again reflected or absorbed, and the part that is absorbed warms the Earth even more. The result is that Earth is a warmer place than it would be without its atmosphere.

Exactly how much warmer depends on the precise concentrations of the chemicals in the atmosphere because they determine what proportion of the radiation in which precise wavelengths gets blocked. In fact, most of the longwave radiation does get blocked. What avoids getting blocked, what escapes out to space directly, falls into a really narrow bandwidth. It is as if there is a hole in the atmosphere at a particular point along the longwave spectrum. There are chemicals that absorb radiation in this narrow band, and they would plug this hole if they were present in the atmosphere in substantial concentrations. But they are not, and that is why the Earth maintains the heat balance that it does.

I think a Gore-Tex ski parka makes the best metaphor for this. It protects you from wind, snow, and freezing rain, just as the atmosphere protects the Earth from harmful UV radiation. But when you move around skiing, you also perspire, giving off water vapor; to stay comfortable and dry, you want the parka to breath. Modern fabrics like Gore-Tex can do this. They are impervious to air, liquid droplets of water, and crystals of ice, but they have tiny holes that are just the right size to let water molecules in vapor form pass through. If you were to clog these holes, such as by smearing them with grease, then the parka would still protect you from the elements, but you would get all hot and clammy inside as you perspired, and your sweat could no longer escape. Earth has a ski parka with just the right sized spectral hole to keep things comfortable for the plants and animals that have adapted to today's climate. Plugging the holes will make things uncomfortable. The more common metaphor, predating Gore-Tex, is that of a greenhouse. The gases that plug the spectral hole are known as *greenhouse gases* (GHGs).

Water is the GHG with the greatest total impact. The concentration of water in the atmosphere varies with the weather, from close to 0 percent to up to about 4 percent. Water enters the atmosphere when it evaporates from oceans, lakes, and rivers and when it transpires from green vegetation. In general, warmer air holds more water, and this can create a feedback: if you raise the temperature, the air holds more water, which plugs the spectral hole and forces the temperature even higher. This is one of the reasons for the large difference in temperature between the tropics and the polar regions, larger than the simple difference between their exposure to sunlight would account for. When the temperature of a given air mass drops, such as when the air mass rises, it is often no longer able to hold all its water in gaseous form, and so clouds form – collections of tiny liquid water droplets. The water in clouds plugs the spectral hole just like water vapor, but the clouds themselves also reflect incoming solar radiation back out to space. Thus, clouds have a net cooling effect on the ground underneath them while also slowing the rate of evaporation at the Earth's surface. The cycling of water into the air from the

Earth's surface, forming clouds if the air gets too cold and then precipitating out of clouds in the form of rain and snow if the clouds get too dense, is among the most complicated and most important determinants of climate.

Carbon dioxide (CO_2) is another chemical that plugs the spectral hole, and, like water, it occurs naturally. Its current concentration in the atmosphere is about 0.04 percent or 400 parts per million (ppm). It enters the atmosphere from volcanic eruptions and when carbon compounds react with oxygen. The major place where carbon reactions occur is in living things, namely the respiration of plants and animals and the decomposition by microbes of dead organic matter in the soil. Carbon dioxide leaves the atmosphere through dissolving in gaseous form into oceans and lakes, through photosynthesis in plants and algae, and through geologic weathering. Geological weathering is a process by which atmospheric CO_2 slowly reacts with rainwater and with minerals founds in exposed rocks to form bicarbonate ions, which then dissolve in water. This whole set of processes – the flux of carbon among the biosphere, the atmosphere, the soil, and the oceans – is known as the *carbon cycle*. Scientists have good estimates of the flux rates, but there is a lot of uncertainty about how some of these rates – such as from the soil to the atmosphere because of plant matter decomposition – vary according to changes in temperature.

There are two other naturally found greenhouse gases: methane and nitrous oxide. Like CO_2, they are emitted and absorbed by natural processes. Each of them is present in far lower concentrations than CO_2, about 2 ppm for methane and 0.3 ppm for nitrous oxide. On the other hand, they are also much more effective than CO_2 at plugging the spectral hole: a molecule of methane is twenty-five times more powerful; a molecule of nitrous oxide is three-hundred times more powerful.

Finally, a whole class of purely manmade chemicals, such as chlorofluorocarbons (CFCs) and other fluorinated compounds used as refrigerants and insulating materials by industry, also act as important GHGs. These so-called *F gases* are present in truly tiny concentrations – less than one part per billion – but are also very potent GHGs, most of them several thousand times as powerful as a molecule of CO_2.

Human activity has been putting more of the nonwater GHGs into the atmosphere. Mainly, this is CO_2 from the burning of fossil fuels. It also includes CO_2 from other industrial processes, such as cement production, as well as from land-use changes, such as

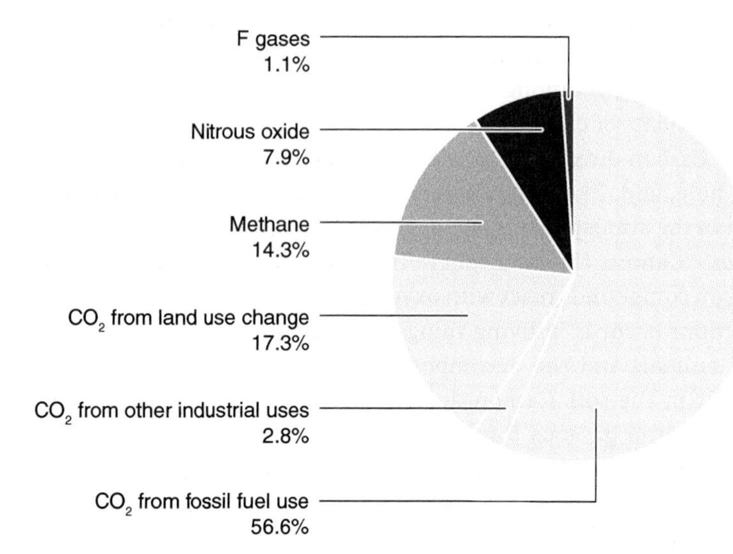

Figure 2.1. *Different greenhouse gases' (GHGs) relative contributions to radiative forcing.*

deforestation, which both releases stored carbon in trees and soils and reduces the rate at which photosynthesis removes CO_2 from the atmosphere. Methane and nitrous oxide have risen from a combination of fossil fuel extraction and burning and from agricultural practices such as the raising of cattle. We also continue to release the F gases, although this has slowed remarkably because of the Montreal Protocol, a treaty signed in 1986 to ban those F gases that damage the stratospheric ozone layer.

To figure out the relative importance of each of these GHGs, you need to calculate the amount of each that people are putting in the air, how long they are likely to stay in the air, and their power at filling the spectral hole. Figure 2.1 shows the commonly accepted result of these calculations, based on current emissions rates, in terms of their relative contributions to long-term radiative forcing.[1] Carbon dioxide is the major culprit, and most of this is from fossil fuel use, although a substantial amount of net CO_2 emissions does result from land-use change. But the non-CO_2 gases are also important. Climate scientists often talk of rising concentrations of "CO_2 equivalent." As I will discuss later, many scientist suggest that an appropriate upper limit for CO_2 equivalent concentration would be 450 ppm, up from CO_2's preindustrial value of 280 ppm. What they really mean is that CO_2 should be at this value if the other GHG concentrations were to be at their historical values. If concentrations of these other gases also rise, that would push up the CO_2 equivalent concentration; then the concentration of CO_2 itself would have to be somewhat less than 450 ppm.

Back to John Tyndall. His work in 1859 was the first demonstration of these processes and suggested even to him that modifying the concentrations of CO_2 in the atmosphere could affect temperatures. Thirty years later, a Swedish chemist named Svante Arrhenius took a first stab at figuring out how big the effect would be, working at a time when the Industrial Revolution was in high gear and coal burning was beginning to change CO_2 concentrations. Realizing that a change in CO_2 would in turn trigger a change in water vapor, he estimated that doubling the concentration of CO_2 in the atmosphere would increase the surface temperature of the Earth by an average of 4–5°C. That is about the difference in average temperature between Berlin and Madrid, or between Boston and San Francisco.

What Arrhenius calculated – the rise in temperature associated with a doubling of GHG concentrations – has come to be known as *climate sensitivity*. The best guess as to its number has changed very little. At the same time, the range of plausible values has increased: it could be as low as 1°C or as high as 10°C, as scientists have discovered complexities and feedbacks in the water and carbon cycles that they never imagined could exist.[2] For example, in a warmer world, there is less snow cover and more bare ground, and bare ground absorbs more sunlight than snow. So this would amplify the temperature change, especially in those places with thin snow cover, where a small initial temperature change could make a large difference in exposed ground. Or it may be that some of the forests will turn to desert sand as things get hotter, and sand reflects more sunlight than do forests. So this would tend to dampen the temperature change, again mainly locally. Moving from calculations of radiative forcing to calculations of climate sensitivity is where the relatively precise laboratory science ends and the much less precise and vastly more difficult planetary science begins. It can be difficult to appreciate how complicated, and at times random, the process of pushing this science forward can be. At the same time, appreciating it is important because only then do you understand why this science moves ahead so slowly and why this doesn't mean that climate change is a hoax. Most of what you read about science tells the success stories, and this can make you think that well-planned success is as easy as putting together a jigsaw puzzle where all the pieces were in the same box. But it's not. It is worth understanding this.

Climate science in the field

My own first exposure to real-world climate science came through a friendship with Daniel Schrag, a professor at Harvard who runs a lab for geochemical oceanography. The business of that lab was to study chemical clues to past and present climate dynamics.

One of the most important sets of clues is to be found in atomic isotopes. The various elements, like hydrogen and carbon and iron, are all defined by the number of protons in their atomic nuclei. Usually, there is also an equal number of neutrons there as well, and the sum of the two determines the mass of the atom. Sometimes, however, this isn't the case, and atoms have a different number of neutrons. The different versions of an element, with different numbers of neutrons and hence different atomic masses, are known as *isotopes*. Carbon atoms, for example, usually have six neutrons to match their six protons, but sometimes have seven or even eight neutrons, raising the atomic mass from twelve to thirteen or fourteen. Some isotopes, like carbon fourteen, are unstable, meaning that some fraction of the carbon fourteen atoms today will shed a neutron to become a carbon thirteen atom tomorrow. The radioactive isotopes, like what you find in a neutron bomb, are all unstable. Most of Schrag's work was with stable isotopes. These are just normal atoms, with nothing like radioactivity to worry about, just a little bit heavier or lighter.

The various stable isotopes of a particular element, like carbon twelve and carbon thirteen, by and large behave the same as each other, but there are fine shades of difference. Sometimes, chemical reactions take place in which the heavier, more sluggish isotopes are less likely to participate. In a leaf, for example, a series of reactions fueled by sunlight combines CO_2 and water to form the sugars and starches that become the wood of the tree. The leaf gets the CO_2 it needs from the air around it, and yet, in grabbing CO_2 molecules, it is slightly more likely to take in CO_2 containing the lighter isotope of carbon. That means that the carbon in the plant's tissue has a higher proportion of carbon twelve to carbon thirteen than is found in the air. How big this difference is depends on a lot of environmental factors, such as the temperature. You could infer something about those environmental factors – the growing conditions throughout the tree's lifetime – by measuring the amount of carbon thirteen in the wood. The machine for measuring this is a *mass spectrometer*. Schrag had two mass spectrometers in his laboratory, each of which was about the size of a large refrigerator and as expensive as a small airplane.

In particular, Schrag used his stable isotope methods to study El Niño, which is a climatic event that takes place every few years in the tropical Pacific Ocean. Most of the time, strong east-to-west trade winds blow over the tropical Pacific. These create a big puddle of warm water at the western end, around Indonesia, and lead to cold water from the ocean's depths rising to the surface at the eastern end, off the coast of Peru. Every few years, however, the trade winds stop blowing for a few months, with corresponding changes in water temperature. Around Indonesia it is colder, and around Peru it is warmer. The name *El Niño* (Little Boy) comes from Peruvian fishermen, because when it happens, it usually starts in late December, and the change in cold water upwelling from the deep ocean temporarily boosts their fish catches, which they originally took as a Christmas present from the baby Jesus. But El Niño also has a noticeable effect on rainfall all around the world. It shuts off the rains in some places, most notably Indonesia and Australia, but also a large swath of Africa. By contrast, the west coast of the Americas, from Chile to California, typically gets drenched.

Around the 1970s, American scientists connected the dots and realized that the fish catches and the rain and snow were somehow related. They convinced a government agency, the National Oceanic and Atmospheric Administration (NOAA), to place buoys all throughout the tropical Pacific and start to collect a temperature record that they could use to figure out what was going on. By the mid-1990s, they had done a good job and were at the point of being able to predict El Niño up to a year in advance – and, with that, something about the weather in many other parts of the world. But the temperature record didn't go far back enough to be able to answer one important question related to climate change: would El Niño grow more or less frequent if the Earth got hotter? That is what Schrag started figuring out by measuring stable isotopes, primarily in very old coral.

When some species of coral grow, they set down visible annual rings, just like the rings in a tree. Some of the isotopes in those rings depend on the temperature and salinity of the water in which the coral was growing at the time. Schrag found large corals growing around the Pacific Ocean that were hundreds of years old. Wearing scuba equipment and wielding a pneumatic drill, he bored into them and extracted plugs. In this way, he was able to construct a longer ocean temperature record than the NOAA buoys could provide. Analyzing that temperature record, he found that, indeed, El Niño had become more frequent over the last hundred years and that the change in frequency had matched the change in temperature on land.[3]

But then Schrag took his question further. If little changes in temperature on land make a little difference in El Niño, would a big change in temperature – several degrees – make a big difference? His hypothesis was that it would, but figuring out if this were indeed correct would require more data. His idea was to find corals that had lived thousands of years ago, at a time when the climate had been several degrees warmer than it is now, and do the same kind of analysis. But where do you find corals that grew at the right time?

Shrag had a friend, Mark Erdmann, who lived in Indonesia. Erdmann was a tropical reef ecologist who had eventually settled into living on a volcanic island that was a short boat ride from Manado, the capital city of North Sulawesi, one of Indonesia's many provinces. His little island was right in the middle of one of the most pristine tropical reefs in the world, where he did most of his work. In 1997, Schrag had been visiting Erdmann, and he had found an old dead coral poking up out of a mangrove swamp not far from Erdmann's house. Schrag had some drilling equipment with him, and he took a sample with him back to his lab at Harvard to analyze. It turned out to be just the right age – from a time when geologists knew the Earth was several degrees warmer than it is today – and yielded data on seawater salinity and temperature, a record that was several decades long.

In early 1998, Erdmann went looking for other good reefs to study. He drove five hours across North Sulawesi to the port town of Gorantalo and then took a ferry to an archipelago a hundred kilometers to the south, the Togian Islands, which lie directly on the equator and at the western end of El Niño. While there, Mark noticed a lot of big round rocks lying around, all at about the same height, a few feet above sea level. They looked to him like they might have once been corals of the species that he knew Schrag liked to drill and sample. This species normally grows just below the water surface. But it made sense that they would be on land if the temperature when they were growing had been warmer because the glaciers and ice sheets would have been smaller then and sea levels higher. And because they were a lot bigger than the old coral in the mangrove swamp near his house, they could potentially provide a data record covering many more years. So he contacted Schrag and volunteered to lead a little expedition back to the Togian Islands to collect core samples if Schrag agreed to pay for it. Schrag had the money, and they planned the trip for September 1998. Schrag took two others along: Konrad Hughen, a postdoctoral scientist in his lab, and me. At the time, I was a PhD student in the school of government at Harvard, researching climate policy; Schrag convinced me that it would be a good investment to spend a year working in his lab, gaining firsthand knowledge of how natural science actually happens. A three-week trip to Indonesia to kick that year off was a very good incentive. I packed my bags.

The bags never made it beyond a basement room in the Singapore airport. After arriving in Manado and discovering the absence of our luggage, we stayed in an upscale hotel until the next plane arrived from Singapore, three days later, and we passed the time scuba diving and shopping for fresh clothes. Our luggage didn't appear on the next plane either, and, after a quarter hour at her keyboard, the woman from the airline admitted that the computer system had lost it. Annoyingly, a new set of drilling equipment was among the vanished items, but Schrag dug out an old set that he had left at Erdmann's house, and, after a time with Erdmann's mechanic and boat driver, Saïd, we managed to get it working again.

The next day, we headed off to Gorontalo. There were eight of us: Schrag, Hughen, an Indonesian geologist who was our local research partner, Erdmann and his wife Arnaz, Saïd the mechanic, Denny the van driver, and myself. We each had a small duffel bag for our clothing and personal effects. Taking up the most room in the back of the van, and requiring most of us to lift, was a spare outboard engine that Saïd had convinced us to take along just in case things weren't working where we were going. We arrived in Gorontalo somewhat thinner from the combination of the local food and the twisty car ride through the mountains. The ferry ride to the Togian Islands left the next evening. It was hot and smelly, twenty hours on a hundred-foot long wooden boat with rusty nails sticking through the cracked paint, filled with people, more people, and chickens. It was particularly full because, until recently, there had been two ferries, but then the other boat sank on its way out to the islands, and the biweekly service was cut in half.[4] Twelve hours into the voyage, the boat's engine started sputtering, the crew did nothing, and we began to worry. But Saïd ran below deck and did some magic, and the problem went away. After stopping at a string of fishing villages to let off passengers and chickens, we reached the last stop. A small fleet of outriggers, puttering along as slowly as the ferry, took our group, our bags, and our spare engine to our final destination.

The Kadadiri Paradise Dive Resort was a collection of grass huts on the beach, and the coral reef started at the end of the dock. Staying there cost $3 a day, excluding diving fees and bottled beverages. Its permanent staff consisted of two cooks, one cleaning lady, a boatman, and a certified diving instructor, originally from Denmark. In addition to our large group, the guests consisted of German couple on their honeymoon, whom we had met on the ferry, and two Australian college students who were spending the year traveling and had been at Kadadiri for a couple of weeks. The diving instructor, however, was unhappy because the engine on his dive boat was broken, and so the diving was limited to the closest of the reefs. Saïd came to the rescue once again, this time with the spare engine that we had been lugging around.

The next morning, the Kadadiri boatman dropped us off at the island where Erdmann had seen all the rocks earlier that year. While Saïd and Denny went off to the main village to source more kerosene for the engine, the rest of us went off exploring. The three geologists had hammers and went around knocking chunks out of the rocks and looking at their inside structure, the way geologists do. By lunchtime, they had seen enough to figure out that the whole trip was a waste of time. The rocks were indeed corals. But they had become fossilized, and this was a sure tipoff that they were a few hundred thousand years older than the ones we were looking for, and any isotopic information that they could offer would be worthless for our purposes.

We had nothing to do but go scuba diving. The beach faced west and offered us a series of perfect sunsets as we feasted on fish and papayas and drank the local beer.[5] Eventually, Saïd and Denny went back to the main village and found a fishing boat that could take us to the mainland, so that we wouldn't have to wait another week for the ferry. We set off at dawn the next morning.

Schrag rebooked his ticket and flew back to Harvard the next day, leaving in the van with most of the others. But Hughen and I stayed in Gorontalo, and we made a last-ditch effort to find some corals that could suit our purposes. We hired a taxi to drive along the coast, and we stopped at any big rocks we saw, Hughen doing his trick with the hammer. On the third day, miraculously, we found a single large piece of coral that was just what we were looking for. We spent two days drilling, then Hughen and I took a bus to Manado, where Saïd came in and picked us up with his boat. We spent a day drilling more holes in the coral near Erdmann's house and another day scuba diving. Despite the fiasco in the Togian Islands, the samples from Gorontalo and Erdmann's island turned out to be a gold mine of information. They provided a hundred-year temperature record at a time when the Earth was a lot warmer, and they revealed an El Niño that had been a lot more frequent, just as Schrag had hypothesized.[6]

Understanding radiative forcing belongs to the domain of laboratory science and, as such, is relatively straightforward, well-established, and clear in its results. Outside the laboratory, climate science is messy and unpredictable. The purpose of the trip to the Togian Islands was to figure out if there was a correlation between temperature and El Niño. The result – that there appeared to be – can then inform us how to expect the distribution of temperature, both geographically and in terms of year-to-year variability, to unfold in the future. This is just one of the countless pieces that we need to complete the puzzle of what the climate of the future will look like. How will forests respond to a change in temperature combined with a change in CO_2 concentration? What happens chemically and biologically to the CO_2 that gets absorbed into the ocean? Will changing weather patterns results in more precipitation, and potentially more snowfall, in the mountains? What are the dynamics of ice sheets, dictating how quickly they respond to warming temperatures by sliding into the ocean? All these pieces need filling in, based on data gathered in the field, under water, or from satellites. Science of this type goes slowly because it is the all-too-random result of people stumbling around in the dark trying to do things that nobody has ever done before.

Two lessons emerge from all of this. First, we should not be surprised that there is a huge amount of uncertainty about the future climate, even when the basic physics of radiative forcing is well established and beyond doubt. Second, reducing the uncertainty that still remains will be slow going, perhaps slower than the pace of climate change itself.[7] If we are successful at preventing it, we may never know exactly what sort of climate change we are preventing.

Climate modeling

But we can get some sort of an idea, even an inexact one, and that idea is reliable enough to suggest eliminating CO_2 emissions altogether, fairly soon. Where this comes from is climate models, which are the computer programs that climatologists have developed to integrate the pieces of knowledge that we do have into a meaningful picture of the future.

All the cause-and-effect relationships that govern the climate can be written as mathematical equations. With just a couple equations essentially on the back of an envelope, Arrhenius was able calculate the approximate amount by which the planet will get warmer as long as nothing unexpected happens. With a little more work, you can put a few more equations onto an Excel spreadsheet, taking into account some feedbacks with the oceans and the biosphere. The next step would be to capture some of the spatial dynamics of the atmosphere and oceans: not just their overall temperature, but also how heat moves around the planet. That is essential to capture some of the known feedbacks in getter detail and to begin to explore the potential for surprise. But it also vastly increases the number of computations that the computer has to do. At this stage, modelers start working in complicated programing languages, which can handle the millions of computations more efficiently than can a spreadsheet, and running the model on a super computer, which is a lot faster than a laptop.

Today's state-of-the-art models are known as General Circulation Models (GCMs) representing the atmosphere, oceans, and biosphere as a beehive of activity. They draw a grid over the whole Earth, with each grid cell being a few kilometers on a side. They divide each grid cell, in turn, into a stack of pancakes, each layer a few hundred meters thick, with air that is thinner and thinner as you move up into the outer atmosphere. At the Earth's surface, the models make note of whether that surface is land, ice, or water. If it is land, the models represent the particular vegetation and the photosynthesis and soil decomposition that goes with it. If it is water, the layers go downward into the depths of the ocean. The models calculate how heat, water, carbon, and other chemicals circulate through all of this according to physical laws as best as they are known. They break up the continuous flow of time into discrete steps, like the ticking forward of a digital clock, with each tick representing a few minutes of real time. The smaller the grid cells, the thinner the pancakes, and the tighter together the ticks of the clock, the more computing power is needed, but the more accurately one would expect the model to represent reality.

When they look out into the future, different GCMs see different things, and this is a result of three sources of uncertainty – and potential error – each of which the various models deal with in slightly different ways. First, there is basic scientific uncertainty about how some climatic processes and drivers operate. El Niño is one example: we understand very well how it works now, and we have a good idea of how this will qualitatively change under conditions of warming temperature, but for the latter we lack a good quantitative understanding. Another example are aerosols, which are basically all of the fine dust in the air: there is a huge amount of uncertainty as to how aerosols affect climate at the global scale. The second basic source of uncertainty lies in the interpolation of present and past data. There are a lot of weather stations and ocean buoys around the world, but not as many as climatologists would like, especially in developing countries. To fill in the gaps of data, they have to make educated guesses, and the different modeling teams guess differently. The third basic source of uncertainty lies in how to represent processes that exist at too fine a scale for the model to capture. Thunderstorms, for example, are small enough to occupy only a small fraction of a single grid cell as represented in the GCM, and yet they give rise to dramatic convections of heat and water compared to more dispersed cloud cover. Modeling teams differ in how they handle the difficulty of guessing where the thunderstorms are.

The main test for any GCM is whether it can do a good job reconstructing the past. This is a process known as *validation*. Researchers use data from one time period – say, 1930 to 1960 – to calibrate the model, adjusting all of the equations so that the model perfectly reconstructs this period. Then they will let it tick forward and see what the model predicts for the next thirty years, 1960 to 1990. The more the model predictions match what actually happened, the better it will probably be at predicting what will really happen in the future. None of the models is perfect during validation, and their imperfections are all different. Some represent particular world regions particularly well or poorly, some are too hot in the summer and too wet in the winter, and others are better over land than sea.

Given this ambiguity, the best way to get a complete picture of the future is to look at an ensemble of projections, namely, the result of all the different models being run many times, each with different sets of assumptions, and all of those assumptions being within the range of what is theoretically possible. What they show is that average global warming will probably be more than 2°C and less than 5°C if the atmospheric concentration of CO_2 and the other GHGs were to double. That is something that will happen this century, if current emission trends continue. Some model runs show higher average warming associated with an atmospheric doubling – up to 10°C – and others show less – perhaps as low as 1°C – but these results rely on assumptions that, although theoretically possible, are extremely unlikely. The models also show a lot of regional variability. In general, the land will warm more than the oceans, the high latitudes north and south will warm more than the equator, and areas with seasonal snow cover will warm more than those that are either permanently snow free or covered in ice. So, even if the total warming associated with an atmospheric doubling happens to be 3°C, some places might warmer by as much as 10°C. And some places may even get a little cooler because of changes in ocean currents and wind patterns.[8]

To examine the climate into the future using a GCM, and indeed to compare the results from different GCMs, it is essential to specify a precise set of anthropogenic GHG emissions running into the future. These are known as *emissions scenarios*. Coordinating the development of a common set of scenarios in order to be able to compare the results of different GCMs is one of the tasks of the Intergovernmental Panel on Climate Change (IPCC). The IPCC was set up in 1988 by the United Nations, as a panel of the leading scientists who could provide useful information to policy makers, then in the first stage of negotiating a global treaty. In 2000, the IPCC published a first set of global GHG scenarios suggesting emissions pathways in the absence of deliberate climate policy but given a plausible range of socioeconomic development and technological change.[9] Since then, the IPCC has added other scenarios, with the inclusion of efforts to reduce GHG emissions.[10] Where the scenarios primarily differ is with respect to the concentration of GHGs in the atmosphere (measured in ppm of CO_2 equivalent) or the corresponding level of greenhouse forcing (measured in watts per square meter of the Earth's surface) that policy makers make a deliberate choice not to exceed. By running GCMs with these different scenarios, it is possible to explore the relationship between climate policy today and the climate of the future.

Among the more interesting of the scenarios is one designed to restrict GHG concentrations in the atmosphere to 450 ppm, higher than its current concentration at 400 ppm. The GCMs suggest that by implementing policy to avoid surpassing 450 ppm – which in the absence of deliberate efforts to control CO_2 emissions would likely happen within about twenty years – the total average warming compared to preindustrial times would be about 2°C, more or less. More precisely, the 450 ppm target would keep warming to under 2°C with about a 50 percent likelihood. To increase the likelihood to 75 percent, policy would have to limit long-term GHG concentrations to 400 ppm, which is just about today's level. By contrast, a 500 ppm scenario would see the chances of staying under 2°C fall to about 25 percent, meaning a 75 percent chance of exceeding 2°C.[11]

There are two respects in which GCMs, even an ensemble of GCMs, fail to offer much information. The first is with respect to particular years. I was at one climate workshop where a modeler showed a projected trend over the European Alps for the next 100 years. The line was squiggly, with high year-to-year variability, but I noted that one year – I think it was 2036 – was projected to be cold and snowy, and I made a mental note to myself to try to still be skiing by then. Of course, the mental note was a joke; no model can tell us what any particular day, month, or even year will look like. All that it can do is to reveal changing averages and the range of variation around those averages. The second is with respect to the long-term future. Climate modelers typically resist projecting climate change out more than about a century, and that is for good reason. The reason does not have to do with time itself, but rather with the effects of increasing radiative forcing under changing conditions. The equations that make up the heart of climate models are based on observed data from the past, a past with a particular climate. The more the climate changes from that which we have observed, the less robust will be the set of equations projecting how the changes will continue.

From cost-benefit analysis to the 2°C target

With some sort of an idea of what the future could hold, it starts to become possible to think clearly about appropriate climate policy targets and pathways. But for any individual to do such thinking, it is important to identify one or more primary concerns. If your primary concern is limiting the changes to ecosystems, changes such as loss of biodiversity, then it is necessary to understand *climate impacts*, the effects of physical changes in climate on ecosystems living in and on the soil and underwater. If your primary concern is limiting the likely or possible effects of climate change to people, then you need to understand how the activities that people engage in are sensitive to climate impacts – both those that are likely to occur and those that are improbable but still possible – and the potential for people to adapt their activities to minimize the negative consequences. If your primary concern is maximizing human welfare, as measured with economic indicators such as gross domestic product (GDP), then you need to understand not just impacts and their effects on human activities, but also the economic costs of making the changes that would reduce GHG emissions. If your primary concern is winning the next election, then you can probably subtract out any careful consideration of climate impacts, since these will unfold over decades, but focus instead on the short-term costs of reducing GHG emissions in different ways and the short-term benefits of those same actions, such as improvements in local air quality, or green job creation. A great deal of the difference of opinion about climate policy targets can be explained simply by looking at what people's primary goals are.

Until quite recently, the goal that tended to dominate climate policy debates was the third one, maximizing human welfare as measured in economic terms. To come up with the right set of climate targets to satisfy this goal, economists began to develop a new type of computer model, known as the *Integrated Assessment Model* (IAM). An IAM took its name because it integrated actual model components, with greater or lesser degrees of complexity, from a number of different sources. First, an IAM includes an industrial or energy system component modeling the economic effects of different pathways for GHG emissions. Second, an IAM contains an impacts component, translating changes in average global temperature into the effects on society. These would ultimately be valued economically, but the models require the intermediary step of being expressed in terms of changes in ecosystems and ecosystem services (like the productivity of agriculture), as well as the direct effect of physical changes like sea level rise on built infrastructure. Third, an IAM contains a climate model linking the reduction in GHG emissions and its costs with the deceleration of temperature rise and its benefits. An IAM can thus serve as the tool for conducting cost-benefit analysis: finding the climate target, or emissions target, for which the total net benefits (benefits minus costs) are the greatest.

The first IAM of this kind to generate a good deal of attention was the Dynamic Integrated Model of Climate and the Economy (the DICE model), developed by William Nordhaus of Yale University. The DICE model itself began with a set of standard equations – standard for neoclassical economists – of how the economy generates wealth, as measured in GDP. Generally speaking, the economy combines capital, labor, and energy to manufacture consumer goods and services, which people buy and then consume. In the process, the economy also produces more capital, which is one of the production inputs, and so the total amount of goods and services produced and the rate at which energy is burned to create it rise over time. The DICE model was simple enough so as not to specify exactly what sort of goods or services are being produced, assuming it will be whatever people want at the time. However, where the model was specific is that the use of energy involves, to a degree determined by the user of the model, the burning of fossil fuels. That, in turn, causes climate change. Climate change causes impacts, which the DICE model translated into economic damages by assuming that they reduce the efficiency with which goods and services can be produced and hence lower GDP. To mitigate climate change, it is necessary to either use less energy or to switch from fossil fuels to another source of energy. Using less energy results in fewer goods and services and lower GDP. Switching from fossil fuels to something else also reduces GDP because the other energy sources generally require more capital and labor to operate, so they eat up resources that could otherwise be going toward final consumption of goods and services. The model assumed that the relative amounts of capital and labor to operate different energy technologies – and hence the economic penalty of switching from fossil fuels to something else – change at a rate that is independent of the policy choices. The parameters for all of the equations, such as the relationship between temperature change and damage costs, were grounded in previous empirical studies.[12]

The DICE model is quite simple, and indeed it is not hard to program it oneself based on the basic set of equations that Nordhaus provides. Nevertheless, the model does allow one to explore how total societal welfare changes as one turns the control knob on deliberate emissions reduction ever higher to find the level of emissions reduction that achieves the highest possible welfare. One can also modify some of the basic parameters that could influence where this optimum point might lie and see the results. The most important of these, it turns out, is the relative value placed on current versus future welfare. People appear to value the present more than the future, a fact that economists typically model by applying an exponential discount function, which makes use of a single discount rate, like an interest rate. The higher the discount rate, the less the future matters relative to the present. Five percent is a typical value for economists to use, and indeed those who use a lower rate, such as 1 or 2 percent, typically have to justify their decision on some rationale lying outside of conventional economic theory.[13] But a 5 percent discount rate also means that changes in welfare in the future, even only a few decades out, have very little influence on the total net present value of all welfare. Welfare fifty years from now, discounted at 5 percent, is worth less than 10 percent of welfare experienced today. Put another way, if the decision to reduce emissions results in lower welfare today but higher welfare fifty years from now, then the gains fifty years from now must be more than ten times the current losses in order for the decision to appear worthwhile if one is using a 5 percent discount rate.

So what does DICE tell us? In the most recent version, published in 2008, the DICE model projects that, in the absence of any climate mitigation policy, CO_2 equivalent emissions will grow over time from their present level of about 8 billion tons per year to 19 billion tons in 2100. This, in turn, would lead to a rise in mean global temperatures of 3.1°C by 2100 and 5.3°C by 2200. The damages associated with these temperature changes would be 3 percent of global GDP by 2100 and 8 percent by 2200. The optimal level of emissions, the level that would maximize the net present value of economic welfare, would be much lower. By 2050, the optimal pathway would reduce emissions by 20 percent from what they would have been without policy in place and by 45 percent in 2100. That 45 percent decrease from the business-as-usual trajectory would mean emissions of about 10.5 billion tons in 2100, only slightly higher than their current value. The optimal emissions path would lead to a temperature rise of 2.6°C by 2100 and 3.4°C by 2200. Qualitatively, the result from the DICE model is this: it makes sense to limit the growth in GHG emissions, but it does not make sense to work on eliminating them. Given tradeoffs between the economy and the environment, the best solution is one where temperatures continue to rise into the future, but by less than they would have in the absence of policy.

This basic result found broad resonance in the economics community, and, indeed, many other economists reached similar results.[14] To many noneconomists, however, the result was deeply disquieting. Limiting climate change to "only" 3.4°C by 2200, instead of 5.3°C, doesn't seem like it does much for the other three possible goals: it doesn't save nature, it doesn't eliminate risks to people, it doesn't help you in elections by appearing to really solve the problem.

So political discussions, more and more, began to ignore the advice of the economists as to what the optimal target should be. They turned instead to a different one, first proposed in 1996 by the European Union, of limiting climate change to no more than 2°C average warming above preindustrial times. As the British historical geographer Samuel Randalls has documented, the basis for the 2°C target lies not in a single goal, but rather in a set of political compromises between multiple goals and objectives.[15] It does not completely protect nature, but it does a reasonably good job. To those concerned about risks to people, it appears to limit climate change to a level at which the risks are manageable. To those concerned about winning elections and making a good show of actually fixing climate change, the 2°C target represents a line in the sand.

There are many who suggest that the 2°C target is unwise because it is impractical. They may be right. Achieving 2°C with equal odds, 50/50 chances, requires limiting CO_2 concentrations to 450 ppm; to achieve greater certainty, the levels would have to be even less, such as 400 ppm. Concentrations are already at about 400 ppm, and rising by a couple ppm every year. Staying at or below 450 ppm would require immediate action to begin reducing emissions to slow the rate of rise of concentration in the atmosphere and ultimately eliminating those emissions altogether by the second half of the century. The cost-effective pathway to a 450 ppm target, energy analysts have suggested, passes through a point at 2050 in which global emissions have been reduced by 50 percent from their levels of a few years ago and in which emissions from industrialized countries have been reduced by 80 percent. To get to 400 ppm, the drop would have to be even faster – perhaps reducing industrialized countries' emissions by 100 percent by 2050 – and there would almost certainly need to be efforts beginning quite soon to actively remove CO_2 from the atmosphere.[16] The 2°C skeptics point to the fact that 2050 is less than forty years away and growing closer every day. A coal-fired power plant built today – and there are many that are being built today – will still be polluting in 2050.

The skeptics notwithstanding, the 2°C target has become the one most adopted by politicians and, to a large extent, by the climate change research community. The Copenhagen Accord, a document agreed to by almost all countries in 2009, states 2°C as the explicit target. Most national governments have made pledges to help achieve the 2°C target. Scientists have started studying what the impacts associated with the 2°C target would be and what the costs of achieving it would mean to society.

One of the interesting things is that as the 2°C target has started to frame political debates and academic research, it has also had the effect of reframing some of the basic assumptions that economists have programed into their IAMs. Nordhaus and DICE, after all, were looking for the optimal target, and they found it at a level of emissions reduction that was fairly modest. One particular assumption that made sense in this context was to treat technological change as something exogenous to the model; that is, something that proceeded at a particular pace, regardless of the level of emissions control. With modest efforts to control emissions, such an assumption stood on firm ground. But more recent models, similar in many ways to DICE, have explored the costs of achieving the sort of deep emissions reductions necessary for 2°C. In this context, treating technological change as exogenous makes no sense; it seems clear that efforts to completely replace fossil fuels with decarbonized sources of energy over the next few decades will have a dramatic effect on the different technologies' relative costs. Indeed, empirical results suggest that efforts to spur the development of renewable sources of energy could quickly make those technologies less expensive than fossil fuels. Building that into their models, economists began to reach the conclusion that the economic costs – ignoring the benefits to the climate – of deep emissions cuts would actually be less than the costs that Nordhaus had estimated for relatively modest emissions cuts.[17] With some additional assumptions, some models even suggested that the costs of eliminating emissions would be negative, meaning that deep emissions cuts would be justified on purely economic grounds, even if climate change were not a problem.[18] Based on many of these findings, the respected British economist Nicholas Stern broke ranks from the Nordhaus crowd and declared that substantial emissions cuts were warranted even on economic welfare grounds.[19]

The 2°C target does not have a firm basis in science or in economics. Achieving it may or may not be practically feasible, and it may or may not be the most attractive thing to do from an economic welfare perspective. But, given the latest economic models, it does appear to be affordable. These models project that the deep emissions cuts will end up reducing long-term GDP by about 2 percent or less;[20] in a world of nearly constant economic growth, that is the equivalent of arriving at a particular wealth level not at the turn of the calendar year to 2020, but rather a few months later. On the scientific side, there are reasons to believe that exceeding 2°C could eventually lead to dramatic effects, such as the rapid dieback of the Amazon rainforest, the spread of deserts, and the melting of the polar ice caps, which would flood most coastal cities.[21] Indeed, there are reasons to believe that some of these impacts could occur given sustained temperatures of less than 2°C above preindustrial averages.[22] But the 2°C does offer clear guidance for what ought to happen today. It implies that society needs to stop building new fossil fuel infrastructure and stop using the fossil fuel infrastructure that is already there as soon as is practically feasible. Society can do so, at very little cost, by building and using something else. The question is simply how government can initiate this process.

3

The solution space and its distractions

A 2°C target makes sense, not because it is necessarily grounded in solid science or economics, but because it represents a good compromise of competing interests. Moreover, the actions needed to achieve the 2°C target are ones that would be needed for even lower targets and would be actions that would make sense even if the target were somewhat higher. The important recognition underlying the 2°C target is that the world needs to eliminate all carbon dioxide (CO_2) emissions as quickly as is practical and feasible. Indeed, it is undertaking these actions that then becomes the immediate problem for policy makers: primarily the replacing of the fossil fuel-based energy system with something else, but also doing something about forest degradation and the other industrial sources of greenhouse gases (GHGs). But before turning to how they can do these things – the transformation of the energy system in particular – it is worth clearing the air of some distractions. The distractions involve protecting ourselves from what we have wrought and include both adapting to a changing climate and coaxing the planet into staying cooler despite the CO_2 that we continue to emit. They are not distractions because they are unimportant; indeed, they may prove to be crucially important for protecting human lives and vast ecosystems. But they can only work as long-term strategies if they complement, rather than substitute for, the core challenge of reducing GHG emissions and CO_2 emissions in particular. They are distractions if they cause us to forget about this imperative.

Adaptation

Humanity has always been protecting itself from nature one way or another and constantly doing a better and better job of it as the technologies at people's disposal have improved and expanded.

The first house I owned, a Vermont farmhouse built in the early 1800s, had window frames that kept out the winter as good as wood pressed against wood could do. My second house, an Austrian one built a century later, had windows that kept out the wind better, with a seal of rubber against wood. My current house, a farmhouse built in the early 1700s but fully renovated ten years ago, does better still, with window seals of rubber against rubber. Adapting can also mean changing the things that people do, taking into account not only climate, but also competing demands on our time. Mediterranean cultures adapted to hot summer days by taking a mid-day siesta. Maintaining that practice has proved challenging as they transformed from largely agrarian societies to urbanized ones, and so the siesta has been all but lost. Instead they have air-conditioned offices, more relaxed dress codes, and more drinking water.

In theory, continued adaptation could erase many of the costs associated with climate impacts. One of the paradigmatic examples is the contrast between a "smart" and "dumb" farmer. A dumb farmer will continue to plant the same crops and suffer declining yields as the conditions for growing these crops become less and less appropriate, whereas the smart farmer will constantly change the crops that she or he is planting according to what makes most sense at the time. Integrated assessment models (IAMs) like Nordhaus's Dynamic Integrated Model of Climate and the Economy (DICE) model implicitly assume smart farmers. The climate damage estimates in the model come from a comparison of farm productivity in a variety of climates: as the climate of southern Canada starts to look more like that of the northern United States, so too will the farmers' productivity, yields, and profits.[1]

Adaptation seems like it ought to be relatively straightforward since it is something that people have always done, and yet, in recent years, academic researchers and policy makers have been seeing increasing challenges associated with doing it well and as quickly as anticipated climate change would seem to demand. The problems boil down to three sets of issues: how to build new infrastructure to protect human settlements and human activities from climate-related hazards; how to do a better job of considering a changing climate in a wide range of planning decisions, a process known as *mainstreaming*; and how to improve the capacity of lesser developed societies to cope with the climate they already have.

Climate-proofing settlements and activities

When Hurricane Katrina hit New Orleans in 2005, people were unprepared and, indeed, had been unprepared for a long time. A great deal of the city lay beneath sea level, and the dikes that had been built to keep the sea out were not designed to withstand a storm surge from a major hurricane. It was not for a lack of information. Previous studies had identified the basic design criteria for a more effective protective barrier, one similar in design to the system off the coast of the Netherlands, and had estimated the cost of such a barrier at $15 billion.[2] But, for a variety of reasons, it had not been built. Katrina flooded the city, killing hundreds of people, and causing direct losses of over $150 billion.[3] Indirect losses, those associated with an economic slowdown, are harder to measure, but one government estimate was that the hurricane reduced the growth of the U.S. gross domestic product (GDP) by between 0.5 and 1.3 percent, costing up to 400,000 jobs.[4]

The example of Katrina is extreme, but it illustrates the mandate to adapt. It is neither possible nor particularly useful to say whether Katrina would or would not have happened without climate change,[5] and indeed there is reason to believe that a rising trend in climate-induced damages is not a result primarily of climate change, but rather of the increasing value of human infrastructure built in highly vulnerable locations.[6] But the research is nevertheless convincing that because of rising sea surface temperatures, the frequency of the strongest classes of hurricanes is increasing.[7] So there is reason to believe that there will be several more events like Hurricane Katrina over the next hundred years or so, not just in the United States, but also elsewhere. And at least a few of these, maybe all of them, would not happen if climate change were not ongoing. It is also worth noting that the upper estimate I quoted of the indirect cost of Katrina, a reduction in GDP growth of 1.3 percent, is not a lot different than mainstream estimates of how much it will cost countries to completely decarbonize: something less than 2 percent.[8] In an ideal world, we would have already decarbonized, prevented climate change, and avoided the increase in costly natural hazards. But the world is not ideal. Even if we stopped emitting GHGs today, lags in the climate system mean that average temperatures will continue to rise for another few decades. We cannot avoid a great deal of climate impacts. If building proper sea walls can not only save lives, but also reduce the associated economic costs by 90 percent, then let's build proper sea walls.

Investing money today to save a lot more tomorrow may be possible in countries like the United States, but it becomes extremely challenging in the context of developing countries. There have been a number of estimates of the cost of building the needed infrastructure and related investments around the world, and these have varied over a wide range depending on the exact assumptions. Among the most recent of these assessments, conducted by the World Bank, put the value at $70–100 billion per year,[9] a number slightly higher but not out of line with other estimates.[10] This is investment that is needed that is additional to what is already being planned and would be undertaken if people were not taking climate change into account. If you believe these numbers, as I do, then they are mind-boggling. Official developmental assistance from wealthy countries to less wealthy ones totals roughly $130 billion.[11] Climate change, then, means that these countries need almost this much additional money if they are to minimize the adverse impacts.

One issue about climate protection concerns what sort of infrastructure to build; whether, for example, a seawall ought to be made of concrete and steel, or if it should better make use of organic solutions, like preserving mangrove swamps to ensure that they buffer incoming storm surges.[12] The primary debate, however, has turned into a debate about finance and financial assistance. A huge amount of the effort in international climate negotiations has been devoted to the issue of how much money will change hands to pay for adaptation and whether this money is to be included in or is additional to the money that is changing hands to finance development projects more generally.[13] Money put to this purposes could pay large dividends, starting immediately. The danger is that it comes at the expense of investments in infrastructure or land-use projects that will stop future CO_2 emissions. Global subsidies for renewable energy in 2010 were $47 billion, well below the lower end of the World Bank's estimate of adaptation costs.[14] It is essential that investments in renewable energy rise, not fall. In the short run, the investments into adaptation may save more lives. In the long-run, far more people will die if adaptation finance crowds out mitigation finance.

Mainstreaming climate information into daily decisions

In 1994, an article appeared in the journal *Nature* that showed that roughly 60 percent of the variance in maize yields in the southern African country of Zimbabwe – essentially whether farmers there would have a good crop or a bad one – could be predicted close to a year in advance simply by observing sea surface temperatures off the coast of Chile and Peru, half a world away.[15] The researchers, led by the oceanographer and climate modeler Mark Cane of Columbia University, were among the earliest to investigate El Niño and its effects around the world. El Niños were statistically associated with snowy winters in North America; warm temperatures in India; wet conditions in northeast Africa and the *pampas* of Argentina; and dry conditions in southern Africa, Australia, and the northeastern region of Brazil.[16] What Cane and his team realized was that if you could predict El Niño in advance, which their ocean simulation models could,[17] you could predict this other stuff with some degree of likelihood, and perhaps you could give people an opportunity to prepare.

The Cane paper set off an avalanche of activity to make El Niño forecasts available to decision makers, especially in developing countries. In 1995, the U.S. National Oceanic and Atmospheric Administration (NOAA) organized an international workshop in the ecotourism resort of Victoria Falls, Zimbabwe, where delegates from around the world designed a set of activities aimed at seasonal climate forecast development and dissemination.[18] Two years later, the first of these activities went on-air; the Southern African Regional Climate Outlook Forum (SARCOF) took place in a conference center near Harare, Zimbabwe. It was a weeklong meeting of climatologists and government decision makers, people coming from ministries dealing with agriculture, disaster management, water resources, finance, transportation, and energy. Hopes were high that if they worked together they could do a lot of good because that very year, 1997, an El Niño was developing that would prove to have the highest sea surface temperatures yet recorded, what one World Bank official declared "the mother of all El Niños."[19] Climatologists at the SARCOF predicted an abnormally high likelihood of drought, especially over Zimbabwe, then the "bread bowl of Africa." Following the SARCOF press conference, the media picked up on the story. Many farmers decided not to plant anything, figuring the money spent on seeds and fertilizer would be wasted if the rains never fell. Banks restricted credit, figuring that money they loaned to farmers for seeds and fertilizer would not be repaid. Emergency meetings were held to prepare for a food security crisis.[20] And then the rains came. Overall they were not enormous, but there were a few heavy downpours, and they certainly were plentiful enough to dispel any fears of drought. Crop yields still suffered, not so much from the rainfall as from the preparations for drought that SARCOF had induced.[21]

The fallout from the forecast of 1997 occurred just as I was coming off my year working with Dan Schrag, traveling to Indonesia and working in his lab, trying to unravel the ancient history of El Niño. I was interested in whether everything might have been better – if the loss of yields that farmers suffered and the loss of credibility that climate forecasters suffered might have been less – if the SARCOF forecast as widely communicated had been more explicitly probabilistic. Of course, it hadn't: the message that most people heard from the SARCOF meeting was that there would be a drought, period. Nobody told them that the probability of drought in 1997 was only somewhat higher than in any other year. As one senior official at the Zimbabwean ministry of agriculture told me, his father was a farmer, and farmers can't understand complicated things like probabilities.[22] I thought he was wrong, so I spent a month driving around the Zimbabwean hinterlands with a homemade roulette wheel, finding farmers, many of them illiterate, who were willing to win some money by playing a gambling game I had put together. The game was designed to test whether they would change their bets when the ratios of different colors on the wheel changed, effectively responding to a change in probabilities. As it turned out, they did pretty well, about as well as the typical American college student who participates in psychology experiments.[23] Interestingly, the women participants did a particularly good job of learning: by the end of the series of gambling games, most of them were displaying a betting strategy that responded perfectly to the changing probabilities on the wheel.[24]

Then, over the next five years, I went on-air myself. I organized annual workshops in four Zimbabwean farming villages, where my local partners and I discussed the forecasts (the ones still being developed each year at the SARCOF) with a group of about fifty randomly chosen farmers. We asked farmers whether they used the forecasts to make different decisions and what factors prevented them from doing so. Many revealed, for example, that it was their relatives who lived in the city who bought them their seeds – seeds were cheaper and the selection greater at the big stores in the city – and yet these people in the city didn't have a good idea of how to change what they bought in response to the forecasts.[25] Ultimately, we compared the yields of those farmers who had changed their decisions because of the forecasts with those who hadn't, and we discovered the former yields to be somewhat higher than the latter, with a high degree of statistical confidence.[26] So, forecasts could be useful after all. We also found that the farmers who had attended the workshops – these were farmers randomly chosen and invited – were five times more likely to do something with the forecast information than were those farmers who had read the forecast in the newspaper or heard it on the radio.[27] At the conclusion of the study, NOAA, which had funded this research, invited me to evaluate results on forecast use from across all of Africa. We found only a few cases where forecasts had been successfully applied, but lots of cases where they could have been but weren't. The difference, the data showed us, was the attention that scientists and government officials had given to the process of communicating the information.[28]

By the time this work was done, I found that I had become ensconced in the community of climate adaptation researchers. This was the case even though nothing about using seasonal climate forecasts has anything directly to do with climate change. Rather, it was about trying to use an improved understanding of the climate as it is today to plan better and make better decisions. But this process has taken on the name of "mainstreaming," since it involves bringing climate information into the mainstream of decision making, and it has become one of the key focal points of government efforts to promote adaptation. In the European Union (EU), for example, mainstreaming stands front and center in the current adaptation strategy.[29] Making it happen consists of two essential elements. The first is to produce climate information in a form that can be useful to people; one of the steps toward this is to establish an online information platform for climate research, to be set up and maintained at the European Environment Agency in Copenhagen.[30] Similar efforts are occurring in the United States, where NOAA has set up and funded a set of regional climate information centers (known as RISAs) to support adaptation. The second element is to try to coax people into using the information. In Europe, in theory, this happens during the process of environmental impact assessment, which is mandated in one form or another across all levels of governmental decision making and major project development. In practice, this does not appear to be happening; efforts are under way to amend the guidance and mandated content of impact assessment to make sure that things improve.[31]

Mainstreaming is the process that can turn dumb farmers into smart ones, the kind that the IAM developers assume are already there. Research has shown that there is still a lot to be done to make this happen as well as it could, and nobody can quarrel with the value of doing so. At the same time, it is important not to overstate the value or to see mainstreaming as in any way more important than reducing CO_2 emissions. The research results that my group obtained in Zimbabwe, for example, showed subsistence farmers improving their yields up to about 17 percent when they started using climate information. Perhaps they could do even better with more practice. But the results also suggested an interannual variance in yields among these farmers – the difference in yields between a wet year and a dry one, – to be more than three times as great.[32] No amount of mainstreaming can make their crops grow if the water isn't there. Unfortunately, increasing frequency of drought is exactly what climate models predict for the villages where I worked.[33]

Building adaptive capacity

Having the right infrastructure in place and making sure to consider the climate when making important planning decisions are two ways of adapting, but there is a third aspect to adaptation, and that is making sure that people have the practical ability to engage in the first two. This is known as building *adaptive capacity*. Especially for developing countries, a growing body of research has started to illustrate how important it is.

In early 2008, I received an email from a Dutch woman, Barbara van Logchem, who was consulting for the Mozambique National Disaster Management Agency (INGC). Given where it lies, Mozambique is a natural disaster waiting to happen: the country suffers frequent droughts (especially in El Niño years), floods along the banks of the four major river deltas located there, and storm surges from the cyclones that hit it after gaining strength over the channel that separates it from Madagascar. INGC wanted to appraise what might happen over the coming decades on account of climate change. But Mozambique also has one of the most quickly growing economies in the world, and that is in turn changing the degree to which its people are vulnerable to climate impacts. In addition to hiring a set of climate modelers and people looking at impacts, van Logchem hired me to do one additional thing: try to understand the role that socioeconomic development could play.

What I did seemed fairly straightforward, although it turned out that nobody else had done it before. I analyzed data on the number of people killed or displaced by climate-related disasters over the past twenty years, seeing if there was a relationship with various national-level indicators that one might think would correlate. Doing so, I built a statistical model based on a limited number of variables: the number of disasters to strike a country, the size of the country, the country's Human Development Index (HDI; a United Nations composite indicator comprising GDP, life expectancy, and educational attainment), degree of urbanization, and female fertility. My climate modeling partners on the INGC project then provided estimates for how the frequency of climate disasters in Mozambique was likely to change, while various socioeconomic forecasters developed scenarios for the other variables in the statistical model. Putting the statistical model and the projections for the future together, I could estimate what the trend in deaths and displacement was likely to be and how sensitive this was to the climate side, relative to the socioeconomic side. I did this for Mozambique, and then, while I was at it, I extended the analysis to a group of twenty-two other least developed countries.

The results? Really, there were two. First, the socioeconomic signal greatly dominated the climate signal. Indeed, there is a very good chance that socioeconomic development will outweigh any increase in climate hazards over the next fifty years, leading to a net decline in the number of people killed or displaced. Second, the relationship between the socioeconomic indicators and vulnerability to these natural hazards is not linear. The least developed countries are vulnerable, but those slightly higher on the development ladder even more so. Only by the time countries get to about the level of South Africa does there then start to be a negative relationship between development and disaster vulnerability: more development, fewer people killed or displaced.[34] Others have noticed this nonlinearity in the relationship between development and disasters, and yet its cause remains something of a mystery.

Something about the level of socioeconomic development has a profound effect on people's ability to protect themselves from the climate, with the strongest factor of all appearing to be the number of people in a country, in particular women, who have made it through secondary school.[35] People who are functionally literate do a better job of protecting themselves, whatever danger they face. Building adaptive capacity means addressing these factors.[36] To some extent, it is synonymous with socioeconomic development, although the results from the Mozambique project suggest that the correspondence is not absolute. Somehow, development assistance needs to focus on those factors, like educational attainment, that appear to reduce vulnerability no matter what the level of development. Efforts to raise adaptive capacity can also focus on specific capacity to cope with climate risks and vulnerability, and not just on the general ability to make and execute good decisions. There are many examples of this, such as training programs for climatologists from developing countries,[37] experimental crop insurance programs for small-scale and subsistence farmers,[38] or refocusing disaster aid from *ex post* reconstruction to *ex ante* risk reduction.[39]

Of all the ways for government policy to promote adaptation, the efforts to improve adaptive capacity are likely to have the greatest impact in terms of reducing human suffering in the face of adverse weather and climate. In wealthy countries, events like Hurricane Katrina stand as the exception to the rule; it is very rare for people with a modicum of education and a financial buffer to be seriously harmed by the weather, even if they do persist in building beach houses on the North Carolina Outer Banks. With attention to raising adaptive capacity, it may well be possible that, globally, the number of people seriously harmed by the climate could decrease, even if the number of harmful events themselves increases. At least for the next few decades. Nobody has looked at this question further into the future. One can speculate that if we manage to curtail CO_2 emissions, the trend of adaptation outpacing climate change could continue. If we fail to curtail emissions, then it is difficult to imagine how this would happen.

Geoengineering our way out of a crisis

The final way that we can protect ourselves from the consequences of persistently increasing GHG emissions is through a set of proposed actions collectively known as *geoengineering*. The conservative side of geoengineering is the removal of CO_2 from the atmosphere, then burying it underground or under the sea floor. There are a number of ways proposed to accomplish this. One way, developed by Klaus Lackner, an engineering professor at Columbia University, is to build free-standing machines to do the trick. His prototype machine acts like a giant car radiator, with air blowing through a lattice with high surface area. But instead of exchanging heat, the air gives off its CO_2 to a fluid coating on the surface of the lattice. The absorption fluid drips down, where a second chemical reaction uses energy to regenerate the absorption fluid and create pure CO_2 that can be compressed and pumped underground.[40] The machines work, but they are expensive, and so far there is nobody willing to pay for them.[41]

Another way of removing CO_2 from the air is by getting plants and algae to do the same job. Researchers have been working on fertilizing the oceans with iron filings, stimulating a bloom in phytoplankton. As the first successful experiment showed, it seems to be possible to do it in such a way that the phytoplankton then die and settle quickly enough to the bottom of the ocean to avoid being gobbled up by the food web and converted back into CO_2.[42] Or, one could grow trees and, instead of letting them decompose, which would have the effect of releasing their carbon back into the atmosphere in the form of CO_2, turn them into charcoal and bury the charcoal. That appears to lock the carbon in place while also stimulating overall soil fertility.[43] The problem appears when you calculate how many trees on how much land area it would actually take to offset all of our emissions. It is a very big number.

The other side of geoengineering is more radical, and for many people far scarier: keeping the planet cool despite rising GHG levels by reflecting vast amounts of sunlight back into space. You could start by painting rooftops and roadways white or putting arrays of mirrors in the desert, pointing straight up, and, indeed, there have been studies quantifying the effects of doing so.[44] But to do it big time, it would be far more effective to cover the whole planet. This has happened in the past, although not as a result of human intervention. The last really major volcanic eruption was in 1991, when Mount Pinatubo in the Philippines erupted and injected 17 billion tons of sulfur dioxide (SO_2) into the stratosphere. Much of the SO_2 stayed up there for more than a year, with the effect of reducing incoming sunlight by 10 percent. In that time, the Earth cooled by about 0.4°C.[45]

So why not create an artificial volcanic eruption, or a series of them, one every year? In theory, it is completely possible, and the price tag would be in the range of affordability.[46] The cheapest way of getting SO_2 up there, it appears, would be simply to carry it by airplane. The annual price tag of creating a Pinatubo-like effect would be $8 billion to fly the airplanes, 300 sorties a day, plus the cost of the sulfur, another $20 billion.[47] In other words, for an annual expenditure of about one-twentieth of the U.S. military budget, humanity could return the temperature to what it once was. It certainly is an attractive proposition.

But there are a huge number of risks. There are indications that the change in incoming solar radiation could have a negative effect on agricultural productivity, and there is evidence from the

Mt. Pinatubo event that it would lead to drought conditions in some places, such as the African Sahel.[48] As leading experts wrote in an opinion article in the journal *Science*, these risks give rise to a need for some sort of international framework for deciding whether and in what form to deploy geoengineering.[49] It would also not solve the nonclimate effects associated with rising CO_2 concentrations, such as ocean acidification; this is almost certain to have a negative impact, likely a catastrophic impact, on coral reefs globally.[50] Ultimately, experts tend to see geoengineering as a last-ditch emergency measure. It could be deployed, they argue, to hold off a disaster with irreversible consequences, such as the rapid melting of the Arctic ice sheet or the loss of the Amazon rainforest.[51] Given the risks, and the fact that SO_2 would have to be deployed in increasing quantities, almost nobody is willing to argue that it is a substitute for reducing and eventually eliminating CO_2 and other GHG emissions.

The essential solution: mitigation

Adaptation is important, and society would be foolish not to invest in it. Immediate money spent, especially to raise adaptive capacity, will bring immediate benefits. If you trade off adaptation with other ways of spending public funds, there is every reason to believe that adaptation will fare quite well. And if you really think about it, adaptation is nothing new and nothing really to do very much with climate change. All of the solutions out there to improve adaptive capacity – things like improving schools and hospitals and sanitation – are things that societies have been doing, or trying to do, anyway. And pretty much all of the other things, from building infrastructure that is climate-proof to mainstreaming climate information in planning decisions, belongs to a long tradition of innovation that make people less vulnerable to bad weather. Indeed, my own experience in the field of adaptation has shown me that nearly all adaptation taking place in the world actually involves the application of new technologies or institutional innovations in order to cope better with the climate of today. The only example I have seen of preparing seriously for the long-term future – the exception to the general rule – is also the example that so many people use to describe adaptation: building taller dikes in the Netherlands. The Dutch are unique. They have been building dikes for hundreds of years, they are good at it, and they are more or less comfortable with dikes dotting the landscape. So, for them, it is a pretty routine, even if expensive, matter of building them tall enough to withstand storm surges a couple decades in the future. Almost everyone else in the world is still trying to figure out how better to cope with the climate of today, much less of the future. Let's not think of adaptation as a response to the climate of tomorrow. Let's think of it as the process of applying innovation to deal with the climate of yesterday and today.

Because adaptation is nothing new, we ought to be reasonably confident that it will take place in the future as it has in the past. Impact studies that project a future climate onto today's way of doing things may project huge losses from climate change. Imagine the money wasted if, thirty years from now, people are still building new ski lifts in places where the snow no longer falls. It seems pretty silly, and, indeed, impact studies like this ought not to be taken seriously. But the point is that you don't need to take them seriously in order to be very concerned about climate change. Even with all the adaptation that it is reasonable to assume, everything short of building giant glass domes to cover all of our big cities and growing all of the world's crops in giant greenhouses, the impacts of climate change are worrisome and potentially catastrophic. So adaptation is important, but it doesn't solve the underlying problem.

Geoengineering, by contrast, is a big question mark. Removing CO_2 from the air is certainly possible, although the costs are, so far, higher than what it would be to avoid putting the CO_2 into the air in the first place. But if we find that we need to start cooling the planet down in a hurry, it may be money well spent. Even faster would be the effects of blocking sunlight by shooting SO_2 up into the stratosphere. On this, we simply don't know enough about the potential consequences to do so ethically. One challenge may actually be to discover those consequences in a manner that is ethical: you can't simulate geoengineering in a laboratory, yet should we do so in the real world, even if it might cause the next big drought in India or the Sahel? Perhaps we need to wait for the next big volcano and be prepared to monitor that even better than we did the last time. (It may be a long wait. Before Pinatubo, the last volcanic eruption of that magnitude was Mt. Krakatoa, in 1883.) Many say that we ought to continue research into geoengineering, short of large-scale atmospheric experiments, because at some point it may be worth the huge risks. Today it is not. And because those risks will always be huge, we ought to do everything we can to make sure we don't need to take them.

Ultimately, neither adaptation nor geoengineering is in any way, shape, or form a viable, ethical, or cost-effective substitute for preventing CO_2 and other GHGs from entering the atmosphere in the first place. The word for that is "mitigation." Let there be no doubt of its importance.

PART 2

Failed strategies to reduce emissions

The need for a set of policies to address climate change by reducing or even eliminating greenhouse gas (GHG) emissions has been recognized for close to thirty years now. A watershed year was 1988, a time when the United States could still claim leadership with respect to environmental policy. An exceptionally hot summer across much of the country, with wildfires raging in the west, appeared to confirm fears that this newly identified phenomenon, global warming, might actually be real and might endanger people. Congress held hearings, inviting the leading climate scientists to testify. It appeared that a domestic solution would soon be found, just as domestic solutions had been found for local air and water pollution or the problems of toxic waste. At the global level, the United Nations General Assembly moved toward the idea of an international treaty to solve the problem in the tradition of acid rain or stratospheric ozone loss.

Jumping forward to the present, one could be forgiven for thinking that none of that happened. When we compare action on climate change with so many other things, the differences are striking. In those years since 1988, we have gone from a world largely without email to one in which the Internet seems universal. The Berlin Wall has fallen, and most of the former communist countries in Eastern Europe have become integrated with the West. Apartheid has ended in South Africa. Global attention to poverty reduction has made a profound difference. But GHG emissions are higher than ever, and, indeed, their rate of growth has increased in the past decade.

Why have the efforts to address climate change apparently been so singularly ineffective? What I argue here is that the strategies that people proposed back in 1988, and which society has tried to follow through on ever since, were the wrong ones for the task. In this section, I describe what those strategies have been, why they made sense at the time, and why they have ultimately proved to be ineffective. In Chapter 4, I deal with the strategy of changing the economy by introducing a large new market for CO_2 and GHG emissions reduction. In Chapter 5, I examine the strategy of implementing a global agreement for GHG emissions reduction. In Chapter 6, I look at the idea of addressing GHG emissions by tackling its underlying drivers: human behavior and economic growth.

In all cases, the objective in these chapters is not to bash a particular body of academic theory. Rather, it is to understand the theory and why it continues to be a motivation for the continued application of the strategy it implies. At the same time, however, it is important to understand how these theories mislead us and why it may not be productive to continue to rely on them. We need to understand both arguments and counterarguments.

4

Getting the prices right

Until I moved to Austria, the idea of climbing a mountain in order to have a cigarette at the top had never occurred to me. By the time I did move there, to the outskirts of Vienna in 2006, I was a father. A popular weekend activity for my family, sometimes with friends, was to go up the mountain closest to our house. At the top of the mountain was an Austrian Alpine Club hut that operated as a restaurant during the day. We typically had lunch there and then hiked back down, playing hide and seek in the forest with our children. In the winter, when there was enough snow, we did it on skis or pulled our kids on sleds. It was lovely. But what surprised me, when we first moved there, was that the hut itself was filled with smoke.

Austria, according to the latest statistics available, ranks first among industrialized countries in a way that few people are proud of: 36.3 percent of adults smoke on a daily basis, just ahead of Greece, at 35 percent, and well ahead of the United States, at 17.5 percent. Indeed, 60, percent of Austrians in the 20–50 age bracket describe themselves as occasional smokers, the highest in the world.[1] Why Austria? I don't know. The country, like most others, has been taxing cigarettes extremely heavily as a way of discouraging smoking. A pack of Marlboros will cost you more than $5 in Austria, comparable with most other Western European countries and most American states (New York and a few other states tax at a much more higher rate) and a lot more than most Eastern European countries (in Poland, a pack of smokes will cost $3 and in the Ukraine $1.50).[2]

But, until recently, Austria didn't have or didn't enforce many other laws for smoking, such as regulations against smoking in indoor public spaces or against cigarette advertising. When I moved there in 2006, people smoked in their offices, and, if others down the hallway didn't like it, they could keep their doors closed and their windows open. Officially, you were not supposed to buy cigarettes if you were under the age of sixteen. But cigarettes could be bought from vending machines, and there was nothing to stop a fourteen-year-old or a twelve-year-old from buying them there.

Things in Austria began to change in 2010. A new law took effect that forced large restaurants and bars to have physically separate smoking and nonsmoking areas, while smaller ones had to choose whether they were entirely smoking or nonsmoking and put a sign on the door accordingly. The Alpine Club hut that we frequented decided that it was going to be nonsmoking; the ashtrays vanished from the tables, and the air became smoke free.[3] The Chamber of Commerce initially opposed the new law and predicted that people would stop eating out altogether if they had to worry about rules concerning where they could and couldn't smoke; now they are comfortable with it, and, after an initial downturn in business, things are back up to normal.[4]

It seems intuitive that if the government places large taxes on cigarettes, making them more expensive, then, all things being equal, people will smoke less. Governments around the world that aim to reduce smoking do just this, just as countries like Sweden and Norway tax alcoholic beverages heavily in order to reduce drinking. But it also seems intuitive that all things are rarely equal. Austria has a culture of smoking. There are plenty of people at my former institute who started smoking within a couple months of arriving simply because everyone around them was smoking every time they went for a beer. I doubt they did so because smoking is cheaper in Austria than in New York.

Unless cigarette taxes are astronomically high, it seems intuitive that their effects are going to be fairly minor compared to all of the other cultural and social incentives that people have to light up or not. If a government is serious about reducing smoking and not simply looking for an easy source of tax revenue, it will address these other factors with a diverse set of policy instruments, as most American states and most other European countries have done, and which Austria started to do a little late. The most common and most noticeable of these, the ban on smoking in restaurants and bars, gave people a practical alternative to socializing in smoke-filled environments.

You might expect this type of common sense about taxes and other policy instruments to dominate the political discourse on fossil fuel burning, which governments are also seeking to curtail. It does not. In the area of fossil fuels, the conventional wisdom has been, for about twenty years, that a tax on carbon – or its virtual equivalent, the tradable permit – is the sine qua non of climate policy. If, and if only, we get the price of carbon right will we solve the climate problem. Indeed, there are respected economists who argue that we should refrain from doing much of anything else to curtail fossil fuel burning because these measures will interfere with the effectiveness of the carbon pricing policy and make our lives worse off.

In this chapter, I explain the logic behind this line of thinking – for it is quite logical – and why it has and will continue to lead us astray. It is the most complicated chapter of the book, with a lot of numbers. The fact is that, with carbon markets, the devil is in the details. The overall structure of the chapter is as follows. First, I describe the underlying theory of carbon taxes and tradable carbon emissions permits, which collectively are known as *market-based instruments*. I describe how, in theory, they correct a failure in energy markets in a way that will achieve economic efficiency: the optimal allocation of society's resources. Second, I review the success of market-based instruments so far. The short story here is that they appear to have worked reasonably well in other environmental policy domains – acid rain and fisheries management – but so far have underperformed expectations in the case of climate mitigation. Third, I address the question of whether it would be worthwhile trying to improve market-based climate policy instruments, knowing what we do about the challenges associated with them. My conclusion is that it would not.

Economic theory: markets and environmental externalities

Today's mainstream economics – the kind that students learn in university – is known as *neo-classical economics* because it represents a modern approach to understanding and expanding on Adam Smith's classic idea of the market's hidden hand. It has split into a number of subdisciplines that focus on how markets function in particular sectors or on particular problems that plague perfect market formation and the maximization of welfare. One of these latter subdisciplines is *environmental economics*. For as long as anyone now alive can remember, environmental economists have looked at the problem of pollution in a particular way, a way that supports the idea of a tax or tradable permit as the exactly right way of dealing with the problem. Paul Krugman has described this as one thing that economists agree on.[5] It is worth understanding how they see it.

All neo-classical economists, environmental economists among them, build on a foundation of supply and demand for different goods. *Supply* refers to the actions of individuals and companies producing goods and services to sell, whereas *demand* refers to the willingness of consumers to pay money to buy these things.

Start with supply. The theory of supply suggests that there is a relationship between how much it costs a factory to produce a particular product, how much of that product it will produce, and what price it will want to charge. The particular number to pay attention to is *marginal cost*. Marginal cost is the cost of adding one additional unit to the firm's existing level of production. It's normal to think that as a factory scales up and begins to mass produce a particular product on a production line, the costs of production fall and hence marginal costs decrease. That is certainly right. Once a firm has achieved mass production, however, and then decides to increase production even further, marginal costs may begin to rise. This is the area that economists focus on.

Let's build up a simplified but illustrative example to see why this may be the case and then extend that example to points relevant for thinking about climate change. Imagine that in your community are several factories that build wooden chairs. Once they have bought and paid off all the machines for the production line, the factory owners need two more things: wood and skilled woodworkers. If they are only building a few chairs each month, the trees and the people they can find locally will suffice. But imagine that business grows. At some point, the factory owners will use up the local resources and will need to look further for their inputs; they will have to ship in their lumber and to recruit and relocate workers from far-away places. That begins to get expensive and makes marginal costs rise.

Next, let's move to demand. Demand is about the behavior of consumers, people like you and me. People are different. Some really like these chairs, and others not so much. Common sense suggests that the higher the price, the fewer the number of customers. Just like with cigarettes. So, while marginal costs rise with higher levels of production, demand falls: the only way to get people to buy more chairs is to charge a lower price.

In Figure 4.1, I begin to put numbers to the example, enabling us to do a little bit of math. The math is annoying but useful, although if you don't want to do the math yourself just trust the results I provide. To capture the increasing marginal cost of supply, let's assume that producing the first chair, using entirely local resources, costs $51 and

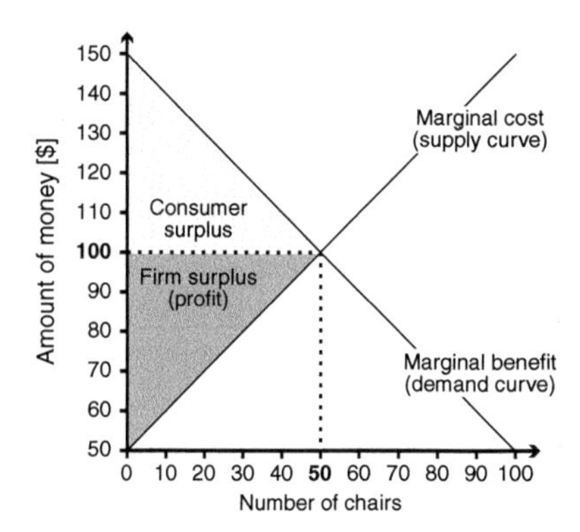

Figure 4.1. *Supply and demand.*

that this rises by $1 for each additional chair produced, as the fraction of imported resources has to increase. The fiftieth chair, then, will cost $100 to produce, and the one-hundredth chair will cost $150. Putting these points together generates a *supply curve*, the relationship between price and the quantity that firms are willing to produce.

On the demand side, capturing the marginal benefit people get from consuming the chairs, the graph portrays heterogeneity in the population. There is nobody who gets $150 of value from a chair, but one person who does get $149 of value, another who gets $148 of value, another who gets $147 of value, and so on. Economists call this value that each individual receives their *willingness to pay*. The fiftieth person has a willingness to pay of exactly $100, and the one-hundredth person a willingness to pay of $50. Putting these points together generates a *demand curve*, the relationship between price and the quantity that consumers will end up buying at that price. In other words, if the price of chairs is $100, there are fifty people who have a willingness to pay at least this high and who will buy a chair. If the price is $50, then there are one hundred purchasers.

The combination of a rising supply curve and a falling demand curve leads to a *market equilibrium*, the price at which the number of chairs supplied and demanded are the same. You can also describe this as *clearing the market*. In this case, the market would clear at fifty chairs, each one sold at a price of $100, as illustrated on Figure 4.1, where the two lines cross. If the example were different, it might be that no market develops. Imagine that the demand curve were the same but that the supply curve started from $250 and moved upward from there, instead of from $50. In that case, the equilibrium number of chairs produced would be zero. This is why we don't yet have much of a functioning market for trips into outer space. The cost of supplying the first unit of trips to outer space exceeds what just about any customer would be willing to pay.

Economics goes further to show that establishing this market for chairs, with the price and quantity at this market equilibrium, adds value to society. Let us start with the value that firms receive. The firms make a total profit equal to the difference between their total revenue and their total cost. The revenue is the product of the number of chairs sold and their price: fifty chairs time $100 makes a total of $5,000. The total costs are the sum of the marginal costs for each chair produced, and this adds up to $3,750. On Figure 4.1, this is the white triangle just below "firm profit," which has an area of $1,250, plus everything that would be below it if the vertical scale were to start at $0 rather than at $50. Hence, the total profit, what economists also call *firm surplus*, ends up being equal to the area of the darker gray triangle and is $1,250. The firms get to distribute this money to their shareholders.

Consumers in the community also receive something like profit, what economists call *consumer surplus*: each of them enjoys the difference between his or her willingness to pay for the particular chair and the price of the chair. The first person to buy a chair, for example, would have been willing to pay up to $149 for it, but only had to pay $100. The second would have been willing to pay up to $148, but also only paid $100. They receive $49 and $48 in consumer surplus, respectively. The total consumer surplus is equal to the area of the lighter gray triangle, and, because the example is symmetrical in terms of the shapes of the two triangles, that is also equal to $1,250. The consumers don't actually see this money in their bank accounts, but they experience it nevertheless. Indeed, as a group they would be indifferent between experiencing this amount of consumer surplus and receiving a check for the same amount in the absence of this market for chairs. The total benefit to the community and to the firm's shareholders, what economists call *total surplus*, is equal to the sum of producer and consumer surplus, in this case $2,500.

The real insight of classical economics, from the famous classical economist Adam Smith, was that this equilibrium happens to be the best possible arrangement for the community as a whole, with the highest possible total surplus. The "invisible hand" of the market means that the aggregate effect that individual people's behavior in the equilibrium state of a free market happens to be exactly the allocation of resources for which the benefits for society are the greatest.

Imagine the alternative. It is unlikely that the quantity of chairs sold would increase beyond fifty because the fifty-first chair, costing $101 to produce, would find no willing buyers at that price. It could be the case, however, that the number of chairs sold would be less than fifty, at a price higher than $100. If forty chairs were produced, for example, the market clearing price could be $110. In this case, firm surplus would be higher (for this example, $1,600 up from $1,250), but consumer surplus lower by a bit more ($800 down from $1,250), and the total surplus somewhat less than in the market equilibrium ($2,400, instead of $2,500). Indeed, this is what happens when one firm is a monopoly and can keep other firms from entering the market. The profit-seeking monopolist restricts supply and boosts its own profits; but, in doing so, society as a whole suffers. That is one reason why countries have laws against monopolies and force firms that do gain monopoly power to split up. Classical and neo-classical economics provide a way to understand why monopolies are bad.

The formation of a monopoly is one event that can lead to a disruption of the perfect market equilibrium, what economists call a *market failure*. In any market failure, the sum total of individuals' selfish behavior does not produce the outcome that is best for the community, and the government has to step in to set things straight. There are other causes of market failure, and one of these is the existence of pollution. It is time to build pollution into the example.

Imagine that the chair factories generate a lot of sawdust, such as 20 kilograms per chair produced. They could pay money to take the sawdust to a landfill – for the purposes of our example let's assume that costs $1.25 per kg – but they can also dump it into the river for free. Scientists recently discovered that the sawdust in the river kills fish, which people in the community like to catch and eat. Each chair produces 20 kg of sawdust, which in turn kills twenty fish. Just about everybody in the community likes to catch the fish to eat, and the surplus derived from each fish is $1. So, for every chair that is produced this is $20 in damage to the community, but not necessarily to the factory shareholders themselves. The market clears with fifty chairs, and so the damages add up to $1,000 in total. Economists call this an *externality*, because it is a real cost to the community, but one that is external to the balance sheet of the firms operating in the market.

Discovering that an externality is present can have a big impact on consumer surplus. Before we knew there was pollution, the consumer surplus in this market was $1,250. Discovering the source of the pollution leads us to revise that number downward. The community now experiences a net consumer surplus of $250, which is equal to the $1,250 consumer surplus from the chairs, minus the $1,000 from the dead fish. The profits of the firm, however, don't change.

Economic theory: policy approaches to externalities

Historically, when governments looked at this problem of externalities, they did one of three things. The first thing they typically did was nothing, which wound up being fine if you were the owner of a factory, but not so fine if you were a fish or someone in the community who liked to catch the fish.

The second thing they did, starting in the 1800s, was to solve the problem in the courts. If the people in the community could prove that the chair factory had killed the fish, they could sue the factory owners and collect their damages, in our example $1,000. The lawsuits weren't so good for the factory owners (the $1,000 would eat away substantially at their profits of $1,250), but it was good for the people in the community receiving the compensatory damages and also good for the lawyers representing the community, who would typically take a 30 percent cut of the damages.

The third thing that governments did, starting around 1970 in the United States with the Clean Water Act, was to command the factory owners to control their polluting, typically forcing them to take the sawdust to the dump. Economists call this a *command-and-control* solution. Taking the sawdust to the dump, at $1.25 per kg, 1,000 kg in total, would cost them $1,250. This would completely eat away their profits. Indeed, the firm might decide to simply close up shop, stop producing chairs, and lay off its employees. So this latter solution works out best for the fish, but terrible for the firms. It could even kill the whole market, which works out badly for the local community as well, both the people working at the chair factory and those who get value out of consuming the chairs.

Beginning around the 1980s, economists began to point out that there was a better way, based on economic theory. To illustrate this, they built on the graph of Figure 4.1 by adding the externality, as can be seen in Figure 4.2. The left-hand graph introduces a new marginal cost curve: the *marginal social cost*, representing the sum of the marginal firm cost and the $20 externality. In the absence of government intervention or with an *ex post* legal solution, the market equilibrium for chairs is likely to continue on as before, with fifty chairs sold at a price of $100, with the firm profits and consumer surplus also being the same.[6] But

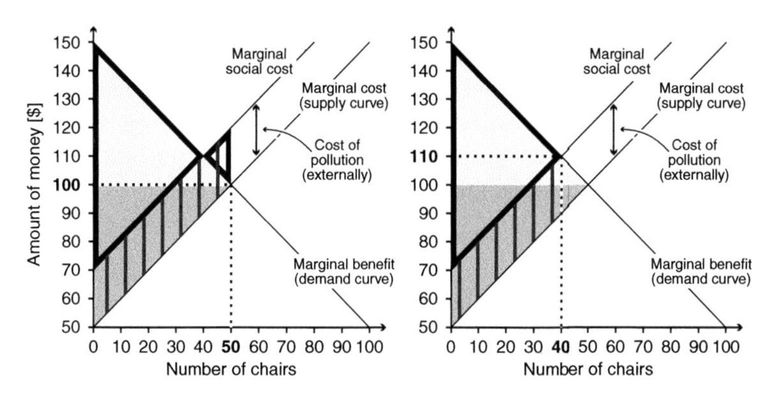

Figure 4.2. *Externalities and their solution.*

this will also kill twenty fish per chair, 1,000 in total, for a loss of value to the community of $1,000. The net surplus is now the original $2,500 (the sum of the two gray-shaded triangles, as before) minus the $1,000 from the dead fish (the parallelogram-shaped area with the vertical hash marks), for a total of $1,500. Graphically, this is also equal to the large fat-edged triangle on the left (which has light or dark shading but no hash marks), minus the area of the small fat-edged triangle on the right (which has hash marks but no shading).

Now, imagine that the government steps in not with a regulation calling for the elimination of the pollution but by changing the market so that the number of chairs sold, and the price, happens to be where the marginal social cost and marginal benefit curves cross. In this example, the regulators would restrict the market to forty chairs, and the market would clear at a price of $110. The effect can be seen in the right-hand graph. With the rise in price, firm profits (shaded darkly) now rise from $1,250 to $1,600, just as in the case where a monopoly restricted output to the same level.[7] Also just as with the monopoly, consumer surplus associated with the chair market falls from $1,250 to $800. So the total surplus from the chair market, as with the monopoly, falls from $2,500 to $2,400, a loss of $100. But the loss in value because of the dead fish now falls even more, from $1,000 to $800. Total surplus including the dead fish is now $1,600, up from $1,500 in the left-hand graph. Graphically, this is now the area of the single fat-edged triangle with shading but no hash marks. By reducing the quantity of chairs, the government made the small fat-edged triangle seen in the left-hand graph disappear from the picture. Economists call this amount, that small triangle, *deadweight loss* associated with the unregulated market in the presence of an externality. Dead weight is stuff that you carry around that doesn't do anything and only weighs you down; that is exactly what the extra level of production was doing to the economy as a whole. Restricting the size of the market to eliminate the deadweight loss, economists have consistently shown, guarantees the best possible outcome in terms of total surplus.

Economic theory: market instruments and their distributional consequences

There are three basic legal mechanisms or *policy instruments* by which the government could restrict the market, each with somewhat different consequences on particular stakeholders and over time. The first and perhaps most obvious way is simply to place a cap on the amount of sawdust that firms are allowed to dump into the river, limiting it to 800 kg. That is what I implicitly suggested in the last paragraph. If there had been four chair factories, each dumping 250 kg into the river, then, under the new rule, each would be entitled to dump 200 kg. This practice is known as *grandfathering*. Even better, the government could issue each firm 200 permits to dump a kg of sawdust and let the firms trade these among themselves. The advantage of this is that it would allow those firms with more flexibility to eliminate more pollution and those firms with less flexibility to do less. For some firms, it might be very expensive to eliminate pollution or scale down production, and so they could buy pollution permits from the firms for whom these changes are easier. Compared to a uniform cap without the possibility of trading, *cap-and-trade* would save the industry as a whole a good bit of money – good for their owners and the pension funds – compared to simply cap without trade. Unless heterogeneity is unexpectedly high, we would expect the price for permits on the secondary market to hover around $1 per kg of pollution.

You might think that the firm owners would complain about this government interference in the market, but you would be wrong. Indeed, this sort of cap-and-trade, grandfathering with permits initially handed out for free, creates the opportunity for a windfall profit. The mandated restriction in the supply of chairs would have the effect of raising the price, just as in the case of the monopoly, and firms' profits rise; in this case, from $1,250 to $1,600. Hurt by this solution would be the factory workers, some of whom would be laid off. Also hurt would be the consumers (i.e., the community as a whole). Given the numbers here, the consumer surplus from consumption of chairs ($800) would exactly match the cost of the dead fish (also $800), which they would still have to bear.[8] Those among them purchasing chairs would continue to do so – the market for chairs wouldn't collapse – but, as a whole, they wouldn't benefit from the market and would be just as happy if the government were to simply ban the pollution altogether.

The second way that the government could modify the market would be to force firms to buy the permits that they need for discharging the pollution. In practice, this would mean holding an auction, with 800 permits up for sale, and also allowing a secondary market in case firms decided that they had bought too many or too few permits. In this case, the price that firms will be willing to pay for a permit ought to equal $1: any more than this and the cost of producing the fortieth chair – $90 plus the cost of purchasing the needed twenty permits, one for each kg of pollution – would exceed the price that this chair will fetch at market, $110. Note that this permit price, as with the permit price on the secondary market in the grandfathering system, is exactly equal to the cost of the externality.[9]

The distributional outcome, and hence the politics of the auctioned permit, is very different from that of the grandfathering system. The same number of factory workers will still lose their jobs as the production of chairs decreases. But the firm profits will also decrease, especially relative to the grandfathering system but also relative to the case where the market was unregulated. The cost of production would now be equal to the marginal social cost, and the firm's profits would decline accordingly, in this example to $800. The consumers, meanwhile, would experience a benefit relative to the unregulated market. They would still suffer the cost of the dead fish, but their government would collect revenues from the auctioning of permits of an exactly equal amount. The government could then use these revenues to reduce other taxes that the consumers would otherwise have to pay, such as the income tax.[10] This would cancel out their loss of fish, and their consumer surplus would again match the size of the lightly shaded triangle, $800.[11]

The third way that the government could influence the market is to place a tax on the pollution, in this example $1 for each kilogram of sawdust dumped into the river. This would have the exact same effect as the auctioned permit system in terms of raising the cost of producing each chair by $20, thus internalizing the externality. Now, the market itself, rather than a cap, would lead to the outcome of reducing chair production from fifty chairs to forty chairs and the number of dead fish from 1,000 to 800. The revenue and distributional effects of the tax would also be exactly the same as the auctioned permit system. The firms and their workers will be unhappy, whereas the rest of the community will gain on average, relative to the unregulated market.

The numbers in the last few paragraphs were complicated, and so I summarize them in Table 4.1. There, I compare all of these effects for the example of the hypothetical chair market, showing each of the possible government responses in the various columns. Each row provides the value for a different indicator of the market's performance and also aligns with the interests of a different interest group.[12] The bold numbers indicate the best possible value to be found in each row and hence a possible preferred outcome for that group.

Consider the different perspectives on the problem. We might expect the chair factory workers to advocate for the policies that keep the market at fifty chairs, so that none of them is laid off. What works for them? The unregulated market or the lawsuit, and potentially the command-and-control solution, unless the command and control causes the factories to shut down. Environmental advocates, by contrast, might focus on the second row, to minimize the number of dead fish. They would advocate for the command-and-control solution. Consumers – the local residents – would care about the values in the third row – net consumer surplus – and would advocate for the lawsuit or the command and control, unless the command and control were to shut down the chair factories entirely. Firms and their shareholders would focus on profits, shown in the fourth row, and advocate for the grandfathered permits. According to economists, the "wise" policy maker ought to focus on the last row. By doing so, they should choose any of the three market solutions, choosing between the grandfathered permits, the auctioned permits, or the tax, all of which maximize total surplus in the presence of the pollution. Which of the three options they choose might be based on their preference for helping the firm owners (shareholders like pension funds) on the one hand (choose the grandfathers permits) or the consumers in the local community on the other (choose the auctioned permits or the tax). Interestingly, the two solutions that generate both the greatest total surplus and the greatest equity – the auctioned permits and the tax – would not be any of the other stakeholder's top choice. It's a sure recipe for political conflict.

By the way, according to economists and their wise policymaking sympathizers, the worst possible government policy in this example is command and control, even if the market for chairs survives. Recall that the firms might decide to close down, in which case

Table 4.1 *Comparing the effects of different policy instruments*

Interest group	Market performance indicator	Unregulated pollution	Lawsuit	Command and control	Grandfathered permits	Auctioned permits	Tax
Factory workers	Chairs produced	**50**	**50**	**50**	40	40	40
Environmental advocates	Dead fish	1,000	1,000	**0**	800	800	800
Consumers/ local residents	Net consumer surplus	$250	**$1,250**	**$1,250**	$0	$800	$800
Factory shareholders	Firm profits	$1,250	$250	$0	**$1,600**	$800	$800
"Wise" policy makers	Total surplus	$1,500	$1,500	$1,250	**$1,600**	**$1,600**	**$1,600**

the market would disappear, and all consumer and firm surplus values would decline to zero. Economists really do dislike command-and-control policies. Indeed, they invented the term "command and control" in order to give them a bad name. Political scientists call these policies by other, more innocuous names, like *technology standards*.[13] Political scientists often suggest that technology standards are a good option, although for very different reasons usually having to do with their ease of enforcement and predictability.

So far, I have just considered the benefits of the market instruments at a single point in time, the moment when they are introduced. But economists are quick to point out that they also have dynamic benefits in terms of being a stimulus for innovation. With either cap-and-trade or a tax, all firms have an incentive to figure out how to reduce some or all of their pollution at a price below the permit price or tax level (in our example, $1 per kg). There are lots of innovations that might be out there: they could invest in stronger but thinner saw blades, which generate less sawdust with every cut, or perhaps they could even figure out something to do with the sawdust instead of throwing it away, like pressing it into wood pellets for people to burn. A firm making any of these innovations would increase its profits by employing this new technology, rather than having to buy permits (or, in a grandfathering system, selling its excess permits) or pay a tax. Society as a whole would benefit as well. Suppose not just some but all of the pollution could be eliminated at an average cost below $1 per kg. In a cap-and-trade system, the government could reduce the cap to zero, and yet the firms would still increase production to something above forty chairs, thus benefiting consumers while killing no fish. In a tax system, the government could leave the tax at $1, but the firms wouldn't pay it; they would use the new technology instead, again increasing production to something above forty chairs and again killing no fish.

The possibility of innovation, however, does illustrate one important difference between the cap-and-trade system – grandfathered or auctioned – and the tax. Ideally, the permit price, just like the tax, should always be equal to the cost of the dead fish, $1. If there has been technological innovation that the government doesn't know about, and the government continues to grandfather or auction the full 800 permits, then the price of those permits will drop (in the extreme, to zero). Then the cap-and-trade system will have no effect on the market, which is less than ideal. Only by continually adjusting the size of the cap, reflecting the changing state of technological development, can the government keep the permit price at its optimal level. In the presence of changing technologies and costs, by contrast, the tax system will continue to achieve the optimal solution because a $1 tax stays a $1 tax, no matter what new technologies come along.[14] If there is substantial uncertainty about the costs of technologies, but it is clear that each kg of pollution kills one fish worth $1, then the tax is the more robust instrument. What will vary with the tax, however, as technologies change, is the total amount of revenue that the tax generates.

Historical experience: successful examples of market-based instruments

That's the theory. What about practice? The first major attempt to put a price on pollution started in the United States in 1990. That was the year that Congress strengthened the Clean Air Act to reduce sulfur dioxide pollution from coal plants to a level that was not only safe for people – as the existing legislation required – but would protect the health of ecosystems from acid rain. Acid rain was especially a problem along the east coast of the United States and up into eastern Canada, in part because the coal from the eastern side of the continent is very high in sulfur. To solve the problem, sulfur emissions from coal-fired power plants east of the Mississippi River needed to fall by about 50 percent. Congress designed a cap-and-trade system that would come into effect over several years, starting in 1995, through a complicated set of steps. Loosely speaking, it would require increasingly more power plants to acquire emissions permits even as the total number of permits declined. These were distributed according to a complicated system that combined auctioning and grandfathering, with a secondary market.

By almost all accounts, the cap-and-trade system to stop acid rain has proved to be a success. First, it achieved the desired level of emissions reduction on schedule, standing in stark contrast to the slow pace of implementation that the rest of the Clean Air Act had become well known for. Second, the cost to industry of reducing emissions eventually proved to be far less than analysts had previously estimated. Before the cap-and-trade system went into place, many people predicted that power plants would have to install flu gas scrubbers to remove the sulfur dioxide. Indeed, the alternative to the cap-and-trade system, the existing processes of the Clean Air Act requiring the installation of best available pollution control technology, would have been to require this. This would have also been the rational thing for plant owners to do because amortizing the installation cost of the scrubbers and adding it to their operating costs was the least expensive way to reduce emissions.

What happened instead was that many of the power plants decided not to make the high initial investment in scrubbers but instead to switch to low-sulfur coal from the western United States. Doing so was expensive because the coal needed be to shipped halfway across the continent and the railways at the time were highly regulated in a manner that restricted competition and pushed up freight transportation costs. Coincidentally with the cap-and-trade system, however, Congress was in the process of deregulating several economic sectors and industries, including the railroads. This led to greater competition and falling costs, exactly over the period of time when the sulfur trading system was coming into effect. Over time, as rail freight volume increased and the industry learned how to operate more efficiently, the costs of western coal declined, and it became by far the cheaper option for reducing emissions. In the end, the market mechanism for sulfur dioxide, combined with deregulation and innovation in the railway system, proved to be far less expensive than the initially proposed technology standard (or command-and-control policy) requiring scrubbers.

There is a stronger success story for market-based instruments, although it is one that is unfolding right now. It has to do with overfishing in the oceans. Over the past fifty years, the growing global fishing fleet has depleted the world's oceans of one species after another, such that, by now, most fisheries are in a state of being overexploited, with populations plummeting. The seriousness of the crisis began to emerge in the 1990s, and that is when policy makers started trying out new policies to halt the trend. There were many different options tried, including placing limits on the fishing season, on the size of fishing boats, or on the ability of foreign vessels to fish in coastal territorial waters. None of these has made much of a difference. What have ultimately proven effective, by contrast, are "rights-based" approaches. Under this system, scientists determine the maximum sustainable annual fish catch for a particular species and allocate this to fishing vessels using a market mechanism. Each vessel has the right to catch up to its allotted amount of fish and can decide when and how to do so. The vessel owners can also sell their rights to other fishing vessels in a system known as *ITQs* or *individual tradable quotas*. The results are encouraging.[15]

Overfishing does not involve the production of an externality in the form of pollution, but it can still be viewed as an externalities problem. When too many fish have been caught already, the overall productivity of the oceans declines, and the number of fish spawned each year and available to be caught falls. At that point, every fish that is caught further undermines the ocean's productive capacity; this is an external cost, one that the owner of the fishing vessel catching that fish does not bear on his or her own balance sheets. So, in this respect, the fisheries problem does not differ so much from the problem of climate change.

But in two other ways it does. The first has to do with innovation. The main effect of innovation in the case of fisheries has been to make the problem worse: larger and faster boats are able to travel farther to find fish and to cast their nets deeper, ultimately turning what had been isolated cases of overfishing close to shore into a global problem facing all the world's oceans. For now, there is no beneficial technology that policy can promote.[16] That is different from pollution, such as carbon emissions, where, through innovation, it can be possible to create the thing of value with less of the polluting externality. The second difference is the size of the cap. The maximum sustainable fish catch for any given species is some number that is greater than zero and that will likely grow in time as the fish stock recovers. Fishing fleet owners have an incentive to support a stringent cap because by doing so they can guarantee their economic future, and so their interests are aligned with those of society more generally. With carbon emissions, things are the other way around. The optimal size cap may be above zero now, but not in the long run. Industry might support a cap if, through grandfathering, it temporarily boosts profits. But one can expect industry to resist a cap at the level that is ultimately needed – zero – which will eventually put them out of business.

Historical experience: market-based climate policies in the United States

Back to air pollution. With the Clean Air Act success in hand, economists turned to the next major environmental problem as a place to use market instruments: climate change. Here, the results have so far been less rosy. We will start by looking at the United States, where there have been two failed political attempts at the federal level to put a price on carbon, and also at several state programs. Then, we will move on to Europe, where carbon markets have figured more prominently but arguably been no more successful.

The first attempt in the United States came from President Bill Clinton in 1993. He proposed a large and broad-based energy tax, justifying it on three grounds: it would reduce America's dependence on imported oil, it would reduce CO_2 emissions and address climate change, and it would be an important revenue source to reduce the federal budget deficit. Clinton's tax proposal initially had widespread support among economists, environmental advocates, and even some energy corporations, and there was a great deal of optimism that it would pass.

But that is when political negotiations began. The first compromise was to base the tax on the energy content of fuels – measured in British Thermal Units (BTUs) – rather than on their carbon content. This request came from Robert Byrd, a senior senator for West Virginia, a coal mining state. Clinton agreed, and the tax became known as the BTU Tax. But then more requests started coming in, from more senators, to exempt certain sources or uses of energy that were economically important to particular states. Clinton kept agreeing, wanting to keep the votes in Congress and pass the legislation. But, as the number of exemptions increased, so too did the opposition to the tax from those industries that would still be negatively affected by it. Based on their lobbying efforts, Senators who had once supported the tax now opposed it. Ultimately, the only part of the tax that became law was a 4.3 cents per gallon tax on gasoline. Everybody agreed it would do little to curb oil consumption, little to stop climate change, and little to generate government revenue.[17] It was a symbolic victory for Clinton at best.

The second attempt at a market mechanism for CO_2 in the United States started a decade later, during the presidency of George W. Bush, and slowly gained momentum in Congress. The bill to establish a comprehensive cap-and-trade system passed the House of Representatives under the name Waxman-Markey, its two proponents. Introduced in the Senate by Republican John McCain and Democrat Joe Lieberman, support for the bill grew over the years. When Barack Obama became president in 2009, passing the bill became a real possibility, and he threw his support behind it. Indeed, it fell onto his agenda right behind healthcare reform and economic recovery, the latter of which included a fiscal stimulus combined with reforming the financial industry. All three of these proved to be prolonged legislative battles, increasingly polarizing Congress and, in particular, the Senate.

In October 2009, 60 percent of Americans supported passage of a cap-and-trade bill in Congress.[18] Nevertheless, opposition to cap-and-trade began to grow, much as it had to the tax under Clinton. Although the bill passed the House of Representatives, it got stuck in the Senate, where a 60 percent supermajority proved essential to passage. This didn't happen, and, by the summer of 2010, it was clear that it would not happen; Senate majority leader Harry Reid pulled the cap-and-trade bill from the Senate agenda. Following Republican gains in the 2010 congressional elections, it became clear that the bill was effectively dead.

One could interpret this story to say that cap-and-trade, like new taxes, is a political nonstarter in the United States. One could also interpret it to say that the United States Senate has become dysfunctional, unable to pass cap-and-trade even though close to 60 percent of its members supported it. At the state level, where politics may be less dysfunctional (or at least differently dysfunctional), cap-and-trade has passed in a number of jurisdictions, most recently California. The best known of these efforts is the Regional Greenhouse Gas Initiative (RGGI, pronounced Reggie).

RGGI is a partnership among the six New England states – Maine, New Hampshire, Vermont, Massachusetts, Rhode Island, and Connecticut – as well as New York, Delaware, and Maryland. Initially, New Jersey was also on board, but pulled out in 2011. Collectively, the states have the goal of reducing CO_2 emissions from fossil fuel-fired power plants by 10 percent by 2018. Each state has grandfathered a number of permits, or *allowances*, based on historical emissions from its power sector and then quarterly auctions 90 percent of those allowances off to the power generators within its borders, grandfathering the remainder. Each allowance covers one ton of CO_2 emissions. The revenues from the auctions are intended to be used to support energy conservation and renewable energy programs, although a number of the states have used the revenues to balance their budgets, thus reducing the need to raise other taxes to do so.[19] A secondary market allows power companies within the region to purchase allowances from anywhere within the region. The secondary market also allows firms to buy and sell carbon market derivatives, such as futures, forwards, and options. The auctioning of RGGI permits began in September 2008, and the market price for allowances started off at $5. Since then, the price has steadily declined. By September 2009, it had fallen to below $2.50.[20] Another year later, the price had declined to $1.90,[21] and it has since remained relatively stable, between $1.87 and $1.89.[22]

It is useful to translate RGGI's carbon price into the additional cost for a unit of energy that people consume using conversion tables that take into account the amount of CO_2 emitted in the process of energy conversion.[23] Doing so, one calculates that a CO_2 allowance price of $1.90 translates into an additional cost of about 0.1 cents per kilowatt hour (kWh) of electricity. For comparison, over the past ten years, the residential price for electricity in New York State has fluctuated between 13 and 20 cents per kWh and, according to the most recent data, was 17.8 cents.[24] Let's put that another way. If the RGGI carbon price were to be added not to electricity but to the price of gasoline, it would add just one-twentieth of a cent per liter, or roughly four times this – less than one-fifth of a cent – per gallon. So I think it is safe to say that the RGGI permit price is trivial.[25] Even the total amount of revenue raised, $952 million dollars over the life of the program,[26] four years so far, sounds impressive, but this translates to roughly $5 per person per year in the participating states. That is tiny, although researchers have shown a net positive economic benefit associated with this revenue, mainly associated with how the participating states spent the money to finance new investments.[27]

Why has the RGGI price fallen so low? The reason has been simply that the total number of allowances being auctioned exceeds the number that firms actually need.[28] In other words, whereas the RGGI will have restricted the emissions from power plants by 10 percent, technological progress alone would have done this anyway. There is no evidence that it was the RGGI itself that caused this technological change.[29]

Historical experience: the European Emissions Trading System

Let's move on to Europe, where there have been both carbon taxes introduced at the national level and a continent-wide cap-and-trade system, the European Emissions Trading System (ETS). Let's start with the ETS, which went into effect in 2005. As with the sulfur dioxide trading system and RGGI, the ETS places a cap on total emissions and has distributed permits through an evolving but complicated mix of auction and free distribution. Like RGGI, the ETS has not been a shining success. Indeed, because the ambition of the ETS has been higher – to reduce power plant and other industrial emissions by roughly 10 percent by 2012, and 30 percent by 2020 – there has been even greater room for unintended consequences. Arguably, its main effect so far has been to land power companies with windfall profits due to the initial grandfathering rules, much of which was reinvested in new fossil fuel infrastructure.

How did this happen? The first phase of the ETS covered the period 2005–2007 and was seen as a trial run before the Kyoto Protocol, which (as I will describe in detail in the next chapter) began counting emissions in 2008. The ETS required about 12,000 industrial installations to participate, together accounting for 40 percent of European CO_2 emissions. Because it was a trial run, the European Union (EU) gave each national government almost all of the authority to determine how many permits to make available for emitters within their borders and to set their own rules for allocation. Under pressure from industry, most governments chose to grandfather the permits, giving them away for free while enabling the companies to sell any excess permits to other companies anywhere in Europe. Moreover, industry lobbied national governments to issue more permits, rather than fewer, than there already were emissions. Most national governments listened because by giving out more permits to their own industries, they could make sure that their country would sell more permits to other countries' firms and make substantial profits from doing so. Every country had an incentive to issue too many permits, although if enough countries did this, the permits themselves would become worthless. Which is what happened. As with RGGI, this wasn't obvious at first, and many firms hedged for the future by buying extra permits, such that, by 2006, the permit price reached $40 per ton of CO_2. But then people started noticing the overabundance of permits, and the price began to drop. By mid-2007, it had fallen to close to zero.[30] During the first phase of the ETS, total emissions rose by 1.9 percent.[31]

While the permit price was high, however, it affected firms' profits, just as theory would predict. The grandfathered permits pushed up the wholesale cost of electricity to the detriment of consumers and to the benefit of firms, providing them with windfall profits. For many of these firms, it made more sense to self-finance new infrastructure projects rather than distribute the excess profits as dividends or keep them as cash on hand.

A PhD student named Michael Pahle, working at the Potsdam Institute for Climate Impact Research (known as PIK), studied this phenomenon.[32] He was interested in why it was that, in Germany, a country that at first glance appears exceptionally committed environmentally, there appeared to be a construction boom for coal-fired power plants. In 2009, when he studied the problem, there were ten new plants with a total capacity of 11.3 Gigawatts (GW) in construction and another group of plants with a combined capacity of nearly 20 GW in the planning stage.[33] Altogether, the new plants would generate electricity equal to 40 percent of the country's peak demand.[34] What Pahle found was that there was a large anticipated demand for new base-load power plants, a result of the anticipated decommissioning of the country's fleet of nuclear power plants. The power companies had obtained billions of euros in windfall profits from the ETS, which allowed them to self-finance a large share of the coal plants and pushed down their total cost substantially. Given this self-financing, coal remained the most attractive technology as long as ETS permit prices remained below $40 per ton of CO_2. In fact, firms did not anticipate the ETS price rising above $40 during the new plants' lifetimes. First, the ETS regulators in Brussels had sent clear signals that they desired the permit price to go that high, but no higher. Second, they believed that carbon capture and storage (CCS) would become economically viable at a carbon price of $40, and this would stop the permit price from rising any higher: if the permit price were to rise above $40, firms would simply install and operate CCS instead, making the higher price untenable.[35] At prices of $40 or less, coal was still the more attractive technology than gas, and so that is what they planned on building. The early ETS, then, was a major factor in Germany's "dash for coal."[36]

For the second phase of the ETS, things went better, mainly because the European Commission recognized the perverse incentives and took more responsibility away from national governments for allocating both permit quantities and the means of distributing them. Countries still proposed their intentions for both, but the bureaucrats in Brussels modified many of them to bring the total number of permit allocations down, and they required a greater share of auctioned, rather than grandfathered, permits. It still appears that too many permits were allocated: the price today remains above $10 per ton because firms will have the opportunity to save any extra permits and use them in a later phase. They are betting, most likely, that the next phase will be more restrictive than the last.

An ETS price of $10 is certainly higher than the RGGI price of $1.90, but only slightly less unimportant. It is possible to detect a contribution that the ETS made to innovation in energy transformation and use, although the size of the contribution is very small compared to the effects of other factors happening simultaneously.[37] There are signs that the ETS is improving as policy makers learn from their past mistakes. In its energy strategy, the European Commission has declared the intent to further limit the supply of permits so that the price will rise to $40 by 2020 and something like $70 by mid-century. These goals reflect a tremendous amount of political negotiation. Whether the European Commission has the ability to foresee the potential loopholes that can lead permit prices to tumble as they have in the past remains to be seen. But, even if they are successful, permit prices in the range of $40–70 will have little effect, as I will soon show.

Historical experience: carbon taxes in Europe and Canada

Finally, let's consider the experience with carbon taxes. Five European countries – Denmark, Finland, the Netherlands, Norway, and Switzerland – have modest carbon taxes in place, set in the range of $10–30 per ton of CO_2. One country – Sweden – has a much more substantial tax in place, which now stands at close to $130 per ton. All of the taxes are complicated, including or exempting certain economic sectors or users of energy, and many of them place an additional tax on transportation fuels such as gasoline and diesel fuel. The Swedish tax covers the use of fossil fuels in transportation, space heating, and power generation, although it excludes power generation that also provides space heating (known as *combined heat and power* [CHP]).

Many people believe that these taxes have made a large difference, particularly in Sweden. As Professor Thomas Johansson from the University of Lund has explained, the tax "increased the use of bioenergy [and] had a major impact in particular on heating. Every city in Sweden uses district heating. Before, coal or oil were used for district heating. Now biomass is used, usually waste from forests and forest industries."[38] The Swedish environment minister, Andreas Carlgren, in the same newspaper article, claimed that Sweden's carbon emissions would be 20 percent higher if it were not for the tax.[39] Researchers at Harvard University made a similar but more modest claim, namely, that the northern European countries "have demonstrated that carbon taxes can deliver greenhouse gas emission reductions and raise revenues to finance government spending and lower income tax rates."[40]

But, behind these opinions it is hard to draw a concrete causal link. District heating, which Johansson cites in particular, makes economic sense with the tax, but it would also make economic sense without the tax. The tax may have helped to create the additional stimulus to push district heating onto the agenda of cities and towns that had not yet considered it. It may have worked because the political landscape needed just a little bit of a push to make a change.

These are all qualitative insights. The only quantitative study to be found has been far more guarded in its conclusions than either Johansson or Carlgren. Two Chinese researchers analyzed the effects of the carbon tax in Denmark, Finland, the Netherlands, Norway, and Sweden, using the well-established "difference in difference" statistical method.[41] That is, they compared the five countries with the tax with another group of countries, a control group, which had not put a carbon tax into place.[42] Correcting the data to take into account other predictors of carbon emissions, such as economic growth and investments in research and development (R&D), they compared how CO_2 emissions had changed in those countries that put carbon taxes into place with how the CO_2 emissions had changed in the control group countries. What they found was surprising. In only one country – Finland – did the researchers find an effect of the tax that was statistically significant at the 10 percent confidence level. There, the tax appears to have decreased CO_2 emissions growth by 1.69 percent, compared to what that emissions growth would have been without the tax. In three countries, the tax also may have caused a decline in emissions, by 0.42 percent in Denmark, 0.52 percent in the Netherlands, and 1.16 percent in Sweden, but the results appeared as statistically insignificant. What appeared to lead to a larger effect in Finland than in the other countries had nothing to do with the size of the tax, but rather with its degree of inclusion: Finland exempted fewer industries and fuel users from the tax than did the other countries. Also statistically insignificant was the result from Norway, where the tax appeared to have *increased* CO_2 emissions by a tiny amount, 0.12 percent. So, rather than the effect of 20 percent for Sweden, as Carlgren had claimed, these results suggest an effect of less than 2 percent, if there has been any effect at all.[43] A single scientific study generally does not settle a matter like this, so perhaps one should hesitate to put too much faith in its results, even if there doesn't appear to be anything wrong with it.[44] Still, it calls into question the claims that the effects of the tax have been substantial, claims that have less methodologically sound empirical support.

One other political jurisdiction – British Columbia – has instituted a carbon tax, the centerpiece of its policy to reduce CO_2 emissions by 33 percent by 2020. The tax rate was comparable to the European cases, starting at $10 per ton of CO_2, then rising to $30 by 2012 and staying at that rate. In its first two years, the tax raised $846 million in revenue, and this allowed the provincial government to reduce other taxes accordingly.[45] Based on theory-driven economic models (as with the European countries), the BC government expects the tax to reduce the provinces' CO_2 emissions by 3 million tons annually by 2020.[46] Given that existing emissions from the BC energy sector are roughly 54 million tons,[47] the tax itself would lead to about 6 percent of the total 33 percent emissions reductions that the government seeks. There does not appear to have yet been an empirical analysis of its effects, however.

Historical experience: conclusion

So, let's sum up what has been observed of market instruments in practice. There was good reason to be optimistic about what a cap-and-trade system could do based on experience in the United States with cap-and-trade to control sulfur dioxide emissions. In that example, industry was able to meet a stringent cap at a cost much lower than anticipated because of the opportunity to switch to lower sulfur coal shipped in by rail, rather than investing in costly new capital equipment to reduce emissions. The two main examples of cap-and-trade in practice for carbon emissions show results that are less promising. So far, neither has managed to set a cap that was ambitious enough to maintain a carbon price high enough to do much of anything with respect to reducing emissions. In both cases, the main effects of the policy was in terms of generating revenue for government in the case of the RGGI and the ETS, and, in the case of the ETS, for those firms that benefited from grandfathered permits. Cap-and-trade has failed politically at the U.S. federal level (i.e., in Washington).

Taxes have also largely failed in Washington DC, but have succeeded elsewhere: in Canada and in Europe. In one country, Sweden, the tax is high enough to suggest that it would make a major difference in influencing investment behavior, and there are many people who believe that it is having this effect. The empirical evidence, however, is very thin. The one methodologically robust study looking at the effects of European countries' carbon taxes has found them to be very modest, reducing emissions by less than 2 percent, and only in the case of Finland is this statistically significant. In all other countries, it was not possible to distinguish the effects of the tax from other factors that might be driving emissions.

Economic theory suggests that market instruments such as carbon pricing are the way to go, but the results from practice suggest that they have had very little effect. The reasons is that, in nearly all cases (with Sweden being the one contrary example), they have not been ambitious enough. The RGGI states picked a target – reducing emissions by 10 percent – that was really just what would have happened anyway. As a result, the number of permits available for auction was more than industry actually needed or wanted to buy, and so their market price fell to close to zero. The ETS, so far, has not been that much different, even though Europe has had a more ambitious target for greenhouse gas reduction than do the RGGI states. But the European countries have also been doing a lot more other things, like promoting efficiency improvements in households and industry or renewable energy in the electric power industry, that have also brought down emissions from sectors covered by the ETS. So here, as well, the ETS price has been very low, and the main effects have been the indirect ones from the revenue. Unfortunately, with the ETS, the revenue effects themselves have been mixed, in some cases leading to increased investment in long-lasting fossil fuel infrastructure.

Leaving Sweden aside for a moment, the taxes have been in the range of $10 to $30 per ton. That sounds like something meaningful in the abstract, but it sounds like a lot less when put in practical terms. If we convert the carbon tax into how much it would add to the cost of gasoline, a $10 tax would add just over 2 American cents per liter, or about 9 cents per gallon, whereas a $30 tax would add 7 cents per liter, or 26 cents per gallon. This is simply very small when put into the context of background price volatility. At the time of writing, for example, the average price of gasoline in the United States is $3.48 per gallon, whereas over the past 6 months – six months in which rising and falling gas prices have not been out of the ordinary and have not made headline news – it has varied between $3.34 and $3.92, more than twice the magnitude of difference that would come from a $30 per ton CO_2 tax.[48] Over the past several years, the price fluctuation has been much larger and at times more rapid, between $1.61 and $4.12 per gallon,[49] which is comparable to varying a CO_2 tax by $285 per ton. That did make headline news, and the especially rapid rise in 2007 led to a change in driving and car buying habits in the United States, although less so in Europe where the relative change in fuel prices was less dramatic. A tax of $285 per ton of CO_2 would be ambitious: $10 or even $30 per ton is not.

The lack of ambition with market-based instruments reflects a political reality: carbon taxes and markets create very clearly defined direct economic effects, almost all of which are negative, whereas the positive effects are a lot harder to pin down.

The way that markets operate, a price on CO_2 – whether from a tax or a cap-and-trade system – will translate directly and immediately into higher energy prices. As climate policy expert Roger Pielke Jr. suggests, this runs up against the "Golden Rule" of climate policy: lots of people support doing something about climate change, but almost nobody supports doing something that will cost them anything substantial in the short term.[50] Carbon taxes and auctioned permits also create another class of immediate economic losers, and these are the firms operating in the fossil energy sector whose profits will fall. Grandfathered permits, by contrast, turn some of these firms into winners, which may explain why the ETS was at all able to make it through politically. But grandfathered permits, every bit as much as taxes or auctioned permits, push up energy prices, making the firms in energy-intensive industries less competitive compared to those in countries where energy prices are lower.

Sweden is the one country that has instituted a tax that is high enough to make a major difference, at least in theory. I can only speculate that it has something to do with the country's already having the second highest taxes in the world, combined with some of the highest provision of social services financed by that money. Apparently, a lot of Swedes are comfortable with high taxes and see them as improving their quality of life,[51] a quality of life that happens to be ranked among the highest in the world. But the high carbon price in Sweden appears to make very little difference in emissions levels, and this appears to have to do with politics there as well: the carbon tax contained a number of exemptions, meaning that many industrial sectors did not have to pay in.[52]

Examining potential effects: the efficient allocation of resources

Given the lack of ambition so far, an important question is whether it is reasonable to pin one's hopes on an adequate level of ambition in the future. To think clearly about this question, I break apart the issue of what an adequate level of ambition actually is, which in turn depends on the specific problems that we want climate policy to solve. The overall argument for market-based instruments in general is that they allow society to achieve an efficient level of pollution, and this is where I start. From there, I move on to other problems that we might want a market-based instrument to solve.

The economic theory that I presented earlier suggests that one goal of policy should be to achieve an economically efficient level of pollution, an efficient allocation of society's scarce resources. This is the level at which the sum of firm profits and consumer surplus is maximized, where the marginal social cost curve crosses the demand curve. Policy makers can achieve this by putting a pollution tax in place that is equal to the marginal social cost of pollution; when added to the marginal cost of production, it will result in consumers having to pay the full marginal social cost of the good in question, which, in the case of climate change, is energy. With such a tax in place, then whatever quantity of energy that people actually consume must be, ipso facto, the efficient amount. Alternatively, policy makers can institute an emissions cap with a market for tradable permits. If the cap is set at the efficient level, then the market price for the permits will come to match the marginal social cost of pollution. This then defines the appropriate level of ambition for carbon taxes and markets. In the case of a tax, then the appropriate level of ambition is to set the tax equal to the marginal social cost of carbon. In the case of a tradable permit system, the appropriate level of ambition is to set the cap at whatever level it is that leads to a market price for permits that is equal to the marginal social cost of carbon.

What is the marginal social cost of carbon? A lot of people have tried to assess this, and their estimates vary widely.[53] The variance in part depends on the difficulty of predicting what climate impacts actually will result from a ton of carbon put into the air (things like floods and droughts), as well as the economic costs associated with those impacts. The difference in cost estimates also depends on how much you discount future costs: the estimate will be a lot lower if you assume a 5 percent discount rate than if you assume a 0 percent or 1 percent discount rate. Finally, the difference depends on your attitude toward risk. There is a lot of uncertainty about future climate impacts and how much they will cost. If you are risk neutral, then your estimate for the social cost of carbon will simply be the average of all those possible future costs. If you are risk averse, then your estimate for the social cost of carbon will be greater than the average, placing a disproportionate weight on the more serious, and more expensive, possible outcomes.

Robert Mendelsohn is a professor at Yale University, and his estimate for the social cost of carbon is on the low side of the spectrum at about $1.35 per ton.[54] To reach that number, he has estimated the range of possible climate impacts, estimated their costs to society in a manner that implicitly assumes "smart" farmers and other decision makers (i.e., successful adaptation), used a pretty standard interest rate, and assumed something close to risk neutrality. This estimate is low compared to others, but not by as much as you might think. Richard Tol, a professor at the University of Sussex, conducted a thorough survey of the literature. He found estimates ranging from below $0 (which means that climate change will benefit us) to above $500. Within that range, most of the estimates were closer to $0, and Tol observed a mean value of around $5.[55] Martin Weitzman is a professor at Harvard University, and he estimated a social cost of carbon that is infinite.[56] To get to that estimate, he noted that the upper bound of the possible economic losses from climate change is effectively limitless because, with some probability that is greater than zero, it could result in the complete destruction of humanity. If you take an average of a group of numbers, and any of those numbers is infinity, then the average will be infinity as well.

What do these number imply about an appropriate level of ambition? If we believe that the estimates at the lower end of the range are correct, then that would suggest that policy makers have already achieved something close to the appropriate level of ambition for market-based instruments. The RGGI price of carbon, $1.90 per ton, already exceeds the Mendelsohn estimate. The ETS price, as well as the carbon taxes in Northern Europe and Canada, do as well, by an even greater margin. None of these instruments is having much of an effect, but then that must be the efficient thing. With a social cost of carbon down around $5 per ton, having much of an effect would be a waste of resources. Put another way, it just doesn't make economic sense to stop climate change, at least not now.

If you believe the higher estimates of the social cost of carbon, say up to the $500 per ton estimate that Tol notes, then you would reach a very different conclusion: the appropriate level of ambition is far higher. If the Weitzman line of argument convinces you, by contrast, then the whole question is nonsense. A social cost of carbon that is infinite implies an efficient carbon price that is infinite as well, or at least beyond what anybody on the planet could afford. This is the same thing as a complete prohibition on CO_2 emissions, of not being able to buy the right to emit CO_2 at any price.

Essentially, this is what the 2°C target represents. The 2°C target corresponds to what will happen if, all around the planet, people reduce

CO_2 emissions about as fast as politicians consider to be technically, politically, and socially possible – down to a level of zero – and then never emit any CO_2 ever again. I have already argued that the 2°C target is a reasonable one, and so essentially I am agreeing with the Weitzman reasoning. Climate change could cause catastrophic damages, and so, for ethical reasons, we ought to stop it as quickly as we can. Economics does not offer a useful set of insights into the optimal level of policy ambition.

But even if we have decided on a climate policy target for non-economic reasons, it still makes sense to get there in a cost-effective manner. And for this it is worth considering the things we need to change and how well market-based instruments can influence them. I will look at three things: consuming less energy, investing in market-ready clean technologies, and developing new clean technologies for the future.

Examining potential effects: reducing energy use

At first glance, market-based instruments would seem to be a good way to get people to use less energy. This is what we saw in the theory illustrating example of the chair factory: the price rose, the quantity demanded fell, and the pollution dropped to its optimal level. In that example, as it was set up, the optimal level of pollution was 20 percent below its equilibrium amount and achieved through a 20 percent contraction of the market. In the case of climate change, almost all of the energy that we consume is currently from fossil fuels, and the optimal decline in fossil fuel burning – or at least the decline that is consistent with the 2°C target – is 100 percent. Of course, a 100 percent contraction in the energy market is not something that policy makers would ever suggest, but they will suggest that people can become substantially more efficient in their use of energy. And, indeed, tremendous gains in efficiency are possible and have been well documented. The problem is that those efficiency gains are not very responsive to the price of energy.

In economics, *price elasticity* refers to the amount by which the quantity of a good consumed changes in response to a change in price. Elasticities are in almost all cases expressed as negative numbers, capturing the inverse relationship of rising prices for a particular good leading to falling consumption for that good. An elasticity of −1 means that proportions are exactly inverse: if prices were to go up by 1 percent, the quantity consumed would fall by 1 percent. Goods that are known as elastic have ratios more negative than −1, whereas those known as inelastic have ratios between −1 and 0. Elastic goods are to be

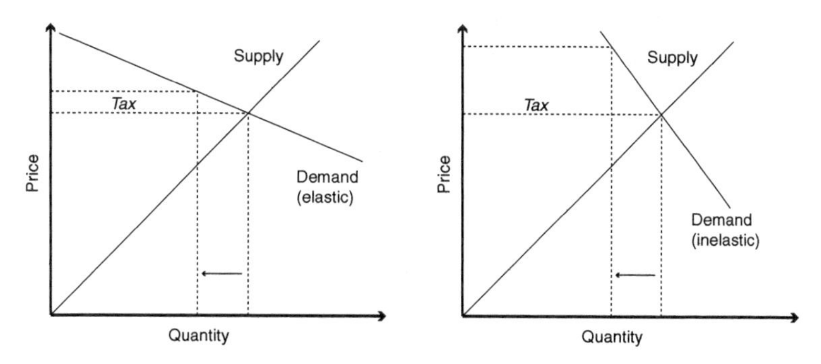

Figure 4.3. *Elasticity and the effect of taxes.*

expected where there exists another good that is a ready substitute. For example, researchers have calculated that the demand for a particular brand of automobile has an elasticity of −2.8.[57] If a Ford becomes 1 percent more expensive, demand will fall by 2.8 percent because lots of consumers will decide to buy Toyotas or Chevrolets instead. Energy, by contrast, appears to be quite inelastic, precisely because of the lack of an easy substitute.[58] Reducing energy use involves more of a lifestyle change than, for example, switching from a Ford to a Toyota, and so the process is stickier.

Figure 4.3 illustrates the effect of price elasticity on the use of a price instrument like a carbon tax to affect consumption behavior. The supply curve is the same in both graphs, but, in the left-hand graph, the demand curve is flatter, with an elasticity of about −2. In the right-hand graph, demand is inelastic, about −0.5. Moving from the left graph to the right, from elastic demand to inelastic, it takes a much larger tax to achieve the same drop in the quantity consumed.

Elasticity is something you have to measure because it depends on the fickle tastes of consumers. Lots of people have measured it, examining how the quantity consumed changes in response to a change in price. In 2012, a team of researchers from California and the Czech Republic published a meta-analysis of the many empirical estimates of the price elasticity of gasoline.[59] They looked at forty-one separate studies that had been undertaken between 1978 and 2010, and which collectively had 202 estimates of price elasticity using different datasets. They found that, on average, the short-run price elasticity of gasoline was −0.09, whereas the long-run price elasticity was −0.31. That the short-run elasticity is closer to 0 than the long-run elasticity is not surprising. When prices jump in the short run, the only thing that consumers can do is to drive less.[60] If prices rise and stay high, they can also drive less, but they can, when the time comes, buy a more fuel-efficient car or even move to a home closer to the other places they need to go, such as work or shopping.

But, in either case, the numbers reveal the demand for gasoline to be very inelastic. Let's put the numbers into perspective.

Imagine that the price of gasoline is $3.50 per gallon in the United States, and a short-term price shock occurs that pushes it up by 36 cents, to $3.86. This is the amount that would correspond to a CO_2 tax of $40 per ton, which is about the most we can hope to see, outside of Sweden. It also happens to be about a 10 percent increase in the price of gasoline, which is a big deal for the overall economy, something that could push it into recession. With a short-term elasticity of −0.09, demand would fall by just about 1 percent. A person who typically drives 100 miles on a typical day would instead drive 99 miles. Of course, a tax is not a short-term event; it would stay in place for some time, and consumers would gradually and increasingly adjust to it, perhaps buying a more fuel-efficient car the next time around. With a long-term price elasticity of −0.31, demand would fall by 3 percent. To get demand to fall by 10 percent, the price would have to rise by $1.40, from $3.50 to $4.90. That corresponds to a carbon tax of $157 per ton, somewhat more than what the Swedes were able to put into place and much greater than the Swedish tax in relative terms. If the tax were high enough to make gasoline as expensive in the United States as it is in Europe, where it costs roughly $7.00 per gallon (€1.40 per liter), then one could expect long-run demand to fall by about 20 percent. In fact, the average European uses about one-half as much gasoline as the average American; this reflects factors even longer term than those captured by long-run price elasticity, such as urban planning and a culture of using public transportation, as well as the fact that Europeans on average are not quite as wealthy and for that reason alone tend to drive smaller cars.

Other studies have run similar analyses of short- and long-run price elasticities for other energy carriers, such as electricity, and found similar results.[61] Given an energy market dominated by fossil fuels, and carbon price levels as high as most politicians are discussing – $40 per ton of CO_2 – we should not expect a reduction in energy use of more than about 3 percent. To get even that, the market would have to be very inclusive, without the various exemptions seen in the Northern European countries. Even with carbon prices much higher, such as the $130 seen in Sweden, we shouldn't expect to see drops in energy use of more than about 10 percent.[62] Ten percent is something, but if the target is to reduce CO_2 emissions from energy by a lot more, like the 80 percent envisioned for 2050, then it is essential that the larger reduction in emissions will need to come from a switch in energy supply from fossil fuels to something else, like renewables.

As Amory Lovins documents, end-use efficiency gains well over 10 percent, often more than 50 percent, are possible across the economy and already make sense from an economic perspective.[63] People are adopting them, but they are doing so slowly, and the reason for the slow pace is not the economics of the situation. Rather, it is the institutional and organizational dynamics that lie behind the making of decisions and the flow of finance. Companies fail to invest in energy efficiency, even when those investments could yield 25–30 percent returns on investment, simply because these kinds of investments are not in their ordinary business practices, and they lack the analytical skills to substantiate the returns and the risks. Informed consumers often make choices toward more efficient products that save them money over time, but typically only when doing so does not involve a large up-front investment or a change in behavior. Carbon prices will, in almost all cases, increase even more the long-term financial incentive to adopt new technologies and practices that already make good economic sense. But carbon prices – unless they are so high as to grab people by their pocketbooks and wallets, as the increase in American gasoline prices in 2007 did – will not overcome the primary barriers to improved energy efficiency.

Examining potential effects: investment in clean energy

The fact that reductions in energy demand driven by carbon prices will be small illustrates the importance of a second thing we want market instruments to accomplish, which is to give energy suppliers the incentive to shift their portfolios toward an ever-increasing mix of carbon-neutral energy sources. Even if people were to make very little change in their driving behavior, it would make a big difference if they were filling their tank up with fuel derived from ocean algae.

All things being equal, relative prices do matter a great deal: in the context of any market, investors will be more likely to choose the low-carbon or carbon-neutral technology option if it costs less than the fossil fuel alternative. Research has shown that it doesn't matter very much how much less.[64] It just has to be lower. There needs to be some economic justification.

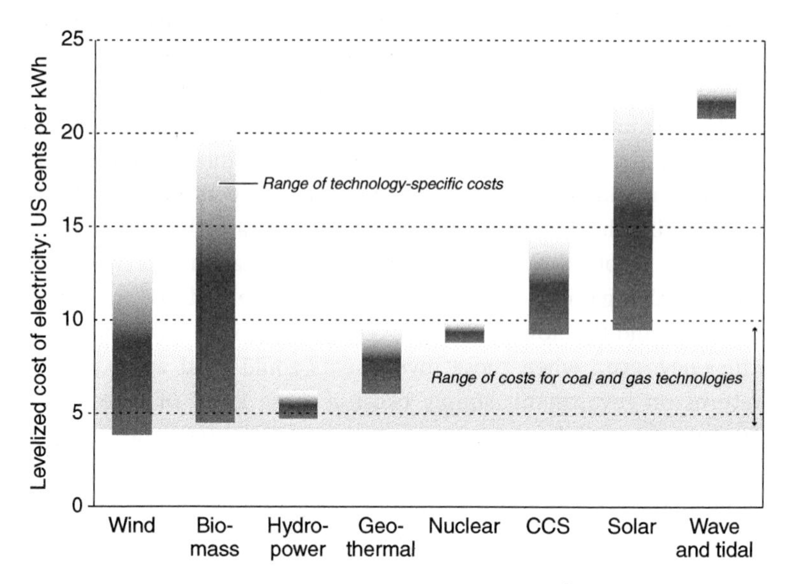

Figure 4.4. *Relative costs of different sources of energy.*[65]

On first glance, the picture here seems quite promising for market instruments because the cost differential between fossil fuels and other options is not large and is growing smaller every day. Figure 4.4 is based primarily on data appearing in a 2011 report of Intergovernmental Panel on Climate Change (IPCC) and illustrates the costs of different sources of electricity. It shows the role that carbon prices could play, comparing the range of costs associated with the different renewable technologies to the costs of using fossil fuels instead. The darker region within the bar for solar electricity, for example, covers the range of median estimates for commercial-scale photovoltaics (PV), homeowner roof-mounted PV, and concentrated solar power (CSP). The background band for the non-renewable sources of energy also has some width to it, and this reflects the fact that it, too, encompasses different technologies, that the technologies have different costs in different places, and that the price of fuel is constantly changing.

What carbon pricing would do to Figure 4.4 is to push the band of fossil fuel prices upward. Where the fossil fuel is coal, it would push it upward by almost exactly 1 cent per kWh for every $10 of tax on each ton of CO_2. In the case of natural gas, it would push it upward by 1 cent per kWh for every $16 of tax. This allows us to see what a carbon price of a given value would accomplish. A $40 price (close to what the EU has set as a target for the ETS) would increase the chances of the lower cost renewable technologies being competitive. A $130 price (the Swedish tax) would make these always competitive and start to make it possible for the high-cost renewables (wave power, for example) to be competitive. If one were to look at transportation or heating, the results would be comparable. All things being equal, then, a $40 carbon price could make a big difference in companies' investment behavior, and a $130 price an even bigger one.

The problem is that things are rarely equal. Whether or not the average costs of producing energy in a CO_2 neutral manner are low enough is implicit in having a successful business plan, a prerequisite in most cases to getting financing. At that point, the bank is going to ask a set of questions related to the business plan, something like the following:

- Does the project make sense economically: at the precise times when the energy is to be sold, will this energy be able to be sold at market prices high enough to cover all of the fixed and variable costs?
- Does the project make sense logistically: is there an infrastructure network, such as a transmission grid or pipeline, that can deliver the energy to consumers?
- Does the project make sense from a risk management perspective: how certain is it that this project will continue making money long enough for the project developer to be able to pay off the debt?

Carbon prices clearly address the first of these three questions, although even here things get complicated, for reasons that I will discuss in Chapter 9. But they do very little to address the second and third questions.

Let's examine the second question. This is a bit of a chicken-and-egg issue, a coordination problem. Imagine that you are building new renewable energy generating capacity in some out of the way place, like windmills in west Texas or on an island off the coast of Denmark.[66] If there are enough people putting up windmills, the market may well respond, and a new transmission line may appear on the map. But still this can take a while, since whoever is building the power line needs to negotiate with landowners over a long distance.[67] Until it is certain that they will succeed, you aren't going to get financing for your windmills. Until it is certain you and others will get financing for your windmills, however, they probably won't build it. If carbon prices are high enough, creating a gold rush of new investment, then the network problem will probably take care of itself.[68] If carbon prices are lower – which, after all, is better for consumers – then it becomes less certain whether the grid map will change.[69] As I describe in Chapter 9, changing the grid map would require a very different kind of policy instrument.

In practice, it is the third question, about risk, that is often the trickiest. Not only do you need to have a good business proposition now, but you need a good business proposition for the lifetime of the loan, perhaps ten or twenty years. Across the entire energy sector, investors need to consider whether the price of oil in five years will be at $50 a barrel or $250. Investors in renewable energy, however, need to worry about a lot more, especially because renewable energy is so capital intensive. Costs of renewable energy are evolving rapidly: if a PV plant built this year will be competing against one built three years from now, such as for the California air conditioning market, then project developers and financers need to consider whether that could make the current investment untenable. If it is a carbon price that is making their investment attractive, they also need to worry about how the carbon price will evolve, to convince the bank that the carbon price will stay at or above its current level. That is a hard argument to make, especially if the price comes from cap-and-trade rather than a tax. As experience with RGGI and the ETS so clearly demonstrates, regulators need to pay close attention to the market in order to achieve a particular target price. If they mess up and the price falls too low, environmentalists will complain quietly, but if the price goes too high, industry and consumers will scream loudly. Prices have fallen too low.

As carbon markets have developed over the past decade, a growing number of researchers have started to study this last issue, the problem of risk and uncertainty with respect to carbon markets themselves. The results are unambiguous. The dynamics of carbon markets make them intrinsically volatile.[70] Their volatility makes the profits from investments in carbon-neutral energy supply highly uncertain.[71] This reduces the amount of investment, especially when investors are at all risk averse.[72] To maintain high levels of investment, the average carbon price needs to be substantially higher, but this comes at a cost to consumers and social welfare: estimates range from 16 to 30 percent, even ignoring possible risk aversion.[73] In all fairness, economists have started to study ways to cure these problems, such as by incorporating "floors" on the market price of carbon; in theory, these can reduce the effects of carbon price volatility, although it is often difficult to predict how they will affect investment behavior.[74] At the same time, however, other policy instruments have proved far more effective at dealing with the risk issue. Chief among them is the *feed-in tariff*, which I will describe in Chapter 9. As I will show there, it has proved to be remarkably effective at stimulating new investment and doing so in a manner that keeps costs to a minimum.

So let's sum up the issue of stimulating investment. Carbon prices can address one of the important prerequisites for investment, namely, eliminating the cost advantage that fossil fuel technologies hold over carbon-neutral ones. They don't directly address the infrastructure and network issues, although if they were sufficiently high, creating a gold rush environment, those investors will probably come too. Carbon prices don't really address at all the riskiness of investment, and, indeed, they create a new class of risk – regulatory risk, the risk of carbon prices changing – that can hinder their effectiveness. Again, setting carbon prices really high can overcome the risk issue, but this comes at the expense of consumers and will be politically unattractive. Because of these latter two difficulties, we should not expect a $40 per ton CO_2 price to make much of a difference. A $130 price potentially could.

Examining potential effects: stimulating innovation

A big benefit of market-based instruments in general is that they give firms the flexibility to use the least expensive available ways of reducing pollution. This showed up in the case of acid rain as firms reduced their sulfur emissions not by installing costly end-of-pipe technologies but by switching to lower sulfur coal. Part of that switch involved innovation in the rail industry. For our purposes here, it is useful to break this effect into the two separate elements. The first is innovation in terms of searching for low-hanging fruit that already exist. The second is innovation in terms of developing new technologies to bring to market.

Start with low-hanging fruit. If it costs $1 per ton to cut CO_2 emissions through improvements in energy efficiency, but $50 per ton to do so by putting up solar panels, then certainly it makes sense to do more of the former and less of the latter. A carbon price somewhere between the two values makes sure that happens. In the EU, for example, the emissions reductions that have come about because of the ETS have been almost exclusively in the form of improvements in the efficiency of fossil fuel power plants.[75] It has also stimulated investment in the Clean Development Mechanism (CDM), which I will describe in the next chapter in greater detail. To summarize the CDM, it allows carbon credits to be purchased from projects in developing countries that offer inexpensive ways to reduce carbon emissions. Here, too, most of the investment has been in energy efficiency or reduction in F gases; very little has been into clean technologies for producing energy in the first place.[76]

But how beneficial is this search for low-hanging fruit? If you take a long-term perspective and take the 2°C target seriously, then it turns out not very much at all. The fact is that improvements in the efficiency of fossil fuel technologies could reduce emissions by a few percentage points. To achieve the 2°C target, we need to reduce emissions by 100 percent. For that, we will need to stop using these technologies altogether and move toward completely different ways of producing and using energy. Finding ways to improve efficiency in fossil fuel technologies will save us a little bit of money now, but will make virtually no difference and not save us any money in achieving our long-run objectives. As one group of researchers put it, if you are setting out to pick all the apples from a tree, there is no particular advantage to hunting for low-hanging fruit. Pick them, sure, but also get started on building a ladder to get the apples higher up since these need to be picked, too.[77] Building that ladder means innovating in non-fossil fuel alternatives.

It would be good if market-based policy instruments were to stimulate innovation. In theory, they should. If there are any technological options that can reduce CO_2 emissions for less than the market price for carbon, firms will have an incentive to use them. If the carbon price is $40 per ton, for example, and a CCS module, fully installed and operating, could reduce emissions at a net cost of less than $40 per ton, then firms would buy it. If there are several kinds of modules meeting these criteria, then the market is there for the technology developer that can manufacture the least expensive version. If there is no such CCS module yet available, then the race is on for technology developers to invent one. It isn't just CCS modules. Any technology that can allow people to get the same or better services from energy – turning on a light bulb, driving to work, heating a building – while reducing CO_2 emissions, as long as the net cost is below $40 per ton reduced (or whatever people expect the carbon price to be in the future), should have a market. The least expensive options, on a per-ton saved basis, ought to have the best market. And so everyone has an incentive to invent and improve.

This basic intuition, however, puts innovation into a black box: in response to market incentive, people do things that result in technological change. The intuition does not describe what those things are. In recent years, however, people have started to look inside that black box and at the actual processes that lead to better technologies at lower prices. What they have found is that there are several different processes at work, each of which responds to a different set of incentives.

One of the processes behind innovation is basic R&D in science and engineering. Although large corporations have historically sponsored a great deal of such research – Bell Labs and HP labs are the best-known examples – today, most such research happens at universities and government-funded research labs. Examples of the latter include the National Renewable Energy Laboratory (NREL) in the United States or the Forschungszentrum Jülich (Juelich Research Center) in Germany. In the United States, for example, recent years have seen about $4.5 billion per year going into energy R&D, with roughly 75 percent of this funding coming from the government.[78] Two things bring results in these research environments: having a good system in place to identify the smartest people and allow them to be productive, and feeding that system with plenty of money so that more smart people can be on board with the equipment and networking resources that they need. More smart people working productively really does equal more innovation. When it comes to the money side, universities and research institutes aren't in a position to take out loans to support potentially promising lines of research; their biggest asset is their human capital, and this can't be offered to banks as collateral. Rather, they need research funding in the here and now. Carbon prices don't offer that. Government research funding does.

A second process behind innovation is known as *demonstration*: showing how great ideas from the laboratory can be turned into things that actually work out in the environment where they will be used. They do not have to be cost competitive for everybody, but their application does have to make sense somewhere. The first customer for PV panels, for example, was the U.S. space program (NASA). Similarly, lithium ion batteries may not yet be financially viable as a power source for automobiles, but they are for telephones and computers, which don't have the option of running on an internal combustion engine. Government laboratories can do some of this work, but more often, today, it happens in the private sector, financed by venture capital. *Venture capitalists* are people who are willing to take big risks on new technologies, knowing that they will win big if the technology is a hit.

So, do venture capitalists respond to carbon prices? A number of studies have looked at this, and the answer appears to be no. Researchers in Switzerland, for example, collected data on how venture capitalists ranked the attractiveness of alternative investments where those investments differed according to whether they would benefit from a carbon price or some other regulatory instrument, like a subsidy. They found the subsidy to be the hands-down winner, with very little effect of the carbon price itself. The problem with the carbon price, for these investors, was that it was too speculative in terms of the short- to medium-term payoff. A subsidy on the new technology would guarantee a market if the demonstration was successful, whereas a carbon price would lead to profitability only if the demonstration were effective, firms potentially buying the new technology decided to go for that way of reducing emissions (rather than something else entirely), and people believed that the carbon price would remain high enough to matter for several years into the future.[79]

Researchers in Germany came to a similar finding: the best policy instrument to support venture capital is that which provides well-defined benefits to well-defined classes of technologies, rather than throwing a wide set of different technologies for reducing CO_2 emissions into competition with each other. In the latter case, the rewards from investment are just too speculative.[80] A researcher from Oxford University came to the same conclusion, examining flows of money into venture capital in new energy technologies: the flows were higher where the policy instruments created clearly winning technologies.[81] Finally, researchers from the United States and France reached a similar conclusion, counting the number of new patents that went along with the incremental improvements of demonstration. They compared the number of patents induced by policies to promote a reduction in CO_2 emissions (i.e., carbon prices) with those that were aimed at promoting particular technologies, such as solar or wind power. They found that the carbon prices did induce some patents, but these were for those technologies (primarily onshore wind) that were already close to being competitive with fossil fuels and not for others. The policies that supported particular technologies, by contrast, induced new patents for those technologies that they were aimed at. By specifically supporting a number of different technologies, the government could induce innovation across a wider portfolio of technological options.[82]

A third process (or actually a set of processes) happens in the course of deploying technologies in increasingly greater amounts, with the effects of improving the quality of the technology while also pushing down its cost. One process is the creation of economies of scale. The more of a technology that is produced, the greater the number of components for that technology that can be mass-produced rather than fabricated piece by piece. Even the production lines themselves can be mass produced. First Solar, the leading manufacturer of thin-film PV panels, copied its production line in factories around the world, with thirty-six parallel lines running by 2011, pushing down average costs with each new production line put into operation. The second process is called *learning-by-doing*. As companies manufacture, install, and operate new technologies, they figure out an uncountable number of incremental improvements, improving quality while reducing costs. One empirical estimate, taking solar panels as a case study, suggests that the two processes – achieving economies of scale and learning-by-doing – account for roughly equal shares of the cost reduction in new technologies.[83]

To get this third piece of the innovation pipeline to work, it is essential to stimulate investment in technologies that are already market ready but that are still more expensive than the fossil fuel equivalent. In the last section, I already described how market-based instruments address one of the issues needed for new investment of this sort to take place but fail to address two others: network infrastructure and risk. The result is that they don't stimulate this kind of investment very well. There are far better instruments available.

Carbon is like cigarettes

I started this chapter by describing the use of taxes to discourage smoking. It is clear that cigarette taxes have some effect. The price elasticity of smoking has been estimated to be about −0.5,[84] which qualifies as inelastic, although more elastic than most forms of energy consumption. And yet it is also clear that taxes are but one of many policy instruments that could be used to discourage smoking. Austria has substantial cigarette taxes, but it is also a country where you will find ashtrays at public swimming pools and other sports facilities.[85] If Austria wants to lose its status as having the highest proportion of smokers, it might want to lose some of those ashtrays.

The same logic ought to dominate the climate discourse. By and large, it does not. If you tune in to the environmental media on any given day, you are likely to find an expert or a pundit suggesting the need for a carbon tax or a cap-and-trade system, first and foremost, as what is needed for climate change. As I wrote the first draft of this paragraph, on a Monday morning, I turned to the *New York Times* website to see what environmental articles it recommended to me. The pick of the day, an otherwise excellent opinion piece by Thomas Friedman, closes with the usual, a call for a carbon tax to fix climate change.[86] Whether or not policy makers are willing to put a meaningful price on carbon has become a litmus test for their commitment to solving climate change. The fact that they have by and large failed in this respect drives a lot of people to conclude that policy makers just don't care enough.

There are good reasons to put a price on carbon. Perhaps the strongest of these is as a revenue raising tool. Even a modest tax, say $5 per ton of CO_2, could raise enough money to really fund R&D on new technologies, an idea suggested by Roger Pielke Jr. and several other climate policy experts.[87] Alternatively, a tax could be used simply to augment the government's core budget and allow the reduction of other taxes that otherwise are a drain on the economy. There have been hundreds of scientific papers examining this so-called *double dividend* hypothesis, and most of them have found it to be important and real. Indeed, there is some reason to believe that a so-called strong double dividend could occur: even ignoring the environmental gains from the pollution tax, the economic gains from reducing some other tax (such as the income tax) could outweigh the immediate economic costs of the pollution tax itself.[88] Put into the context of climate change, this would mean that the added economic growth gained from reducing income taxes would more than compensate for the reduction in growth caused by the carbon price: carbon taxes or auctioned permits would get us more growth now and less climate change later. Whether this is still possible when a carbon tax is designed in such a way as to not disproportionately disadvantage the poor, as was done in British Columbia,[89] is not clear.[90] But, in any event, the revenue effects of a carbon tax or an auctioned permit system suggest one good reason to implement it.

But the other reasons are mostly lacking. Given that we have decided on a 2°C target, the purpose of any climate policy instrument is not to achieve the economically efficient level of CO_2 emissions, which is where the whole economic theory underlying market instruments had its starting point. Rather, the purpose is explicitly to drive change in the energy sector. One change could be toward reduced consumption of energy – or at least energy derived from fossil fuels. Even a carbon tax well in excess of $100 per ton would have very little effect, perhaps reducing demand by 10–20 percent. Other policy instruments can achieve far more. A second change could be to stimulate investment in new technologies; carbon prices in the range of $40 per ton accomplish very little of this, although prices in excess of $100 could certainly do more. Other instruments perform better yet. A third change would be to stimulate innovation in new technologies. Carbon prices have proved to be very ineffective in this respect. Other government policies achieve more.

If we were to have carbon prices in excess of $100 per ton, it is reasonable to expect that they could achieve something, although still less than other policy instruments have been shown to accomplish. So far, outside of Sweden, policy makers have shown themselves to be quite unwilling to consider carbon taxes or market prices anything close to this level. There are good political reasons for this, and these are not changing. A carbon price of this magnitude would have large distributional consequences within a single national economy. Moreover, it would place energy-intensive firms subject to the tax at a huge competitive disadvantage relative to those in other countries with no such taxes. Indeed, perhaps the only way it is possible to imagine a meaningful carbon price, one that could shift investment around, is if it were part of a global agreement, requiring carbon markets or taxes to be everywhere. The challenge of that is where we turn next.

5

Striking a global bargain

Martin Grosjean is a professor of paleoclimatology at the University of Bern in Switzerland and former director of the country's National Center for Competence in Research on Climate Change (NCCR Climate). Grosjean and his colleagues at NCCR Climate engage not only in research, but also in education, and one of the main activities is a summer school each year at which a group of leading PhD students from around the world converge to delve into a particular climate-related topic. On several occasions, the summer school has taken place in the town of Grindelwald, a picture-perfect ski and summer resort at the foot of some of the Alps' most impressive mountains, in a valley where glaciers converge. The Grindelwald valley has an amazing natural history, one that is tied to climate change, something that the students in the summer school have the opportunity to learn about. One afternoon during the week, Grosjean guides the students on a hike up the side of one of the mountains, showing them how to see the signs of the last Ice Age and the fast glacial retreat of the past several decades. Indeed, Grosjean and his team have turned the walk into a guided tour that anybody can enjoy simply by downloading an app onto your iPhone: at various points along the trail, your iPhone will tell you what to look for, to see the signs of climate change before your eyes.

But Grindelwald also holds an important human history, something that Grosjean explains to the students but that doesn't appear on the iPhone app. It concerns the history of farming, the town's primary economic base for all those centuries before the tourists arrived. With cowbells tinkling on the hillsides during the spring and summer months and with locally produced cheese to be found in the mountain restaurants, agriculture is still one of the reasons that the tourists come, at least for the half of the year when the grass is green.

Grindelwald is home to what may be the longest standing agreement concerning the rights to use communal land, forged more than 600 years ago, in 1404. In a document known as the *Taleinungsbrief*, or "valley letter of agreement," the people of the valley agreed on a system of rules for livestock management that survives to this day, most recently amended in 2002.[1]

Tragedy of the commons

The farmers in Grindelwald practice a form of management that can still be found across much of Switzerland, Austria, and northern Italy. In the summer, the cows from all the farms go up to communally owned mountain pastures to graze; such a pasture is known in Swiss German as an *Alp* and, of course, is the source of the name for the mountain range. There, the cows intermingle, with a few people paid by the community to look after them, milk them, and make cheese. At the end of the summer, the cows come back down to the valley and with them come precious wheels of cheese. Since the wheels often differ in quality, a lottery system determines the order in which farmers pick them out, with each farmer's total take determined by the number of cows that had been out on the grass.[2] The cows spend the winter in the valley, munching on hay that was grown there and harvested during the summer months.

The concern in 1404 was that there were too many cows: in summer, they were overgrazing the Alps. The production of cheese was suffering. To deal with the problem, the farmers of Grindelwald did two things. First, they decided to limit the Alp grazing rights to animals from their own valley, thus excluding animals from neighboring valleys. Second, they limited the total number of cows or cow equivalents (one cow was equal to four pigs or eight goats) that farmers in the valley could own, with each farmer's allocation defined by the size of his or her lowland property. Rights to graze the Alps were made contingent on owning farmland in the valley, and these rights, in and of themselves, could not be bought or sold.

There are many different systems of property rights around the world, with different allocations of who gets to do what on or with the land. In many parts of sub-Saharan Africa, farmers are able to live on and till the land and keep the profits, but they are not entitled to sell it. The right of somebody to buy and sell land, by contrast, is a common feature of Western property rights systems, but political jurisdictions differ according to whether the right to develop land – to change its use – belongs to the owner or to the state. In the latter type of system, the landowner can apply for a development permit but has no automatic entitlement to one and little recourse if the permit is denied. In much of the western United States, both development rights and the right of exclusive use automatically belong to the owner: in parts of the eastern United States, exclusive use is not assumed, and other people are allowed to use your land for traditional activities, such as walking or hunting. In Scandinavia, there is the right to camp on others' land for a night, as long as you respect the landowner's privacy and clean up after yourself.

Common across all of the different property rights systems is a concern for maximizing the value of the land to the individual owners and to society, but the systems differ in terms of which they prioritize and how they believe the best value can be attained. As the political scientist John Dryzek suggests, one view is that the maximum value accrues both to individuals and to the society when the greatest possible number of rights belong to the individual landowner.[3] Such a view is consistent with classical economics on the one hand – the idea that what is good for individuals, as they themselves decide it, is good for society – and also with the belief that people will take better care of what is theirs and what they can sell. The *Taleinungsbrief* reflects a more community-centered idea about what is good for society and a unique practical situation. In Grindelwald, it seemed clear that the community benefited from the quantity and quality of its cheese, and the property rights system needed to promote this. It also would have been far less efficient to divide up the mountain land into separate fenced off parcels, rather than letting a few herders guide the cattle over the entire mountainside over the course of the summer. For both reasons, community ownership made sense, and the *Taleinungsbrief* was there to make sure that the community did as well as possible and that each farmer within the community got a fair piece of the pie.

More recently, the basic problem of how to govern community property or *common pool resources* has enthralled economists, political scientists, mathematicians, and legal scholars. In the 1950s, the mathematician John Nash developed the idea into what would become the foundation of *game theory*, later winning the Nobel Prize in economics for this work.[4] In 1968, the ecologist Garrett Hardin put another name to it, calling it the "tragedy of the commons"; he showed it to be a widespread problem in environmental management at various scales of governance and suggested that it was often intractable.[5] In 2009, the political scientist Elinor Ostrom won the Nobel Prize in economics for a body of work that showed how people had, as in Grindelwald, managed to develop the necessary institutions to manage common pool resource problems, identifying some of the factors associated with success and failure.[6]

The *tragedy of the commons* is actually a special case of an externalities problem, one in which all the people are in the same situation of being both polluter and polluted upon. With the cows, it isn't exactly pollution that is the externality, but rather the effects of their grazing. Every cow that eats bit of grass makes that grass unavailable to everyone else. By setting a limit on the number of cows, the Grindelwald villagers reached an agreement that made life better for all of them, maximizing the aggregate surplus of the villagers. The case stands out in terms of having found a good solution that still involved communal ownership. The far more prevalent way to solve the problem, at least in Europe and its legal descendants, has been simply to carve up the common land and give it to the farmers, for each of them to own as private property. In this case, each farmer then faces the incentive to graze the right number of cows for his or her land. But, just as it was impractical to carve up the land above Grindelwald into individual plots belonging to individual farmers, so too is it impossible to carve the global atmosphere into separate parcels. The global atmosphere is our common resource, and we need a way to protect it.

The United Nations Framework on Climate Change

Because the solution settled on in Grindelwald works so well, you may be happy to know that its basic ideas are embodied, heart and soul, in the global treaty covering greenhouse gases (GHGs).[7] That treaty is the United Nations Framework Convention on Climate Change (UNFCCC), and, for twenty years, almost every country on the planet has agreed to be bound by its basic terms. The UNFCCC also has an important spin-off agreement, the Kyoto Protocol, which is where all the action is, and yet which somewhat fewer countries have agreed to.

The time it took to travel the road from starting to identify climate change as a problem to having a global treaty to manage it was remarkably short. Scientists first started becoming alarmed by the possibility of climate change in the 1970s, and they eventually convinced several specialized agencies of the United Nations – most importantly the United Nations Environment Program (UNEP) and the World

Meteorological Organization (WMO) – to organize two meetings that would pull together the leading scientists from around the world. Following the second of these, held in Villach, Austria, in 1985, the WMO and UNEP decided to join forces and establish an independent scientific body to assess whether climate change was indeed a serious concern. This was the Intergovernmental Panel on Climate Change (IPCC), and it took on the task of delivering an assessment report to the UN General Assembly in New York by 1990. The IPCC, drawing on a pool of leading climate scientists nominated to serve on it by national governments, finished its report on time. On the basis of its findings, the UN General Assembly set up an international committee to negotiate a climate treaty. The goal was to have something ready to sign by 1992, when heads of state would converge on the city of Rio de Janeiro for the long-planned UN Conference on Environment and Development. The negotiation committee finished its work on time, too, and the UNFCCC signing ceremony took place exactly as anticipated.

Following the fashion of UN treaties, the UNFCCC is really just an agreement to agree. Parties to the convention made pledges to take steps to limit GHG emissions to safe levels, although the steps themselves were not spelled out. What was spelled out was the process for future negotiations. As soon as the requisite number of countries had ratified the treaty – that is, submitted their participation to formal vote within their own national government – meetings would start taking place every year. These meetings were called the Conferences of the Parties (COPs). The UNFCCC established a secretariat in Bonn, Germany, to organize the COPs. It also dictated requirements for national governments to collect data on their domestic GHG emissions, which would aid future negotiations, and established a set of committees to take this information and do something with it. Finally, it set up a dichotomy between the rich countries and all the others. A sheet of paper stapled to the back of the agreement listed the rich countries by name, in alphabetical order, and was called Annex 1. Several paragraphs within the main document made it clear that these countries listed in Annex 1 would shoulder a disproportionate share of the responsibility and cost for reducing GHG emissions.

The first of the annual UNFCCC meetings, COP 1, took place in 1995, and, by then, it had become even clearer, following the publications of a second comprehensive IPCC assessment report, that countries needed to start taking real steps to reduce their emissions. The national negotiators at COP 1, held in Berlin, established a negotiation committee, with the charged task of finalizing an agreement on this, with a deadline for COP 3, which was to be held in the Japanese city of Kyoto. The terms of reference for this group, coming out of Berlin, specifically framed the agreement in Grindelwald terms: they had to establish an overall global limit for GHG emissions, and they had to divide up that limit in a reasonable fashion, just like all the farmers with their cows. The only twist on this was that they had to respect a basic difference between Annex 1 countries and all the others, imposing the national emissions limits only on the former. For the third time in a row, another major deadline was met. Negotiators at COP 3, in 1997, signed what became the Kyoto Protocol, the first really binding agreement for GHG emissions.

The Kyoto Protocol

The core of the Kyoto Protocol is an agreement that Annex 1 countries would halt the growth in their aggregate emissions and, indeed, bring them down to a level 5 percent lower than what they had been in 1990.[8] That is an amount not unlike the reductions in herd size to which the Grindelwald farmers agreed, and it arguably achieves a greater level of economic efficiency without requiring any major change in the economic production system. The deadline for Kyoto was a five-year window, 2008–12, known as the *first commitment period*. Every Annex 1 country agreed to a particular national target, with the general pattern that the more committed countries agreed to do more, and the other countries agreed to do less. The European countries, for example, negotiated as a block and, as a block, agreed to reduce their national emissions by 8 percent. To average out to 5 percent, other countries had limits that were less strict, and indeed some had limits that represented a net growth in emissions – albeit a limited one – from 1990 to the first commitment period.

From this simple core, the Kyoto Protocol then became progressively more complicated for a lot of reasons that ultimately make a lot of sense – or at least did so at the time. Many of these reasons reflect the logic of the externalities model of climate change and the idea that establishing markets for carbon would be the way to go. Kyoto became, in many ways, a vehicle for establishing such markets, with all the institutions that such markets would require in order to function well. Indeed, given the fact that even in 1997 most people recognized that a 5 percent reduction in emissions, limited to the rich countries of the world, was pretty meaningless as a cure for climate change, the fact that Kyoto would create the institutions associated with global carbon markets became one of the strongest arguments in its favor. Once the markets were established, the argument went, it would be a relatively easy matter to turn down the screws, push up the price of carbon, and draw emissions down to levels that would actually solve the problem.

Kyoto sets up three basic markets. The first is a market between countries. If one country fails to meet its own national target, it could buy excess emissions credits from another country that would otherwise cut its emissions by more than it had obligated itself to. In fact, there were almost certainly going to be several of these latter countries, and this was no accident. The base year, 1990, coincided with exactly just before the economic collapse in Eastern Europe and the Soviet Union, a collapse associated with their rough (some would say disastrous) transition from communist to free market systems. By 1997, all of these countries were recording annual CO_2 emissions far lower than they had been in 1990, and it was almost certain that their targets for the first commitment period would turn out to be very far from a meaningful constraint. So they would have a lot of excess emissions to sell. Experts called this "hot air." Right from the start, it was clear that the 5 percent target was going to take very little effort because most of it had already happened as a result of the fall of communism. It was also clear that the market for national-level emissions was money in the bank for Russia and its neighbors.

The second and third markets are a little more complicated, in that both are tied to the complicated world of development project finance. Let's say that you are a company that builds things, things like electric power plants, in Asia. You are planning on building a coal-fired power plant, which will generate a lot of CO_2 emissions. For a little extra money, you could build it in a way that would reduce the emissions, say by making efficiency improvements or by switching the fuel from coal to natural gas. Of course, you won't do this because there is nothing in it for you. But what if one of those countries that failed to meet its first commitment period target and needed to buy emissions credits made a deal with you? It would pay the cost of the CO_2 reducing upgrade and, in turn, would get the credits for the reduced emissions.

The Kyoto Protocol makes these deals possible. The first instrument is known as joint implementation(JI) and covers development projects taking place within Annex 1 countries, which for all practical purposes means the countries of the former Soviet Union. In this case, in exchange for project finance coming from another country, that other country would get to count the emissions savings, rather than the country in which those savings actually took place. Really, this is just an alternative way to structure an emissions trade of the first variety – between Annex 1 countries – although in a way that lets the money bypass the national government of the country selling emissions and receiving money, putting money instead into the pockets of project developers.

The second instrument is known as the Clean Development Mechanism (CDM) and covers development projects located outside of Annex 1 countries. As with JI, the country providing the financing for a low-carbon development project get to count the emissions savings as if they were its own. But with JI, the project's host country could not then count the emissions savings as its own, toward its own Kyoto target. With the CDM, the host country has no Kyoto target. Here, it is important to make sure that CDM-based finance is going for development projects that really do reduce CO_2 emissions and that would not have taken place without the finance. If both are the case, then the effect of the CDM is to move money from rich country to poor and the credit for the corresponding emissions reductions from poor country to rich. To make sure that both *are* the case, the Kyoto Protocol established a new bureaucracy, the CDM Executive Board, based in Bonn. If the Executive Board fails in either respect and the emissions reductions are either not genuine or not contingent on the additional finance, then the CDM still moves money, but its net effect on CO_2 emissions is simply to relax the overall 5 percent target. The evidence so far is that the Executive Board has failed in both respects.[9] They didn't fail because they slept on the job. They failed because the task proved far more difficult than anticipated. The details of this story are messy and too long to describe here.

Closely coupled with the three market mechanisms, another way that the Kyoto Protocol became complicated is on the issue of land use. Following the initial agreement in 1997, two points quickly became evident. First, there were large and relatively low-cost gains to be made in net GHG emissions by improving land use and land management practices, actions having nothing to do with transformations to the energy system. Second, the countries where these actions could take place were the same ones as those benefiting from JI and the CDM: countries in transition and developing countries. Given the fact that Kyoto was becoming as much as anything else a vehicle to shift money from western European countries to these others, why not make it possible to do so in the area of land-use change as well? This has happened, coupled to JI and the CDM, but requiring an additional bureaucracy, one based in Bonn, to come up with accounting rules for the carbon that is in trees and the soil.

By now, the first commitment period has come and gone – what are the results? At time of writing, the final accounting has not been complete, but several points are reasonably clear. First, Annex 1 countries did achieve their aggregate target of 5 percent emissions reductions. Second, they were able to do so primarily because of the excess emissions allowances in Eastern European and former Soviet countries, the so-called *hot air*, with very little in the way of actual emissions reduction taking place in the remaining Annex 1 countries. Third, together, the CDM and JI have resulted in billions of dollars of project finance flowing from wealthy to less wealthy countries. Coupled with this has been the establishment of institutions for the management of these markets.

By all accounts, then, the Kyoto Protocol delivered everything it promised. Included in that promise was very little in the way of actual emissions reductions. Most of the promise was in terms of laying a solid foundation for the future, both in terms of setting a precedent for countries making and honoring national GHG emissions commitments and in terms of establishing the institutions – the bureaucracy and all the associated rules – needed to manage a global carbon market.

A solid foundation is great – as long as you build something on it. Otherwise, it's just a chunk of unpainted cement, and only serious skateboarders would consider it an improvement on the landscape. To evaluate the success of Kyoto, we need to see what has come after it.

The road from Montreal to a new Kyoto

In December 2005, I was an assistant professor teaching a master's level class on climate change at Boston University, and the opportunity came up for the students to go on a field trip of historic importance. COP 11 was scheduled to take place in Montreal, just a few hours' drive away. What was more, this was to be the first COP since the Kyoto Protocol formally went into effect and, with Kyoto out of the way, would mark the starting point for negotiations on a set of targets to follow those of the first commitment period. Even better, whereas the two-week COP started on a Monday with a set of meetings concerning the implementation of Kyoto, the negotiations on post-Kyoto targets were scheduled to commence on the following Saturday morning. So, after I organized the needed credentials to gain access to the COP as official "observers," my students and I piled into a large Ford van on a Friday afternoon and drove to Canada for the weekend.

Well-behaved and wearing "business casual" attire, my students and I showed up early to the meeting room on Saturday morning and took seats at the back. The rows filled up in front of us. Another class – we all assumed from the University of Vermont – was in tie-dye. Soon all the seats were taken up by observers, including the seats around the rectangular table at the front of the room. Then some of the national negotiators arrived and stood there, having no place to sit down. Minutes passed, more and more of them were standing there, and then, eventually, someone fetched another person to ask those observers sitting at the front of the room and around the table to move to the back of the room, where they would have to stand, so that the negotiators could sit. The official representative of the Canadian delegation – the hosts – brought the meeting to order, and all of the national representative introduced themselves. In the middle of this, someone asked if it would be possible to open the doors to get some air into the room because it was getting stuffy. That happened, thank goodness. Then the Canadian outlined the seriousness of the negotiating task and what it was all about, building on the foundation of the existing Kyoto Protocol to come up with a real solution to one of the biggest challenges mankind has face, and how important this was. That is when a hand went up in the second row of chairs. It was a leader of one of the national delegations. If this was so important, he asked, then why didn't all countries have a seat at the table. The table was too small. The Canadian started to explain that he hadn't picked the table and that he would personally make sure that everybody got a chance to weigh in, even if they weren't seated at the table but were sitting in one of the front rows. That is when the muttering started. No. It wasn't good enough. The Canadian pleaded. No. Finally, a motion was proposed, and passed, to send someone out to the main office in the conference center to find out whether there was a larger room available, with a larger table. That person went, and it was time for a coffee break.

Returning from the coffee break, the messenger returned to say that there were some rooms that could be made available – furniture would have to be moved – but that it was important to know how large the table needed to be. So then there was discussion to figure this out, and then the messenger went off again, and people started checking their email using the still novel technology of wireless Internet. The messenger returned to say that a suitable room could be made available by the evening. Any other day of the COP, negotiations would have reconvened that evening, but not on the Saturday in the middle because on the Saturday night in the middle there is always a big party, hosted by all the observer organizations in honor of themselves and the national negotiating teams. They had reserved a big nightclub featuring some of the acrobatic performers that Montreal is famous for, being the original home of Cirque du Soleil. Nobody was going to miss it. Sunday was already reserved for scientific events and presentations, and so the Canadian at the head of the room informed everyone that negotiations would have to wait until Monday morning.

My students and I had to leave Montreal on Sunday afternoon. We never got to watch real negotiations take place, other than over the size of the table. But we didn't miss anything important. Because, starting on that Monday and proceeding for the rest of that week, negotiations never moved beyond anything equivalent to that table. And essentially that is how it has been ever since, now going on ten years.

What they did agree on in Montreal, which didn't surprise anyone, was to set a deadline for the next agreement. That deadline was two years later, for COP 13, to be held on the Indonesian island of Bali. By Bali, they agreed in principle to an ambitious global target – one that would limit total average warming to no more than 2°C above preindustrial times – but they got stuck on the basic framework within which national commitments would take place. Some wanted simply a second commitment period, with new targets for Annex 1 countries, tacked onto the back of the existing Kyoto Protocol. Others wanted a fundamentally different type of agreement, in one way or another, perhaps with the important step of establishing commitments for non-Annex 1 countries. Unable to reach an agreement on this fundamental point, the main outcome in Bali was to establish separate negotiating committees for both approaches. Within two years, by COP 15 in Copenhagen in December 2009, both committees were to come up with an agreement, and then a final step would be to resolve how the two agreements would work together.

Copenhagen was the big disaster. As leaders of the negotiations, the local Danish hosts made some tactical mistakes, not unlike the Canadian and his too-small table, resulting in a large block of countries walking out in protest. Neither negotiating track came up with anything close to an agreement. To save face, on the last day, countries agreed to a document known as the Copenhagen Accord. The Copenhagen Accord restated countries' collective commitment to prevent climate change beyond 2°C and established a collective commitment by the wealthy countries to provide hundreds of billions of dollars in financial assistance to developing countries in order to help them with mitigation and adaptation. But in terms of specific national commitments with respect to their share of this finance, there were none. In terms of national commitments with respect to emissions reductions, a sheet of paper was tacked onto the back on which countries wrote down what they would like to do. As with the commitment on finance, there was no agreement as to whether these actions would be enforceable. Analysts wondered what the meaning of it all was.

The COPs have improved since Copenhagen, but mainly because expectations have been lowered. A meaningful agreement finally did come together at COP 18 in Doha, Qatar.[10] First, countries agreed that there will be a second commitment period to the Kyoto Protocol, to start in 2013 and run through 2020. The various flexibility mechanisms, most importantly the CDM, will continue on. Unlike the Kyoto Protocol and like the Copenhagen Accord, countries' national targets will simply be what they volunteer, not what they negotiate in the context of reaching a predefined aggregate Annex 1 total. If one adds up what countries have volunteered to do, it is quite clear that it is far too little to achieve anything close to the 2°C target.[11] Second, countries have agreed to negotiate a new protocol, having explicitly decided that the Kyoto Protocol will end. The new protocol will cover emissions starting in 2020. To agree on whatever it is, negotiators have set themselves the deadline of 2015 for the COP scheduled to take place in Paris. Maybe the negotiators are serious this time, coming out of Doha. I think they were serious the previous times, too. I also don't think that any of the things that came between those serious intentions and actual results have changed. The most important of these is that now, unlike in Kyoto and unlike in Grindelwald, they are not negotiating an incremental change that is clearly economically efficient, one that also leaves production systems essentially unchanged. Rather, they are now negotiating a pathway toward much more strenuous emissions reductions, one that will require a fundamental redesign of the global energy system. Nobody really knows how this fundamental redesign will take place, not to mention its immediate economic and social costs.

What's wrong with the UNFCCC and Kyoto

For years now, social scientists have been writing about how the basic architecture of the UNFCCC and Kyoto are flawed, and, because of those flaws, it is extremely unlikely that further negotiations at the COPs will produce anything meaningful.[12]

The first of such articles to catch my eye appeared in the journal *Nature*, shortly before the COP in Bali.[13] While most of the world was hoping that negotiators would agree on a strong second commitment period for Kyoto, the *Nature* paper's two authors – Gwin Prins of the London School of Economics and Steve Rayner of Oxford University – suggested that it was time to abandon this effort. Their basic argument was that the original authors of the UNFCCC had vastly underestimated the complexity of the climate problem and hence the willingness of countries to agree to massive emissions cuts. The UNFCCC had come into being, Prins and Rayner noted, shortly after another global agreement had gone into effect, the Montreal Protocol.

The Montreal Protocol was signed in 1987 and constitutes the main agreement to limit substances that harm the ozone layer, like certain kinds of coolants for refrigerators and propellants for spray cans. As a model for what was to come for the UNFCCC and the Kyoto Protocol, the Montreal Protocol codifies binding national commitments to limit the production and use of these substances. Most people, Prins and Rayner included, view the Montreal Protocol as a success, and there is reason to believe that over time it will close the ozone hole.[14] But, Prins and Rayner argue, the very success of the Montreal Protocol gave a false sense of hope about the potential to negotiate a set of binding national targets. Phasing out certain kinds of coolants was something that most countries were quite willing to agree to; there were already other chemicals available as substitutes at only slightly higher cost. Before countries will do the same for CO_2 – agree to phase it out of the energy sector – there had better be alternatives available that everybody knows will work and that don't cost much or any more. That, however, is years away. What we need, Prins and Rayner argue, are policies at the national and international level to stimulate the development of those technologies.

The Prins and Rayner article came under immediate attack from a number of mainstream climate policy experts. At the time, 2007, I was working in a very large European Union-funded research project, known by its acronym ADAM, which had drawn together a *Who's Who* list of Europe's top climate policy researchers from the leading universities and institutes. We had a project meeting shortly after the Prins and Rayner article appeared, and it was the topic of frequent discussion. Most people criticized Prins and Rayner for forgetting that Kyoto was and is meant to be iterative in nature, with early iterations driving the technological progress that is needed for the later ones. A stronger Kyoto, they argued, would do two things. First, it would de facto force more countries to put in place carbon markets and turn down the screws on those markets, pushing the price of carbon higher. Second, it would send a signal to industry that world leaders were serious, that the days of fossil fuels are numbered. Together, the carbon price and the signal accompanying a visible political commitment would stimulate technological change. And, with that technological change, the next iteration of Kyoto could be even stronger.

This argument against Prins and Rayner, and for Kyoto, makes sense if you believe in carbon markets and if you believe in the power of signals. Both are overrated. For carbon markets to make a difference, as the evidence presented in the previous chapter suggests, the carbon price would have to be very high, high enough to do some damage to the economy. This is exactly the problem that Prins and Rayner pointed out. What about signals? Recently, people have put a lot of stock in another signal, the idea that countries' commitments to balance their budget would signal good economic times ahead, stimulating investment and leading to economic growth and an end to recession. Paul Krugman calls this idea the "confidence fairy" and builds a convincing argument, based on solid numbers, for why it doesn't work.[15] I would suggest that there is a "Kyoto fairy" of similar importance in climate debates. The idea is that countries' signing of a treaty like the Kyoto Protocol becomes a signal that they are serious about climate change and that this alone will convince private actors to change their investment behavior, confident that future government policies will reward this. Unfortunately, the evidence suggests otherwise. Venture capitalists, for example, have shown that they need clear and immediate government commitments to particular technologies, rather than far more vague expectations of future policies.

Prins and Rayner have not been the only ones to suggest that Kyoto had some major problems. One of the leading political scientists working on international climate policy is David Victor, a professor at the University of California San Diego. In a book published in 2011, *Global Warming Gridlock*, he reached essentially the same conclusion as Prins and Rayner's, but he took their ideas to a deeper level of political analysis.[16] To understand why the architecture of the UNFCCC and Kyoto is faulty, he focused on three main aspects.

First, what Victor calls the *geometry* of the agreement is flawed. By geometry, he means the number of countries and their relative power. The UNFCCC is structured such that all countries are members. It is a commonly understood goal that any protocol agreed to under the UNFCCC should also have universal membership or as close to universal as possible. This means that every country has the power to block a universal agreement, which means that every country has, essentially, a veto. Victor suggests that this is a guarantee to a weak agreement. It doesn't have to be this way, he suggests. First, you could do a great deal among a much smaller group of countries with disproportionate power to influence both emissions and technological change, such as the G20. An example of a previous treaty that followed this path is the European treaty to limit acid rain deposition. In this case, a small number of highly committed countries agreed to significantly reduce their emissions of chemicals causing acid rain. As they proceeded to make the reductions, demonstrating that the economic costs were minor, leaders in more and more other countries came under domestic political pressure to join them. Eventually, just about all countries had signed up. The pathway was to go deep and then broad, and it worked. Kyoto's strategy of going broad and then deep is a mistake.

Second, Victor argues, the basic policy instrument is inappropriate for a situation in which there is a high degree of uncertainty. The instrument he points to is the legally binding emissions reduction target. The uncertainty lurks in a couple of important places. First, it is very difficult for countries to actually predict what their emissions are going to be several years down the road. With an energy sector being slow to change but tied to economic output, the greatest determinant of emissions is in fact the state of the economy itself. If there is a recession, emissions go down. If there is a boom, emissions rise. Countries would like to control this, but they can't. In turn, they are hesitant to agree on too binding a target, one that they might miss for reasons essentially beyond their control. Second, the economic costs of achieving emissions reductions are uncertain. Indeed, if some new technologies, like renewable energy sources, don't work as hoped for, then countries would have to take draconian measures to rein in their emissions, such as restricting total energy use, with a clearly negative effect on the economy. Again, countries are hesitant to agree to a binding target that could ultimately require this. It doesn't have to be this way, Victor again suggests. Countries could agree to take particular action to reduce their emissions, the effects of which on the economy would be easier to predict. This could include agreeing to phase out coal combustion plants, for example. Countries could also agree to take specific steps to promote new technologies through spending on research and development (R&D) or on international cooperation.

With less uncertainty about whether they would meet their obligations and what the costs would be, countries would be more willing to agree to more ambitious commitments. The problem with Kyoto is that it forces countries to commit to results, rather than actions, in an environment where results are difficult to anticipate.

Third, Victor argues that the incentives that states have to comply with their Kyoto obligations simply don't match up with the nature of those obligations in the first place. UNFCCC negotiators, in coming up with Kyoto and in all of their work since, have aimed for legally binding targets. That makes sense, or at least it sounds good. After all, if the treaty is to be a serious one, it ought to have teeth to it, in terms of punishing those states that don't comply. The problem is that the actual incentives that states have to comply, directly tied to their power to punish noncomplying states, are quite limited. Formally, there is an enforcement mechanism written into Kyoto: a state that fails to meet its first commitment period target pays a penalty in the second commitment period in terms of a higher target. It also is not allowed to make use of the flexibility mechanisms such as emissions trading. But the enforcement mechanism is weak for two reasons. First, it supposes a set of second commitment period targets that is strong and that had already been negotiated prior to states' being able to predict their compliance with the first commitment period targets. Neither has happened. Second, and more importantly, a state that decides it doesn't like its Kyoto commitment can simply repudiate the treaty and walk away from it. This is exactly what the United States did by refusing to ratify it; while plenty of people didn't like it, there was nothing they could do about it. Victor argues that it doesn't have to be this way. To make the commitments more enforceable, one could tie them to some other international institution, like the World Trade Organization (WTO), that states do not want to walk away from. One could imagine a world in which compliance with a climate obligation would be a prerequisite to enjoying access to national markets free of trade barriers. That would be a substantial carrot. Alternatively, one could structure a climate treaty around voluntary commitments. Victor cites substantial evidence that these typically prove to be just as effective as binding international commitments, contrary to common intuition.

Consistent with one prevailing view in the field of international relations, one founded in empirical study of states' behavior, Victor argues that states rarely if ever commit themselves to take a particular action by way of an international treaty unless they are already

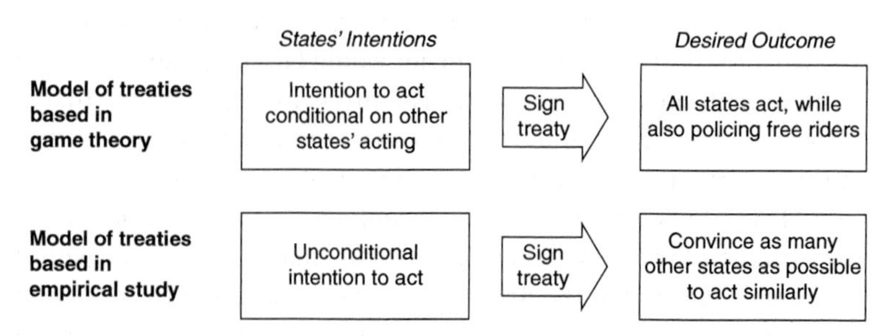

Figure 5.1. *Comparing game theoretic and empirically founded models of treaties.*

prepared to take that action anyway.[17] This is a very different model of how international treaties operate and what one can expect from them.

I draw this distinction out in Figure 5.1. As Figure 5.1 illustrates, the basic game theoretic model of treaty formation starts with an intention to take a costly action that is conditional on other parties taking the same action. This idea is at the heart of the collective action solution to the tragedy of the commons, the kind that the farmers in Grindelwald all participated in. By signing a treaty, states commit themselves to take an action that they otherwise would not have taken. What they hope to get out of a treaty is the commitment that all states will do the same. Moreover, since all states will have an incentive to be a free rider – benefiting from other states' costly actions without taking those actions itself – states hope to benefit from the participants in the treaty collectively taking on a policing function that punishes free riders.

But this is theory, and theory depends on the idea that the thing that they are agreeing to is clearly economically enhancing given the current production system. Those conditions no longer apply to the climate agreement that people are hoping for. Given that theory may no longer apply, it makes sense to base a model of treaty negotiation based on actual observation. In the empirically driven model, by contrast, states actually start with an unconditional intention to act. They would be willing to act unilaterally, and the presence of a treaty neither adds nor detracts from that willingness. They would, however, prefer if their actions were not unilateral and that other states would take similar actions as well. By signing the treaty, they hope to raise the issue on the global agenda, increasing domestic political pressure in those other states and leading them to become willing to act as well.

An example of an issue area where this makes intuitive sense is human rights. Many states had enshrined a commitment to uphold basic human rights within their own borders, within their own constitutions or national legislation. At the same time, they also desired to see that these rights were universal, in part because it seemed like the morally correct position and because universality would benefit their own citizens while traveling abroad. Signing on to the Universal Declaration of Human Rights, then, was an action taken with the goal of universality. At the same time, it is hard to imagine states agreeing to uphold human rights within their own borders only because other states had done the same.[18]

According to the theory of international relations based on empirical findings, then, this should be our basic model of global treaties. There are a lot of explanations given for why this is the case, but the common denominator appears to be that there is no true global policing power of any significance.[19] States know that they can free ride and that others can free ride, and, in most circumstances, there is nothing that anybody can do about it. The only good exception to this appears to be in the area of international trade, enforced by the WTO, as already discussed. Here, the benefits of belonging to the WTO are immense, and so policing is possible because it is possible to eject a state from the WTO, whereby it would lose its benefits.[20] Thus, free ridership is less of a problem. But the WTO is an exceptional case.

Switching away from game theory to the model based on evidence implies two things. First, it implies that we should design treaties with their limitations in mind. There may be value in states signing on, in the context of a treaty, to undertake the full extent of what they were prepared to do anyway. If a state were prepared, unilaterally, to reduce its emissions by 20 percent, then a well-designed treaty would be one in which that state was to do exactly that: make a 20 percent reduction. In fact, Victor essentially argues, Kyoto's architecture, built around binding targets and timetables, has led states to commit to less than this, say, 15 percent or even 10 percent. The ambition gap is especially to be seen in the context of developing countries. Countries like China, India, Brazil, and South Africa have all taken ambitious steps domestically to spur the development and implementation of low-carbon technology. Brazil, for example, has switched almost entirely to burning biofuels in its vehicle fleet, rather than oil out of the ground.[21] South Africa and India have major programs to push renewable energy, solar in the case of South Africa and wind in the case of India.[22] China, too, is pushing renewable energy, as well as nuclear energy, and even though its CO_2 emissions have mushroomed with its economic growth, there are arguments to be made that its leaders are taking positive steps to try to transform the country into a low-carbon economy.[23] Yet all of these countries have, at one point or another, vigorously opposed suggestions to impose binding emissions constraints on them in the context of the UNFCCC and Kyoto Protocol. The same can be said for the United States. Domestically, the United States at both the federal and especially state level has taken positive steps to limit CO_2 emissions and promote low-carbon energy sources, most recently by limiting the construction of new coal-burning power plants. And yet the United States has flatly refused to commit to emissions reductions in the context of the Kyoto Protocol.

Second, and more importantly, it implies that we need to adjust our expectations. We should not count on global treaties to solve true collective action problems.

This may be a truth just as inconvenient as climate change itself. The reality is that we live in a world in which there is no effective government at the global level. The UN is not a global government. It is a negotiation forum for sovereign states and also an institutional channel for them to cooperate on issues of economic, cultural, and political development. But it has the power neither to make nor to enforce laws.

We may at some point get a global government. Personally, I both hope and believe that we will. My favorite author on this theme is the popular science writer Richard Wright, and my favorite quote of his is the following: "In 1500 B.C., there were around 600,000 autonomous polities on the planet. Today, after many mergers and acquisitions, there are 193 autonomous polities. At this rate, the planet should have a single government any day now."[24] What is any day now? I think it is within a few centuries. The point, as he and countless others have illustrated, is that the process of shifting sovereign power to a more aggregated geographical level makes good sense, but it also takes time, especially if one rules out war as the means for effecting a merger. The European Union is probably the best current example of where this is happening, and it shows how it goes in fits and starts. Before they are willing to give up power, sovereign states have to trust that the new international entity will do a good job, and, for that, they need evidence. Arguably, people's very identities have to change. I come from Massachusetts, and a century ago, I would likely have felt residence in that state as my primary cultural and political identity. Now, it is as an American. Perhaps in several generations, more people from Greece or Germany will feel their primary political and cultural identity to be European. It seems that few do now, and the leaders in Berlin and Athens display some lack of comfort in handing over control of their money to the other. If there is one thing that we cannot do, it is to decide that we need a global government, and so we are going to create one right now, one that works. It would be like me consciously deciding not to be afraid of snakes or spiders, and then not being afraid. It just doesn't work that way. My heart still skips a beat.

Why people remain so committed to a global treaty

There is every reason to believe that negotiating a global treaty to solve climate change just won't work. The first reason is the evidence from the past twenty-five years. Governments around the world have repeatedly expressed their desire to end climate change, to limit temperature rise to no more than 2°C. But they have simply been unable to agree on a set of emissions targets consistent with that. The targets agreed to in Kyoto were feeble, really just a warm-up act, and everyone knew so at the time. Every single deadline, goal, and negotiation target that policy makers have set since then, they have missed. The second reason derives from thinking carefully about how politics works. The world lacks the institutions that would be necessary for an effective climate treaty to work. Those institutions are built on a foundation of mutual trust and a common identity. That foundation will take generations to build, let's say 100 years at a minimum. But the science of climate change doesn't give us 100 years to reach the international agreement to start things moving. Really, it gives us less than five.[25]

So, here is an important question: if there is every reason to believe that a global treaty cannot and will not stop climate change, then why do so many people continue to suggest otherwise, and why do so many resources, and so much hope, go into the continuing negotiations? My colleagues in the climate policy research community continue to talk and act and behave as if the UNFCCC negotiations toward binding international targets matter. So do writers in the media and political leaders.

One reason is probably that people don't see any alternative. The idea that a global problem like climate change demands a global solution seems intuitively right. Personally, I think it is wrong, and will demonstrate why in the second half of this book. But I think there are also other reasons that people remain committed to negotiating a global treaty. It is important to understand these reasons because they are valid. Indeed, they suggest that there may be reason to continue negotiating in the UNFCCC, even if it is clear these negotiations will be relatively meaningless from a climate perspective. They may serve other goals.

First, people may be committed to the UNFCCC process because they are committed to the idea of a global government and see climate change as a vehicle for making progress in this direction. Mike Hulme, a leading climate policy expert from England, argued this in his 2010 book *Why We Disagree about Climate Change*.[26] If climate change did not actually exist, he suggested, we would invent it precisely because it is a concrete problem that appears to demand a global solution: solving climate change would necessarily lead to the creation of those global institutions that we need to solve other problems. Personally, I noticed this sentiment particularly strongly in Europe, several years ago, back when George W. Bush was the American president. Europeans, in my experience, hated Bush. They saw him as doing whatever he wanted to serve his own political ends and (in their eyes) screwed-up worldview. And they felt largely powerless to stop him. My impression was that a strong undertone in climate negotiations at the time was that we needed something to stop the United States doing whatever it wanted. A strong global government could do that. Hence, we needed the UNFCCC to be effective, as a step toward strengthening global governance.

Second, people may be committed to the UNFCCC process because it represents an essential prerequisite to getting the full benefits from carbon markets.[27] Without a global treaty, those countries that do put in place a carbon tax or auction carbon permits will give industry an incentive to relocate to jurisdictions that do not impose this additional burden.[28] One way to fix this would be with a duty on imports from these latter jurisdictions, but this could easily violate principles of free trade.[29] Without a global treaty, as well, one of the main benefits of carbon markets, relative to other policy instruments, is lost. That is the flexibility, guided by the market, to locate emissions reductions where they are the least expensive to achieve. Global carbon markets also represent an opportunity for developing countries to earn a lot of money through the sale of emissions permits to developed countries.[30]

People support the creation of carbon markets for a variety of reasons. Some do so because they believe that these markets are the most efficient way to reduce emissions. In other words, we need carbon markets – and hence a UNFCCC agreement – not because this is necessarily the most effective way politically to limit emissions, but because it is the way that will do the least harm to the global economy.[31] Other people support the creation of carbon markets because they stand to gain a disproportionate benefit from them. This includes the people working in consulting and banking firms in London and other similar cities whose jobs are all about managing the flow of finance and corresponding carbon permits. It also includes representatives from developing countries and countries in transition, which can sell to the highest bidder the cost savings associated with locating low-carbon investment in their countries.[32] There are estimates that the establishment of a truly global carbon permit market would generate billions of dollars annually in financial flows from wealthy countries to poorer ones, but this depends on a global treaty of the right kind.[33] In other words, we need carbon markets, and hence a UNFCCC agreement, not because this is necessarily the most effective way politically to limit emissions, but because this will ensure that money moves into the pockets of those who deserve it.

Third, people may be committed to the UNFCCC process because a strong climate treaty would serve other environmental and social goals. One of the most important of these, or at least the one most heavily publicized, is biodiversity. A few years ago, I signed up to teach a class on global governance, which would be one of the electives in a master's program on global change ecology. As I was planning the syllabus, one of the things I realized was that I knew quite a bit about the global climate governance regime but next to nothing about the global regime for ecological protection. Indeed, there is a regime, embodied in the Convention on Biodiversity (CBD), which was negotiated and signed the very same day as the UNFCCC, back in Rio de Janeiro in 2002. During this time of preparing to teach the course, I found myself one day sitting in the front row of seats at a major conference; sitting next to me was the former environment minister of Costa Rica, a country that has been at the forefront of ecosystem protection. This was an excuse to talk to him. I asked him what he thought a group of master's students ought to know about the CBD. And this is what he said: if we are going to preserve biodiversity, it will *actually* happen because of the UNFCCC, and the measures being taken there to influence land use. So teach them about the UNFCCC.

A few pages ago, I described how efforts to reduce emissions from land-use practices came to be included in the Kyoto Protocol. Technically, the program that deals with this is known by its unpronounceable acronym, LULUCF, standing for Land Use, Land-Use Change and Forestry. Under Kyoto, countries can count the effects of afforestation and reforestation programs on their net emissions, and afforestation and reforestation programs in developing countries can be eligible for CDM financing. This provides a strong incentive to plant trees in places where trees are not now standing. What people soon realized, however, was that it missed a much larger problem; namely, the fact that people are still busy turning forests into other things, like pastures. On average, the world loses 5.2 million hectares of forest annually, which leads to a huge flux of carbon from the trees and the soil into the atmosphere.[34] Often the economic benefits people receive when they cut forests down are exceedingly small; if there were a way to pay them just a little bit of money *not* to cut their trees down, they would gladly leave them standing. If you take the amount of money you would need to pay them per hectare of forest and divide this by the carbon emissions caused by deforesting that hectare, you arrive at a cost of avoided emissions, and the number is very small, roughly $10 per ton of carbon.[35] That makes avoided deforestation about the least expensive ways to reduce CO_2 emissions, at least right now. So, people started figuring out ways to channel money in this direction based on this apparent win-win outcome: saving money, at least in the short-term, on CO_2 emissions reduction and protecting vast areas of forests and all of their associated biodiversity. They came up with a new acronym, this one pronounceable – REDD – standing for Reduced Emissions through Deforestation and forest Degradation.

If you go to a COP these days, you will be forgiven if you think it is actually a forum for negotiating REDD. The interests lined up behind REDD are huge, precisely because of its win-win character. The negotiations have also turned out to be very challenging because it turns out that there are also a lot of problems with the idea, similar to the problems with the CDM. How do you tell if the reduction in deforestation is something that would not have happened without the REDD financing? This requires a realistic model to predict which land would have been deforested were it not for the financing. How do you tell if the reduction in emissions are permanent, that the person taking your REDD financing this year does not turn around and chop down the forest next year? That turns out to be difficult, especially in the case of changes in land management that don't actually destroy the forest but degrade it in less visible ways that nevertheless release carbon from soils and other biomass. In addition to private landowners and developers, should governments be able to get REDD financing if they change their laws and regulations to make deforestation practices less attractive? With all these issues, negotiators have been extremely busy for the past five years making deals, and there is a huge collection of people who have become invested in the idea and would be extremely disappointed if it did not come to anything. They believe that it is most likely to come to something in the context of a global climate treaty, particularly one that enshrines carbon markets.[36]

The other important goal that the negotiation of a strong global climate treaty could advance is that of climate adaptation, which in turn is closely linked to sustainable development and poverty eradication. Most people agree that climate adaptation and development are closely linked. One of our own studies found that effects of improvements in living standards in developing countries could more than compensate those of climate change: wealthier, better educated, better fed people are simply less vulnerable to climate change and climate hazards.[37] A follow-up study focused on the role of education and found this to be the most important factor of all the development indicators, again with the message that if you want to protect people from climate change, the best investments may be not in things like sea walls and irrigation systems, but in schools.[38] At the same time, the potential cost of adaptation measures is large. The World Bank has estimated on the order of $100 billion per year.[39] That is an amount roughly equivalent to all the formal development aid flowing from wealthy countries to less wealthy ones.

When it comes to climate adaptation, developing countries have both a strong moral argument and a strong negotiating chip to get that kind of additional money. The moral argument is that it is primarily the wealthy countries that caused the problem of climate change and hence are burdening the poorer countries with the cost of adapting. The negotiating chip is the fact that if climate change is to be solved through an international treaty, ultimately, not just Annex 1 countries, but rather all countries will have to be on board. If you want us on board, the developing countries say, then you pay for our adaptation. Or you pay for our mitigation – the direct costs to us of being on board.

And, indeed, this was the amount vaguely promised in Copenhagen, enshrined in the horribly vague Copenhagen Accord, for activities related to both adaptation and mitigation. Since then, negotiators have been trying to devaguify it (if that is even a word), stipulating who will pay into the $100 billion annually, who will receive from it, and who will administer and enforce it all. As with the money that might flow on account of REDD, which also has a strong moral argument behind it, it's all resting on reaching an international agreement.

Carbon is not like cows

I opened the previous chapter by writing about smoking and ultimately argued that we should think about carbon much the way we think about cigarettes. For both, a heavy tax is one way to discourage their use, but only one way among many and maybe not a very good way at that. In this chapter, my intention is the opposite. I opened with a story of cows. I want to argue that we must not let the story of those cows guide our thinking too strongly when it comes to CO_2 and climate change.

The cow story is quite simple and embodies the elements of the game theoretical model of the tragedy of the commons. Every farmer in the village of Grindelwald had an incentive to graze more cows, rather than fewer, on the mountain pastures. All of them would have been better if they could negotiate an agreement to limit the total number of cows and come up with a fair system for deciding how many cows each farmer was entitled to have. This is exactly what they did. The agreement led to 600 years of relative prosperity and avoided the environmental damage that comes with overgrazing.

But imagine that the Grindelwalders had discovered that they could not have any cows at all, that having any cows at all was fundamentally unsustainable. Suddenly, the effects of an agreement to get rid of cows would be altogether different. An agreement to eliminate the cows would not enhance their immediate economic productivity. Zero cows cannot be more productive than lots of cows. The fundamental problem in this case would not be enhancing the productivity of an economic system based on grazing, but rather making the transition to an economic system based on something else. The game theory that suggests that every farmer has an incentive to participate, as long as free riding could be eliminated, no longer applies. From a game theoretic perspective, this is a completely different problem. And this is where we stand with climate change, if we take the 2°C target seriously. That is why signing a strong global treaty for climate change is not the necessary prerequisite to meaningful action at the national level and also why signing an ambitious treaty is so difficult.

There are two elements of the Grindelwald story that I omitted and that everybody else omits as well. The first is the question of how long it took the farmers there to reach their agreement. Indeed, the issue only occurred to me recently. Had it taken the people of Grindelwald months, or years, or centuries to negotiate their *Taleinungsbrief*? I sent an email to Martin Grosjean, the Swiss professor who had told me the Grindelwald story in the first place, asking him. His reply: "I have no idea, and I doubt that anybody would know."

This timing issue really matters. Maybe it took several generations for the people of Grindelwald to come up with their agreement, and maybe, over that time, they did a fair job of destroying the beauty of their valley. I am sure it was an issue to them then, even if now, 600 years later, we forget to ask the question. For us, now, dealing with climate change, we are in a hurry. If we can come up with a strong and effective global treaty in the next few years, such as by the December 2015 deadline that UNFCCC negotiators are currently working against, then perhaps such an agreement can be the backbone of an effective solution to climate change. If it will take much longer than that, then it will not. In fact, there is every reason to believe that it will take much longer. If we want to solve climate change, then we had better find some way to do so that does not depend on a global treaty.

As David Victor and numerous other political scientists and international relations scholars have demonstrated, you can devote a huge amount of time and energy to trying to devise the basic architecture for an international agreement that would garner the maximum commitment from the maximum number of states and ultimately do the most good.[40] But, at the end of the day, what matters most is whether the sum total of their commitments would be enough to really solve the problem.

To be enough, the treaty would need to effectively hand over sovereignty on the most important energy policy decisions to an international governing body. This is only going to work out well if there is an international governing body that is really up to this task. One that people trust, that is accountable, and that has the power to enforce its will on free-riding states. The world lacks such a body. Because political leaders aren't stupid, they see its creation as a precondition to an effective treaty. Creating such a thing isn't easy. The European Union tried to do something similar, at the regional level, when it introduced the euro, creating a body out of thin air to administer and govern it. The recent result has been to seriously harm the economies of Ireland, Spain, and Greece, leaving millions out of work and without hope for their children's future. The hard fact is that you cannot create such bodies overnight, not at the regional level and certainly not at the global level. At the global level, we should plan on decades at best. A century would be more realistic.

There are a lot of people who continue to support the negotiation of a strong global treaty. I am one of them, and, if a strong treaty were signed, against what I see as all odds, I would celebrate. But I also think that the reasons for supporting such a treaty, right now, largely have to do with concerns other than solving climate change as quickly as possible. One reason is that through a strong treaty we will accelerate the development of global governance institutions. These won't be in time to solve climate change, but they may be in time for other global problems. A second reason is as a necessary precondition for effective carbon markets. I don't think this is particularly important, but other people do. If successfully implemented, these offer the promise of reducing the net cost of solving climate change while also transferring money to worthy individuals and groups. A third reason is to funnel money into biodiversity protection and climate adaptation, as well as sustainable development more generally. All of these other reasons are to varying extents legitimate, and, perhaps on account of them, there is reason to continue pressing forward for the next climate agreement.

The important thing, however, is not to equate the political will to solve climate policy with negotiators' success at signing a global agreement. Signing a meaningful global agreement structured anything like Kyoto would require the political will not just to solve climate change, but also to fundamentally alter the balance of power between sovereign nation states and global governance institutions. Likewise – and that is where this book comes in – the failure to agree on a strong global treaty should not lead us to believe that solving climate change is itself impossible.

I have suggested that because of the time required to develop strong global governance institutions, the time it will take to develop a global agreement to meaningfully limit CO_2 emissions ought to be measured in decades, not years. Perhaps, ironically, we may in fact get a meaningful global climate agreement sooner than that, but only because it is no longer needed. There is reason to believe that states would agree collectively to eliminate their CO_2 emissions as soon as they are willing to do so unilaterally, unconditional on other states' behavior. As I will begin to describe in Chapter 8, this state of the world – in which states see decarbonization as in being their own interests – is increasingly within our grasp.

This relates, too, to the second untold part of the Grindelwald story. The *Taleinungsbrief* accomplished its objectives of limiting the number of cows. Today, however, Grindelwald no longer has a problem of too many cows, but rather of too few. The cows grazing on the Alps, their bells gently ringing, create a set of sights and sounds that draws tourists from all of the world. Today, it is these tourists, not the cows, who form the basis for Grindelwald's local economy. The incentive to free ride now goes the other way; namely, for farmers to convert their barns into hotels. They can make money off summer tourism but that tourism itself depends on the fact that some other farmer is working hard, getting up at 5:00 every morning, to take care of the livestock. When the economic production system changes, so too do all of the incentives.

6

Changing the way we live

As my forty-fifth birthday was approaching, my wife asked me what I would like for a present. Like many of us, I am pretty sick of owning a lot of stuff, clutter that eventually finds its way up to the attic, and so I didn't want to get a physical object. Instead, I asked her if she could give me one of those experiences I wanted to try at least once in life. In this case, it was skydiving. She organized the whole thing, and, on a sunny late-summer Sunday morning, we found ourselves at a little airport an hour's drive from our house. I met the instructor to whose belly I would be strapped, and he explained what would happen from when we climbed into the back of the plane with one other tandem team and four solo jumpers, all the way to touching down again on a patch of grass next to the hangar. Then it all started, and the main feeling I was experiencing was one of complete terror, even though I knew, intellectually, that everything would be fine. Sitting in the plane next to a curtain covering the opening through which I was destined to exit once we had reached the appropriate altitude, I tried to distract myself from my fear by asking the instructor lots of questions. Then we got there, the curtain was opened, the solo jumpers tumbled out without hesitation, and the other tandem team took a few moments to exit as well. The state of fear continued as my instructor slid us to the opening, and I sat there, my legs dangling out, and my feet resting against a metal bar that I couldn't see. Then he pushed us, we tumbled once, I caught a glimpse of the airplane receding into the sky, and then we stabilized, looking down with arms and legs splayed out. At that moment, everything changed. All fear vanished. It felt like we were suspended motionless in the sky and could stay that way forever. Intellectually, I knew that things were otherwise and that if this state of affairs lasted for more than about seventy-three seconds, we would slam into the ground. Of course, the instructor pulled the ripcord a good twenty seconds before then, and we floated slowly down the last bit to land gently in front of my wife and children. That little adventure safely out of the way, we went out for brunch at our local Chinese restaurant in Vienna.

Let's be clear about the little adventure that our generation and the one before us have been enjoying. For most of human history, tens of thousands of years, growth in both population and material consumption had been modest and irregular. Perhaps people simply didn't want to consume that much, but more likely they simply lacked the means. Then, around 1950, everything changed. Collectively and individually, people figured out how to sustain high growth rates year-to-year, decade-to-decade. On every graph charting out indicators of human development and prosperity, lines that had been more or less flat or squiggly took a sudden turn upward and continued in that direction.

Will Steffen, former executive director of the International Geosphere-Biosphere Programme, calls this the "great acceleration" and attributes it to a set of financial, institutional, and technological factors.[1] In 1950, Gross World Product – the value of all economic services – was $7 trillion (measured in 2010 dollars); by 2011, this value had risen to $ 77.2 trillion.[2] Think about it. Less than 10 percent of the growth in economic activity that has taken place since the dawn of humanity happened before 1950, and the other 90 percent in the sixty-four years since.

Like me, however, you were probably born after 1950. Even if you were alive before then, you have no memory of what things looked like through the eyes of an adult. The world that seems normal to us is not that of the thousands of years of human existence, but rather that of the few decades that we have been around to witness, which happen to have been very unusual. Thinking about our time on Earth is like comparing the forty-five years and six hours that I had lived until I got launched out of an airplane to the fifty-three seconds of free fall that occurred next. Fifty-two seconds in, the ground was comfortably far away and there was nothing in my life's history to indicate that I was in any danger. There are reasons to believe that humanity is in the same general position, one that is in very many ways unsustainable. It's time to deploy the parachute and slow things down.

One reason to slow down has to do with the climate. Economic growth is a driving force for climate change, as the most recent assessment report of the Intergovernmental Panel on Climate Change (IPCC) has made clear.[3] In the decades between 1970 and 2000, global greenhouse gas (GHG) emissions rose at an average rate of 1.3 percent, whereas between 2000 and 2010 that rate increased to 2.2 percent. The IPCC authors attempted to figure out what factors accounted for this change, and, to do so, they disaggregated the emissions geographically and according to a set of underlying societal factors. In terms of geography, it became clear that virtually all of the growth in emissions took place in Asia, whereas in places like Europe and North America emissions had pretty much flattened out. In terms of underlying trends, they divided the change in the emissions into four underlying factors: population, gross domestic product (GDP) per capita, energy use per unit of GDP, and CO_2 emissions per unit of energy. What they found was that the first two factors – population and per capita GDP – accounted for essentially all the growth in emissions. Of these two, GDP was the larger driver. This story fits with the geographical one because it is in the years since 2000 that China and, to a lesser extent, the rest of Asia as well have experienced an unprecedented economic boom.

But climate is not the only area of unsustainability. Several years ago, a group of leading scientists published a paper in the journal *Nature* in which they tried to describe the magnitude of this problem.[4] They described a set of ten planetary boundaries associated with different human impacts on the environment. They suggested that in two areas – biodiversity loss and changes in the nitrogen cycle – the Earth had already progressed beyond a point of no return, with potentially disastrous consequences. Climate change was an area where the margin of safety had been infringed upon, and a point of no return was close. Ocean acidification, changes to the phosphorous cycle, freshwater use, and land-use change were areas where global impacts were noticeable, but the margin of safety was still intact. Two other areas – atmospheric aerosol loading and chemical pollution – were not yet quantified. For all of these, one could probably attribute the causes of the problem being primarily economic growth, just as with climate change. So it makes sense that one could do a lot of good by halting economic growth. That is the subject of this chapter.

Consumption and growth

Today, the idea that caring about the environment is synonymous with altering one's consumption patterns is widespread. A good example is from a recent German romantic comedy, *The Hearts of Men*, in which one of the characters was defending his own credentials as an environmentalist. "The environment," he exclaimed, "is precious. And I care about it. Especially global warming. Nobody cares about it more than I do. Why," he claimed, "I am just trying to do my part. And I am doing my part, everything I can. Why," and then he screamed, "I RIDE A BICYCLE!"

The fact is that many of us try to do our part to stop climate change by altering our lifestyles, things like riding bicycles or turning down our thermostats or eating locally grown vegetables. Within the social sciences, and within the field of ecological economics in particular, the assumption is widespread that fixing environmental problems such as climate change demands a change in lifestyles toward reduced consumption.[5]

One of the first to rigorously study the effects on the environment of the growth in consumption was Jay Forrester, a professor in the business school at the Massachusetts Institute of Technology (MIT) in the late 1950s. Strangely, it grew out of work he was doing with the management of the General Electric Corporation, trying to help them figure out why they seemed to always be hiring and firing employees instead of being able to maintain a stable labor force. Forrester's insight was to treat the corporation as an ecological system having stocks and flows of various things, including people. If management needed to increase inventories, then it needed to increase the flow of production, and, for that, it needed a greater stock of employees in the manufacturing department. This in turn created the need for a flow of newly hired people, for which they needed a stock of people in the hiring department. When the inventory became large enough, everything would go in reverse. But all of the processes took time, and Forrester suspected that the time lags generated cyclical patterns of employment that were of larger magnitude than the business cycles affecting their inventories in the first place. To figure out if this were the case, he developed a computer program to simulate the stocks and flows of people in the corporation. Computers were still new, but MIT was a good place to be if you needed to find people who could program them. Forrester's model of General Electric demonstrated his hypothesis and gave GE a tool with which to find solutions.[6]

The idea that a computer program, what he called a *system dynamics model*, could solve a social problem proved to be an immediate hit. Forrester's next big project was developing a systems dynamics model of a city to examine the effects of the urban renewal policies that the U.S. government had put in place. Surprisingly, his model showed how the construction of low-cost urban housing, part of that urban renewal program, displaced denser job creation industries and actually led to greater urban unemployment. This created the need for even more low-cost housing in a never-ending spiral that was killing the very cities people wanted to save.[7]

The results drew attention beyond the United States. In 1970, a group of civic-minded wealthy industrialists known as the Club of Rome approached Forrester and asked him if he could scale his model up from a single city to the entire world. He did, programming a model he called World2, which colleagues of his then improved and renamed World3. World3 revealed big problems on the horizon. They published the results in 1972, in a book entitled *The Limits to Growth.*[8] The book, by its very terms, was crude and exploratory. The modelers looked at five variables: human population, industrialization, pollution, food production, and resource depletion. What they did was to explore the consequences of growth in the first two on the states of the final three, taking into account feedback loops within the system. Then they considered three basic scenarios of future development, all of which were reasonable expectations of what might happen. Two of these scenarios led to a rapid collapse of food supply and resources sufficiency by the latter half of the twenty-first century. The other appeared not to lead to such a collapse. The take-home message, then, was that humanity had better be careful about how it chooses its future.

The Limits to Growth and other books like it, dating back 200 years to Thomas Malthus, have fallen under a great deal of criticism. So far, the cataclysms they have predicted have failed to pass, headed off by human ingenuity, a factor these books have largely ignored. The Green Revolution, for example, helped to avoid mass starvation in the face of steady population growth. I had the chance to meet the lead author of *The Limits to Growth*, Donella Meadows, a few years before her untimely death in 2001. She explained to me that the real value of her work, of the computer model, was to reveal how fast exponential growth can sneak up on you. Her example was of a pond with some lily pads. The lily pads double in number every day and, over the course of weeks, come to cover half the water surface. That is exactly one day before they cover everything, blocking light entirely and killing everything beneath. Our consumption is increasing exponentially. The rate of innovation may be as well, and it may head off collapse at a global scale, even as it has often failed to prevent collapse at the local to regional scales.[9] The problem is that we just don't know.

Although *The Limits to Growth* may not have had much of an effect on global politics, the earlier findings – at the scale of the city – did make a difference. It has also been at the scale of the city that policy makers have attempted to promote efficiency- and lifestyle-based changes in order to solve climate change.

In 1993, the Toronto-based International Council for Local Environmental Initiatives (ICLEI) launched a program called Cities for Climate Protection (CCP). ICLEI's goal was to recruit municipalities to join whose collective CO_2 emissions represented at least 10 percent of the world total, a target they have easily met. Having moved their headquarters in the meantime to Bonn, Germany, they have recruited more than 500 city governments to engage in a five-step process: measuring local emissions, committing to an emissions reduction target either for the entire city or for the city government's operations, plan a set of actions to achieve the target, implement their local climate action plan, and monitor the results of the plan.[10]

The main avenue toward reduced urban emissions is improvements in efficiency through better system planning. What can local governments achieve? For the operations of the city government – things like government buildings, police and fire departments, and public buses – they can do quite a lot, including constructing low-energy and passive solar buildings, switching vehicles from gasoline and diesel fuel first to natural gas and then to electricity, improving the efficiency of street lighting, and managing waste disposal facilities to reduce the escape of GHGs such as methane. To reduce the emissions of all the people living in the city, it gets a little bit harder, but there are still important possibilities. On the efficiency side, the city government can initiate a district heating system, something I describe in Chapter 8. Next, it can provide support to homeowners to insulate their buildings or install rooftop solar panels. In Berkeley, California, for example, there is an innovative program to lend homeowners money for solar panels in a manner that adds the loan payments to their property tax bill over a period of twenty years.[11] The additional tax for the solar panels is more than offset by the reduction in the electricity bill, and there is no need to pay off the loan when they sell the house; it simply passes on to the new owners for the remainder of the twenty years. City governments can change their zoning laws to increase building density and create mixed-use neighborhoods, thus reducing the problem of sprawl and, with it, the need to drive cars long distances. Coupled with this, they can invest in public transportation, reduce downtown parking, and impose a fee for driving in the central business district, all so that people will drive less.

In their first few years, the CCP cities mainly addressed the issue of energy efficiency and, to a large extent, through their own government operations. The results? By 2006, thirteen years into the program, CCP cities had reduced annual CO_2 equivalent emissions from business-as-usual projections by 60 million tons.[12] Most of these savings were in North America, where 290 cities, home to 73.6 million people, saved 27.5 million tons, and in Europe, where 156 cities home to 38.2 million people, saved 23.5 million tons. Sixty-nine Asian cities home to 71.5 million people managed 7.5 million tons of savings. Thirty-one cities in Latin America and Africa, home to 59.8 million people, saved 1.5 million tons. It's worth noting, of course, that per capita emissions in Asia, Latin America, and Africa were a lot lower to start off with.

These numbers seem impressive in absolute terms. But, like ecological modernization more generally, the gains of locally planned efficiency improvements are fleeting when you compare them to the big picture. Global annual CO_2 emissions are about 42 billion tons, with roughly 8 billion of those tons accounted for by the cities in the CCP program.[13] So, the thirteen years of work managed to reduce global emissions by about one-tenth of 1 percent from what they otherwise would have been and emissions in the participating cities by about 0.7 percent.

All of the actions taken by CCP cities are probably great things, adding to the quality of life, the efficiency, and maybe even the economic attractiveness of living in these particular places. But they have made less than barely a dent in CO_2 emissions. All these programs have attempted to improve the energy efficiency of economic activity within the context of a city and yet none of them – for obvious political reasons – has attempted to slow down economic activity and associated consumption. This in turn suggests two questions. First, what are the prospects for reducing consumption at the societal level to an extent that would be meaningful for climate change? Second, is it actually necessary to reduce consumption in order to fix climate change?

Real prosperity

There is a large and growing scientific and policy-oriented literature on reducing consumption and growth that, for the purposes of this book, boils down to three findings. First, within relatively affluent societies, there are really good reasons for individuals to want to limit their own levels of consumption. Second, nobody really knows how to get affluent people to reduce their own personal consumption short of waiting for them to discover this on their own. Third, when too many separate individuals reduce their own consumption, some immediate problems develop at the societal level, making consumption reduction unattractive to policy makers. I will explain each of the points.

Let's start with the individual benefits of reduced consumption. In June 2000, the British government under Tony Blair set up an advisory group called the Sustainable Development Commission to look into these issues in greater detail. Led by the economist Timothy Jackson, it lasted for eleven years, and, during this time, it was among the first government programs to seriously explore the links between rising consumption and human well-being. They produced two large reports on the topics, *Redefining Prosperity* in 2003, and *Prosperity without Growth?* in 2009.[14]

Let's start with the first one. *Redefining Prosperity* addressed the basic question of how much consumption is necessary for human well-being. It gathered evidence that, in many ways, consumption is not contributing to prosperity at an individual and community level and may be undermining it. The strongest predictor of people stating that they are happy and satisfied in life is the strength of their interpersonal connections: whether they are in a stable relationship, whether they have a network of good friends, and whether they are active in community groups and activities. This echoes research in psychology and behavioral economics, which has clearly demonstrated that it is relationships and life experiences that correlate most closely with expressed feelings of life satisfaction and well-being, rather than consumption and achievement.[15] Material consumption is not completely irrelevant, but rather it appears that there is a basic threshold of consumption that is important. Above this threshold, consuming more brings very little additional happiness, whereas differences in other aspects of one's life matter a great deal.

Figure 6.1 shows the results of a national poll taken in the United Kingdom, where people were asked to indicate the factor that

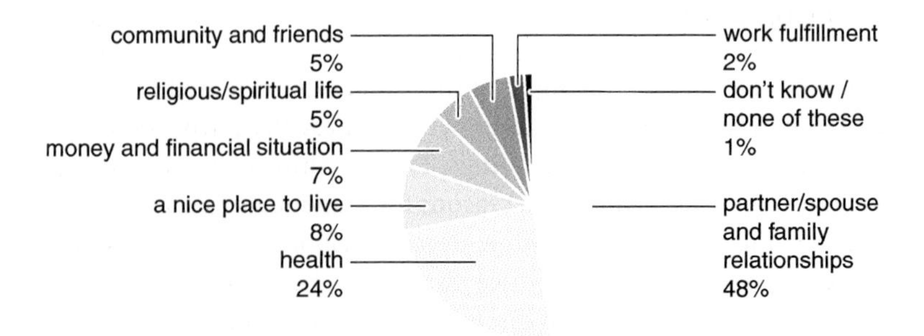

Figure 6.1. *The factors that matter most for happiness and well-being.*[16]

contributed the most to their happiness and well-being.[17] Just about half of the people answered that it was their relationships. Another quarter answered that it was their health. Certainly being in good health has some correlation with income, but the link is one of a threshold and has a lot to do with governance: below a certain income level, you might not be able to afford proper health care, especially if there is no public provision or good things to eat. A nice place to live came in next on the list; this clearly is linked to income, but it can be somewhat of a zero sum game. In Britain, right now, there are only so many nice places to live, and the rich may have already grabbed them. Making everybody twice as wealthy won't change that. Money and finances did come in next on the list. So it isn't completely unimportant.

What the *Redefining Prosperity* report documented was that, for the United Kingdom, a lot of the growth in consumption that has occurred has been at the expense of other factors that appear to influence well-being. Most specifically, societies that have experienced high growth have done so in ways that undermine the strength of communities. The Commission noted that the weakest British communities back in 1971, the ones with the least active community organizations, were stronger than the strongest British communities in 2001, thirty years later. The sociologist Robert Putnam, in his book *Bowling Alone*, has reached similar conclusions for the United States.[18] Economic growth since about 1970 has left people increasingly isolated from each other.

An important insight from this work is that the relationship between material consumption and well-being or happiness is not a simple linear one. For people living in poor communities, increases in wealth and commensurate consumption do bring gains in expressions of happiness and overall well-being. One observes, for example, members of poverty-stricken societies expressing low levels of life satisfaction. Once a society or community has reached a particular level of wealth, however, additional increases in average consumption do not lead to observed increases in average levels of well-being. The residents of Bermuda are not, because of their wealth, on average happier than the residents of Cuba, with both countries being above the threshold. Within any given society or community, by contrast, it is typically the wealthier who express greater life satisfaction. Above a particular threshold, then, it is relative wealth, and not absolute wealth, that matters. But if everyone in a community is seeking to become the wealthier ones within that community, it is a zero sum game.

These findings are also consistent with a more philosophical inquiry into the matter of consumption. Robert and Edward Sikelsky explored this in a recent book, *How Much Is Enough? Money and the Good Life*. They addressed the question of what level of consumption might be best – or at least better than others. To answer this, they proposed a set of seven basic needs that form the basis of the good life: health; security; respect; personality, by which they meant to capture a sense of individuality and autonomy; harmony with nature; friendship; and leisure, by which they mean the opportunity to do something, including nothing at all, simply for its own sake. A good life, they argue, is one that realizes all seven to some extent or another. More importantly, "the lack of one cannot be offset by abundance of another."[19] For those who are relatively lacking in the basic economic conditions underlying health, security, the ability to express oneself, and to engage in interesting pursuits of leisure, growing levels of consumption are a good thing. Beyond a certain point, however, the added benefits become fewer, and the imbalances in life begin to emerge. For the already-wealthy, unbridled consumption adds only to overabundance on the personality side but can leave them impoverished in the other dimensions. This is especially the case when, to achieve that added consumption, they must work harder, sacrificing leisure and perhaps health and friendship along the way.

The Sikelskies also set out to answer the puzzle of why people continue to work more and enjoy less leisure the wealthier they get, given that this seems to be a bad strategy for happiness.[20] They identify a few reasons. Some operate at the level of the individual, such as the fact that while one's absolute level of consumption has little effect on happiness, one's level of consumption relative to one's neighbors does. Others operate at the level of society and the firm, such as the fact that it is simply more efficient to divide a given job among fewer people: those people, working more, learn faster, and this boosts productivity. So, firms have an incentive to try to get their employees to work longer hours, and the employees lack either the bargaining power or willpower to say no.

So, what are the instruments that policy makers could employ to get people to reduce consumption, and what are their prospects for effectiveness?

In their book, the Sikelskies were more concerned with ensuring that people have the "good life" and less concerned with reductions in consumption per se. As to the former, they offer a wealth of ideas, whereas for the latter they have relatively few. To help ensure the good life, government could actively work to see that a broader range of needs is met, rather than tailoring their policies purely around income growth and job creation. They could, for example, guarantee a basic income (to increase security), engage in thoughtful spatial planning (to promote health and the time for leisure), and support communities (to promote friendship) and good environmental legislation (to promote health and harmony with nature). One might notice that, indeed, these are the kinds of things that European governments (at least the center and center-left ones) pride themselves on. And, indeed, that is why my family and I decided to move from the United States to Europe. When it comes to reducing consumption per se, however, the Sikelskies' ideas are thinner. One idea is to impose taxes on consumption. Another is to place limits on advertising, an activity that is specifically designed to get people to spend more money.

Let's turn to Timothy Jackson and the Sustainable Development Commission.[21] They investigated the question of policy interventions in a more quantitative manner and also tied this with some of the possible negative consequences of those interventions. Figuring prominently in their report *Prosperity without Growth* is the work of a Canadian economist, Peter Victor. Victor programmed a computer model of the Canadian economy and used the model to estimate all the interconnected variables: things like output, investment, consumption, public spending, employment, and poverty. By altering a few critical variables over which the government does have some control – things such as investment and savings rates – Victor could generate a variety of future scenarios looking thirty years into the future. Victor started as most economists start, with a so-called business-as-usual scenario. The rates of investment, savings, and other "control variables" stayed at current levels, and, as a result, consumption grew at a healthy clip of a couple percentage points a year, much as it usually does in the real world. Incidentally, the model also calculated CO_2 emissions, and these rose with rising consumption.

Next, Victor experimented with the control variables, to see if he could get the model to stabilize consumption at current levels and explore what this would mean for everything else. The simplest way of doing this is by restricting credit, so that net borrowing by individuals and businesses declines, as does new investment. The results were successful in that consumption stabilized and CO_2 emissions fell slightly, by about 10 percent over the thirty-year period. But the cost of this was enormous. Unemployment began to climb almost immediately, more than tripling over the thirty-year period. Poverty followed soon, more than doubling. Overall debt first declined, as borrowing slowed, but then shot up as the government coped with unemployment and poverty in the face of a stagnating economy.

But Victor kept on experimenting. Eventually, he found a portfolio of interventions that managed to slow down and stop the growth in consumption and in CO_2 emissions as well without the negative effects on employment, poverty, and debt. The differences between this scenario and the previous one included shifting new investment from the private to the public sector, placing limits on the number of hours that individuals can work, and phasing all of the interventions in over a longer time period so that there were no dramatic shocks to the system.

What can be learned from Victor and Jackson is that policies can be developed that would reduce consumption at both the individual and the societal levels, but these risk doing a lot of harm to the things that politicians care about. Victor's model was a deterministic one, meaning that it did not recognize the huge range of uncertainty in a lot of key aspects of how society functions now or in the future. Within that deterministic model, he was able eventually to find a solution that stopped growth without doing economic harm to a lot of people. The real world of the present and future, however, doesn't exactly match his model, or indeed any deterministic model that anyone could build. The more that policy makers try to limit growth, the lower their odds are that they can avoid causing economic harm to large numbers of people in society.

That, of course, is the main explanation for why most policy makers actually try to do the reverse: namely, promote faster and more stable economic growth. Sikelsky and Sikelsky document the main driving force for Western governments' attention to economic growth as a desire to promote full employment. Every year, technical advances make it possible for factories to produce stuff with less labor. A car factory installs a robot arm to screw in headlights. A textiles mill figures out how to run the looms at a faster speed. The only way that countries have figured out to keep people employed, which we know is what keeps them happy, is by promoting steady economic growth. Every year, people keep buying more cars and more fabric, or better cars and better fabric, than they did the year before. If economic growth were to stop, the need for workers would fall, unemployment would rise, and politicians would lose elections. Every time that the economic cycle turns into recession of its own accord and unemployment begins to increase, policy makers take special steps to boost consumption.

Even if there are tools that people have thought of so far to reduce consumption, it seems very unlikely that policy makers would actively promote policies substantially limiting economic growth, enough to make a meaningful contribution to climate change. In Victor's model, in which a careful portfolio of policies managed to slow down and stop economic growth while employment remained high, CO_2 emissions only fell by about 20 percent over the thirty-year period. That is nowhere near enough. The model suggests no way of reducing consumption while maintaining employment sufficient to get CO_2 emissions to fall by the 60–70 percent necessary over this time period.

Decoupling growth from emissions

A few pages ago, I posed two questions. The first, which I have just addressed, was whether the prospects are good for reducing consumption at the societal level. The answer was that they are not good. The second question was whether reducing consumption is necessary to halt climate change. I address this question now. To do so, I am actually going to flip it over into one that is easier to answer. Is it possible to "decouple" CO_2 emissions from economic growth? In other words, how possible is a world in which income and consumption levels continue to rise but CO_2 emissions fall and eventually disappear?

There are three ways to answer this question: from the data on CO_2 emissions, from analogy to other data, and from models. Let's start with the first approach, which is the one that Jackson and the Sustainable Development Commission took[22] and that is consistent with one of the headline messages to come out of the most recent report of the IPCC.[23] Overall, the picture is bleak: rising consumption has consistently gone along with rising energy use and with rising CO_2 emissions associated with the fossil fuel emissions that comprise a large fraction of that energy use.

There are some finer level nuances, but, unfortunately, they do not alter the big picture. One nuance is that, over time and as consumption rises, the ratio of material use to overall consumption falls. Jackson describes this as "relative" decoupling, and it derives from two different factors. First, continuous innovation means that the value associated with a particular material product rises over time, even if the materials embedded in that product do not. I am typing this book on a laptop computer connected to a flat-screen video monitor. Combined, they are, on the one hand, smaller and lighter, and, on the other hand, they give me far more computing power and a better visual display than the computers of twenty or even ten years ago. The rising numbers for consumption – such as GDP – reflect this improvement in value. This trend is also on display in terms of energy use. Cars of today deliver far more performance than those of a decade or two ago without needing to burn any more gasoline. Second, our choices of what to consume have been changing, with services, rather than goods, comprising a rising share of our expenditures. In an industrial economy, people spend most of their income on stuff. In a postindustrial economy, people spend most of their income on experiences.

But the nuance that balances this is that, even as relative decoupling has been occurring, "absolute" decoupling has not. By this, Jackson means that growth in material use has lagged behind growth in total income and expenditures, but it has never gone in the opposite direction. Consider a society that goes from a per capita income of $30,000 to one of $45,000 over a period of time. At the lower level of income, the average family might spend 10 percent of its budget on transportation by owning a car, driving it a certain distance, and burning fifty liters of gasoline a week. At the upper level of income, the family might only spend 8 percent of its budget on transportation, but that would still imply owning a slightly larger car or driving it somewhat more, perhaps burning fifty-five or sixty liters of gasoline per week instead of fifty. Jackson looked hard, but he could not find a single instance where people have, on average, consumed more total value but less actual materials.

The most recent IPCC report presents data from the past forty years that echo this finding using more complicated language. As they write, "all the studies show that reductions in emissions resulting from improvements in emissions intensity and changes in the structure of production and consumption have been offset by significant increases in emissions, resulting from the volume of consumption, resulting in an overall increase in emissions."[24] In plain language, this means that while the CO_2 emissions per dollar have been decreasing, total emissions have still been rising simply because of more dollars spent.

What about looking at other data and drawing an analogy? There are a few examples that offer some room for hope and derive from research into the dynamics of so-called *technological transitions*, something I go into more detail in the next chapter. Here, I offer one of these examples, catalogued by Arnulf Grübler and Nebojsa Nakicenovic, former colleagues of mine at the International Institute for Applied Systems Analysis, and David Victor. They looked at the transition from horses to cars as a means of transportation all around the world, something that happened about a hundred years ago. From the point at which they started looking, 1850, until about 1910, the number of horses in the world roughly doubled, and, within that time frame, was rising at more or less an exponential rate. Then, in a brief period of time, between about 1910 and 1930, the number of horses suddenly declined by 90 percent, and horses were completely replaced by cars. Since then, the number of cars on the road has continued to rise, also at an exponential rate. The overall growth in the number of transportation "units" carried on unabated, but, in the middle of that growth, the units themselves underwent a rapid shift from having four feet to having four wheels.[25]

Super, you say, the transition from horses to cars is why we are in the environmental mess that we are in today. But the hidden story is that the transition from horses to cars got people out of another environmental mess, namely, the problem of urban pollution brought about by ubiquitous, smelly, fly-attracting piles of horse manure. So it is easy to imagine being an analyst in the year 1900 and asking the question: can we ever decouple economic growth from horse manure? The answer, then, might have appeared no. The more the economy grows, the more people use horses for transportation, and there is no escape. But the problem was temporary, solved by a rapid and somewhat deliberate shift in technologies. At least one particular kind of pollution did become absolutely decoupled from economic growth because the technology that generated that pollution was suddenly obsolete.

The data show that it is possible and show us that we need to be careful in how we interpret the pessimistic news of Jackson and the

IPCC authors in evaluating past data. They have been looking at the past fifty years, and these particular fifty years are a period of time in which there has arisen no really competitive alternative to fossil fuels as an energy carrier. An observer of transportation would have seen the same thing in 1900, looking back to 1850. But a surprise was just around the corner because a new technology, the car, was rapidly becoming viable. The critical question for climate change is whether the future could hold similar cards to those seen in the horses-to-cars example. To answer that, one needs to turn to models, and that is the third way to address the decoupling conundrum. For this, we turn again to the IPCC, one chapter later in their report.

What energy system models tell us

In the same IPCC report on climate mitigation, the sixth chapter – roughly the middle of the report – aggregates the results from models of the energy system. In fact, many of the major climate change research institutes have developed their own models, similar in many ways to the DICE model I described in Chapter 2, and the various teams often meet to compare their ideas and compare the results of their different models with each other. There has arisen an annual meeting involving the whole community every summer in Snowmass, Colorado, but they also come together in smaller groups elsewhere, such as via the IPCC or through internationally funded research projects.

Perhaps because of the strength of their community, almost all of the models operate according to a similar structure. That is, they have a fairly detailed set of data about the current energy supply system in terms of the different technologies making up that system, their relative efficiencies (e.g., in converting the energy in coal to energy in electricity), and their relative costs. They assume that there is somebody responsible for planning how to alter that system for the next few years, such as by building a new power plant. The models assume that what this so-called central planner does is to satisfy a set of constraints (e.g., there has to be enough electricity available to satisfy demand, CO_2 emissions need to stay below a particular value) in such a way as to minimize the overall costs of operating the system. Such decisions come along at regular intervals, such as every five or ten years, and the decisions made at one point in time affect the baseline conditions for the decision made in the next period. Such baseline conditions include the existing infrastructure in place and the costs of different technologies. Most of these models are constructed so that the effects of the decisions with respect to energy planning then feed into a macroeconomic model. The latter figures out how changes in the price of energy, as well as other factors like the potential government revenue gained from a carbon tax, affect the demand for different goods and services throughout the economy and, ultimately, economic growth.

How good are the models? It's a tough question to answer. On the one hand, they are based to a large extent on robust theory about how the economy works and robust evidence of the costs and performance of various technologies. On the other hand, they make the bold assumption that there is a central planner who is actually making decisions to minimize costs when all the evidence suggests that decisions are made primarily for other reasons, like protecting particular industries or maintaining needed levels of security. Another problem is that it is impossible to test these models, a process known to modelers as *validation*. With a climate model, you program it to start with the climate as it was observed in 1960, let the model simulate thirty years, and see if it ends up close to what was actually observed in 1990. No energy system model will do this because there are simply too many important and unpredictable events happening along the way, all of which have a profound impact on what policy makers decide to do with the energy system in any given year. Perhaps there is a war in the Middle East. Or a recession. Or someone discovers how to frack for natural gas. Where it *is* possible to validate these models, however, is against such events. In other words, you can introduce some sort of an external shock into the model – such as shutting off all the oil flowing from Iraq – and see if the changes to the energy system that the model predicts qualitatively match what actually happened after such an event in the real world. In sum, the models do not give us a very good indication of what the energy system is likely to look like at any particular point in the future because that will largely depend on unpredictable events. The models do, however, give us a first-order approximation of how a particular policy intervention, such one proposing to reduce CO_2 emissions drastically, is likely to change some key parameters of that system, such as its overall cost.

The simple message to come out of these models is that imposing a very substantial carbon constraint – one that would likely limit climate change to the 2°C target – need not have a substantial impact on economic growth. Indeed, the IPCC authors gathered a database of roughly 1,000 energy system scenarios based on these models with widely varying assumptions about the constraints that a central planner would face. About half of the scenarios were so-called climate policy *baseline scenarios*, in that the constraints the central planner faced did not include a CO_2 limit. Across the baseline scenarios, the projections for overall economic growth varied widely (again, depending on the assumptions embedded within each scenario), with the world economy becoming something between three and twelve times larger over the next 100 years. The other half of the scenarios involved the introduction of a carbon constraint on top of all of the baseline assumptions. In those cases where the carbon constraint was added early and gradually, it stimulated cost reduction in low-carbon technologies. These appear to become the backbone of the energy system later on. The long-term effect of the constraint was to reduce economic total growth over the next hundred years by about 1–2 percent.

A good way of thinking about these numbers is to consider how much the carbon constraint will delay the point at which the world would reach some particular level of economic activity. For example, a given baseline scenario might predict that the global economy will grow to a factor of five times its current level by 2070. Introducing a carbon constraint into that scenario would delay this event by a few months. Instead of being five times wealthier by January 2070, the world will have to wait until June that same year.

The models also show, by the way, that the effects on economic growth would be much more significant – several times greater – if the carbon constraint were added later on, such as thirty years from now. In those cases, the central planner would have to switch over the energy system from old technologies to new ones all at once, and the new technologies would not have benefited from decades of learning-by-doing and associated cost reductions.

Most people cite these results, correctly, to suggest that the overall cost of achieving an ambitious climate mitigation target is minimal. Ottmar Edenhofer, who co-led the IPCC working group within which this result emerged, declared at the news conference following the governments' approval of the IPCC report that "it doesn't cost the world to save the planet." But another way of interpreting the results is that decoupling growth from emissions is not only technically possible but, in a world of climate policy geared toward stimulating low-carbon technologies, highly likely. If policy makers tackle climate change in a way that relies on low-carbon technologies, they can eliminate CO_2 emissions – reducing them by 100 percent – while reducing global GDP by only a couple percentage points. On the other hand, if policy makers were to tackle climate change by taking steps to reduce global GDP while doing nothing to promote new technologies, then the effects on the climate would be minimal. Let's say that policy makers managed to reduce GDP by 20 percent from its baseline amount, leaving technologies unchanged. That would reduce CO_2 emissions by 20 percent.

Of course, a key question is whether the transition to the new energy technologies will harm the world in new and different ways, echoing what happened when the world switched from horses to cars. We fix the climate, but we make other environmental problems even worse. Certainly, there are some reasons to fear this. Photovoltaic panels, for example, contain toxic chemicals, and so it is possible to imagine that manufacturing them, or later disposing of them, could contribute to local pollution problems.

What is critical to do here is to compare the environmental impacts of the new technologies with the environmental impacts of the old. My own research group has started to look at this by focusing on the particular impact of water consumption. We analyzed the effects on water resources of a switch from fossil fuels to solar energy, concentrating on the Middle East and North Africa (MENA) region, where water resources are particularly limited.[26] We compared scenarios in which MENA countries continue to export a lot of oil, to scenarios in which they export electricity generated at concentrated solar power (CSP) plants located in the desert. Indeed, this is a realistic possibility, since CSP offers the potential of highly reliable electricity at a competitive cost.[27] Importantly, CSP is far from being a benign technology when it comes to water: especially in desert regions plagued by sandstorms and dust, it is necessary to wash the mirrors frequently (in some locations every day), and most CSP plants are currently built using cooling systems that evaporate water.[28] Nevertheless, the net effects of a switch from oil exports to CSP-generated electricity exports would be positive. As oil fields begin to be depleted, the oil industry maintains production by pumping groundwater down into the wells, better to displace the oil. We find that the switch from oil extraction to CSP production, for the same amount of energy, would reduce water use by the equivalent of the household water requirements of 6 million people.

As far as I know, the work that my own group has been doing on this issue is cutting edge, although we are not completely alone. We are looking at the issue of water. Others have looked at the issue of local and regional air pollution, such as acid rain, and reached a similar conclusion: the negative environmental consequences of low-carbon technologies are less than the fossil fuels that they would replace.[29] I am not aware of other studies that attempt to aggregate these effects across multiple environmental impacts, and I see this as an important research need. What I suspect that we will find is that the overall environmental effects of a switch from fossil fuels to renewable energy sources will be beneficial simply because fossil fuels are so dirty in so many ways that go beyond climate change. At the same time, there will no doubt be localized consequences that will be negative. For these, it would be very good to have strong local and national environmental legislation to protect people and the environment in these places.

Conclusion: taking a deep breath

In the previous two chapters, I have suggested how two particular policy approaches – carbon pricing and legally binding national commitments – offer the promise of solving climate change and a whole lot more. In the case of carbon pricing, it would be a step in the direction of getting markets to include the effects of environmental externalities. Were this to happen for CO_2, it could happen for other pollutants. In such a world, markets would work even better to allocate resources, and the need for government planners to do so would be reduced. Carbon markets fit a particular political agenda quite nicely. In the case of legally binding national commitments of the Kyoto variety, the solution to climate change would similarly signal the beginning of a brave new world. A meaningful global treaty to tackle climate change, with the needed powers of enforcement, would effectively transfer sovereignty on issues of energy policy from the national level to some international governance body, such as the United Nations. Such a transfer of sovereignty could then be easily replicated in other policy areas, such as criminal justice or health and safety regulation. A strong carbon treaty also fits a particular political agenda very well, although it is a somewhat different agenda than that of the carbon markets. The idea that we can solve climate change by reducing consumption fits this same pattern.

I want to be clear that I have nothing against this particular agenda. On a particular Sunday afternoon two summers ago, my family drove our VW through the wooded hills just east of Cologne, Germany, to the town of Waldbröl, where we would spend a week of summer vacation that year. Just after the first traffic light, we turned left and headed up the hill for a couple hundred meters, until we arrived at our destination: the European Institute of Applied Buddhism. In the blistering afternoon sunlight of a late-summer heat wave, we unloaded our stuff, set up our tent, and then checked into the weeklong family retreat with the title of "Body and Mind Are One." In the presence of monks and nuns and their teacher, the eighty-five-year-old Thich Nhat Hanh, we spent a week devoted to mindfulness. We awoke before sunrise to meditate with the hundreds of other participants in the large tent just downhill from our campsite, starting with attention simply to the act of breathing in and out. We listened to Thich Nhat Hanh's daily Dharma teachings and followed them with discussion groups about peace, environmentalism, and social change. We ate communally, sparingly, and vegan, beginning each meal with a word of thanks and then ten minutes of slow, silent chewing. Our children sang songs, learned techniques of deep relaxation, and went on nature walks through the forest. Finally, the week was over. On Friday, just after lunch, we packed up our tent and the few clothes we had brought, said and hugged our goodbyes, and headed home. Soon, it was back to work and school. But I like to think that the effects of that week are long-lasting.

As far as I can tell, my family is not alone, and the world is witnessing a boom of mindfulness, whether it is embedded in Buddhism or in some other spiritual tradition. Every few days, it seems, I read another newspaper article or see another TED talk that has something to do with the topic. At their heart, these practices emphasize the value in living simply. We can achieve the greatest satisfaction in our lives if we learn to recognize, but ultimately not act on, most material desires and instead concentrate on who we ourselves are, our relationships to other people, and our experience of the wonder of creation that is around us. There is a tremendous amount of psychological literature to show that people who live their lives this way are happier. Moreover, there are reasons to believe that these benefits are even greater when the changes become adopted at a community scale. We do compare our lives to those of our neighbors. I may have chosen not to spend my money on a fancy car, but my neighbor has made a different choice, and there remains a piece of me that is jealous of the luxury his car offers. A simple life is even better when it has good company.

But public policies to promote these kinds of lifestyle changes, although they may make people happier for other reasons, are simply not going to make a meaningful difference for climate change. First, nobody really knows how to get such policies to work. What are the practical things that government can do – really, that anybody can do – to accelerate the pace with which people reduce their own consumption? I don't have a good answer, one that I am confident will work, and neither does anyone else. This is, of course, the kind of question that social scientists like to answer, and maybe someday they will. By the time they do, if they do, it will be too late for the climate. Second, the change in emissions that such policies could stimulate is much less than the change in emissions that effective climate mitigation demands. It would be a major accomplishment if government policies, by stimulating a decline in consumption and lifestyle change, were able to get emissions to fall by even a few percentage points over the coming decades.

Lifestyle change is not a sufficient condition for climate mitigation. At the same time, all the models suggest that it is not a necessary one either. I still think it's a good idea. But let's not confuse it with really solving the problem of climate change.

PART 3

Successful strategies to move us away from fossil fuels

I spent the last three chapters writing a depressing story. None of the things that people thought could solve climate change actually work. They don't work because they fail to stimulate the kind of innovation and the sequence of system change that we need. They don't work because they rely on a set of political institutions that simply don't exist and will not exist for decades. They don't work because they don't go far enough, probably can't go far enough, in reducing energy use and consumption. And they don't work because, at the end of the day, very few people want them.

It is time to turn things around. It is time to identify a pathway that can work, that can mean that the planet will be a good place for our grandchildren and those who follow.

The starting point is to identify a red thread, a red thread that differentiates the solutions that have not worked from those that have and that can continue to work in the future. In Chapter 7, I do that by drawing from theories of technological and social change. These contrast sharply with those theories that have informed climate policy so far. By and large, those theories spoke about a much more static world. The world is not static. More importantly, if we are to solve climate change, we must make the energy sector particularly dynamic.

In Chapter 8, I turn to consider specific technologies. What exactly are the technologies that we need to solve climate change? Do they exist already, or are they still a figment of our imaginations and an artist's rendering? There are plenty of the latter kind, and these may serve us well several decades from now. But there are also enough that exist in the here and now. They could solve climate change right now if there were not social factors standing in their way, such as their being slightly more expensive. But these are factors that will disappear quickly and, in some cases, have already ceased to be important.

In Chapter 9, I dive into a case study: what Germany has done to transform its electric power system by relying on some of the key technologies that currently exist. What has worked well? What has not? On balance, I argue, Germany has proved to be a success story and an incredibly important one at that. German leadership, at a critical point, has led to innovations in both technologies and in policy instruments themselves. Other countries can copy their success. They can also learn from some of the failures – generally the failure to anticipate the consequences of success. By identifying these, I suggest what the major challenges will be if we are to generate all the electricity we need without CO_2 emissions.

In Chapter 10, I move beyond the German case study, which focuses on electricity, to all the other areas where change is necessary. These include the other pieces of the energy system, as well as the other contributors to greenhouse gas emissions, such as changes in land use. I identify the particular problems to be faced in each area and a set of public policies that can overcome them.

Finally, in Chapter 11, I wrap up by considering some of the objections to these arguments. Will the new approach really work? Can it work fast enough? Is the old approach really so wrong? What can we, as individuals, do to help?

7

Theories of transitions

I remember clearly the moment when the Internet became established. Exactly twenty years ago, in 1995, I was a young American in search of life experience, about to fly to Switzerland for a three-month stint working on an organic mountain farm. I had the crazy idea to keep in contact with my parents in Boston with a new technology: email. My mother already sent and received email at her office because she worked at a university with a computer system linked to the Internet. But outside of universities and government research labs, email was almost unheard of. I bought a book (*Internet for Dummies*) that explained to me how I could use my notebook computer and a modem to connect to the Internet anywhere in the world. In Switzerland, after the cows had been milked each evening, I dialed in to a CompuServe access number in Zurich, spent a few minutes exchanging a single set of emails with my family, reading them after I had closed the telephone connection. Within about a year, everyone else I knew was doing the same thing. No tax on postage stamps could ever have killed the handwritten letter as quickly.

More and more people have started to equate climate change with a need for an energy system transition, something like the transition from letters to email or the transition from horses to cars that I alluded to in the previous chapter.[1] Promoting the idea of such a transition has become common sense, and something that nobody would argue against, just like nobody argued against "sustainable development" a decade ago. But, as with sustainable development, the term *energy system transition* can be so imprecise as to be devoid of useful meaning. One crucial task is to give the idea of an energy system transition an actual meaning, something that is well enough defined to be able to say what it would be and what it wouldn't.

Let me start by defining what I mean an energy system transition to be. First, an energy system transition would be a change over from an energy system that had been stable in the past to a very different kind of energy system that is equally stable in the future. Once the transition happens, there would be no reason to go back. Second, an energy system transition would be something that would click into place quite quickly once the necessary pieces of the puzzle were in place.

What would an energy system transition not be? First, it would not be something held in place by the policing power of the state. Imagine if the only reason people were sending email was because the government had forbid the mailing of anything containing only information and was enforcing this by engaging in spot checks on our mailboxes, opening packages to make sure that they contained a new sweater and not just a message. Second, and quite similarly, it would not be something put in place and held in place by a growing and continuing set of government disincentives to use the old technology. In both of these cases, the technologies used would change as a result of government intervention. But neither would represent a true transition, one in accordance with how the last paragraph defined it to be.

The remainder of this book is about whether such a transition in the energy sector is possible and, if so, what kinds of government policies can accelerate it. This chapter sets the stage. First, I describe some of the documented cases of technological transitions in the past, and I identify the key ingredients that were associated with their occurrence. Second, I identify a set of principles to guide the formation of policies, policies found not in theoretical models of the world in equilibrium – such as neo-classical economics – but rather of evolution and change. Third, I suggest that such a core reframing of a problem, a recognition that we have been looking at it 180 degrees wrong, is not actually that unusual. We have reframed other problems with good effects. It is time to do so for climate change.

The multilevel model of technological transitions

In the previous chapter, I described the societal shift from using horses to using cars. It happened quickly and not because government policy put a tax on horse manure. But the simple story fails to convey the complexity of what really happened and the role that government played in making it possible. These are important. The person who has looked at these issues and taken the study of such transitions to a new level is Frank Geels, a professor at the University of Sussex in England. I turn to three of the cases he has examined, including that of horses to cars, which together make up his landmark book on the subject.[2]

Geels's first story is about ships and has its origins in the late 1700s. Britain was the dominant power at sea. It had built a fleet of wide merchant vessels with large cargo holds based on a business model of high-volume transport of raw materials from Africa and Asia to England. Speed was not a factor, and British ships were slow. Indeed, so slow that nobody knew when they would arrive at port. One day, a big ship would show up in England with cargo from India or China, and a host of local merchants would be prepared to buy it. Nobody expected things to be otherwise.

The seeds of change began when the United States declared independence from Britain, and, in retaliation, the King banned American ships from British colonial ports. To survive, American ship builders figured out how to build faster ships, ones that could outrun the Royal Navy. They were smaller and lighter, and they led to a new business model, one concentrating on shipping high-value products such as opium and silk. Soon though, people realized there was value in information traveling faster, and the *Baltimore Clipper* ships that the Americans had developed began transporting mail. From there, they moved to passenger transport, where people also started to place a premium on speed.

About this time, the early 1800s, nautical engineers started to experiment with steam engines. The technology wasn't good. The engines were heavy and inefficient, and they needed so much coal that on wide ocean crossings there was little room left over for cargo. But steam power did offer something that sail power didn't, namely, greater reliability: ships could still move even when the wind wasn't blowing. And so steam-powered ships took over two niche markets. The first was passenger and cargo transport on inland waterways, where winds were less constant but refueling stations were possible. The second was as an auxiliary power source on ocean-crossing vessels. But these were slow to penetrate commercial markets because having two sources of power was so expensive.

British colonial administrators, however, recognized the value that this auxiliary steam power offered them. Just as American ships had improved mail delivery and passenger transport with their greater speed, so could hybrid sail/steam vessels give the British an administrative and military advantage with their greater reliability, also for mail and passenger transport. For the first time, ships could have predictable arrival times, allowing administrators in the ports to plan ahead. Between 1838 and 1862, the British government subsidized the construction and operation of oceanic hybrid steamships, allowing them to gain a foothold despite their higher cost. With time and continued investment by the British government, steam power gradually improved. This included the development of higher power engines and also of propellers to replace paddle wheels.

But the improvements in propulsion brought another problem, namely vibration. Steam power was tearing wooden ships apart. The solution was to shift to iron hulls, another new and expensive technology. So expensive, indeed, that American ship builders decided it wasn't worth it, and, going into the 1850s, they concentrated their efforts on building bigger and faster wooden sailing ships. England, however, was running out of trees, and again the government supported the construction of iron ships in order to maintain British dominance. The investment started to pay off. Both sail and steam saw a period of rapid technological change, but by 1860 iron steamships began to catch up with wooden sailing ships on their speed and cargo capacity. As they gained in market share because of the British subsidies, shipping companies and passengers came to expect and rely on the greater reliability they delivered and were willing to pay a premium for it. When the scales tipped in favor of steam, they did so suddenly. In 1863, the year after British subsidies ended, 45 percent of European emigrants to the United States traveled by steamship. Three years later, in 1866, this figure had jumped to 81 percent. By about 1870, within seventy years of their first introduction, forty years of the first commercialization, and less than a decade of becoming cost competitive, steam engines had completely taken over the high seas, leaving sails as something for sporting competitions.

The second case Geels documented is the transition from horses to cars for urban transportation, which took place over a forty-year period from 1890 to 1930. I already mentioned this historical event in the previous chapter, but Geels provides us with important details that the simple numbers of horses and cars does not reveal. Importantly, one might expect that the invention of the car led to its market dominance, but this turns out not to have been the case. Instead, two other technologies played an important role in transforming people's expectations about road transport and, in so doing, paved the way for the automobile.

The first new technology was the electric tram. Compared to either maintaining a stable of horses or paying a cab driver for a ride, the electric tram offered people a relatively fast transportation option that was also affordable. Trams became immediately popular. A few years later, the modern bicycle, with two equal-sized wheels, appeared on the market. This was as fast as a tram and also gave people a degree of individual autonomy that the trams did not. Bicycles, too, became instantly popular, but their popularity demanded two important changes to the road system. First, city governments had to develop and enforce traffic laws for the first time; otherwise, the hordes of bike riders would continue crashing into each other. Roadways changed from being public meeting places to being transportation arteries, and pedestrians moved onto sidewalks. Second, city governments had to better maintain roadways: lightweight bicycle wheels demanded a smoother surface than either horses' hooves or large carriage wheels had previously needed.

It was these two changes that suddenly made cars more attractive. With smoother roads and well-ordered traffic, people could finally take advantage of the greater speed that cars offered. Indeed, the first cars to gain market entry were based on propulsion technologies that had been around for decades: namely, steam power on the one hand, and electric motors on the other. Gasoline engines, by contrast, were a relatively new technology, and at first they were confined to racecars. When they became cheaper with mass production, however, they instantly pushed steam and electric cars aside. And horses went with them.

The final case Geels has described is the move from propeller planes to jetliners in commercial aviation, a transition that took place between 1940 and 1975. Both World War II and the Cold War led the military to sink vast amounts of money into developing jet engine technology and to sink vast amounts of jet fuel into the engines themselves. Through the 1940s, the combination of fuel and investment costs was so high that jet engines had no market beyond the military. In the United States, things stayed that way. The airline industry was growing fast, benefiting from an influx of planes donated to the private sector by the military at the end of World War II.

Just as British subsidies can take credit for the transition away from steam ships, it was BOAC, Britain's nationally subsidized airline, that developed the first passenger jet, the Comet. It went into service in 1952, in order to transport government officials back and forth between England and its colonies in Asia and Africa. The Comet took more than a day off the journey and was seen as instrumental in maintaining close contact in a turbulent colonial time. It proved immediately popular with wealthy businessmen as well. Unfortunately, it also introduced engineers to the concept of metal fatigue, and Comets began falling out of the sky in 1954. But its early success was a wake-up call to the much larger American industry. Boeing unveiled the 707 in 1955, and, within a year, it was outselling a comparable propeller plane, the DC-8, three to one. There was one final hurdle for jet planes to overcome and that was the danger of midair collisions because they flew so much faster and left little room for evasive maneuvering. This time, the U.S. government stepped in and, in 1958, developed the national air traffic control system. The growth in commercial jet aviation has been unstoppable since then.

Across these different transitions, Geels identified changes taking place at three levels. At the top level was a broad, evolving landscape of society within which particular technologies operate. In the case of shipping in the 1800s, this was the transport first devoted to heavy cargo from British Asian colonies, then lighter cargo and mail, and then mail and passengers across the Atlantic to the United States. The overall size of the market was increasing, and the market itself was moving in the direction of higher value added. Rather than just transport a lot of nonperishable stuff, ships had to transport time-sensitive material. They had to be fast, and they had to be reliable. In the case of horses and cars, the broad societal transition was the rise of large dirty cities with the Industrial Revolution, in turn giving rise to the ring of today's inner suburbs and the desire of a middle class to commute to work and business. In the case of civil aviation, it was the growth of the industry in the wake of World War II, led by the influx of inexpensive former military planes.

At the middle level was a patchwork of social and technological regimes. In the case of shipping, it was the network of shipbuilders and their suppliers of wood and iron, as well as shipping companies, and the navy. Different regimes existed in the United States and Britain, given their different local conditions. In the case of horses and cars, it was the systems of roads, rails, and traffic laws and the expectations of passengers to travel fast or slow, in groups or on their own. In the case of aviation, it was the commercial airlines, the aircraft companies that built these companies their planes, the passengers who flew on them, and the network of airports where they could take off and land, not to mention air traffic control towers to keep them from hitting each other.

At the bottom level were particular technologies operating in narrow niches outside the main market. In the case of shipping, it was steam ships used for British mail and the navy. In the case of cars, it turned out to be racecars, which were the first gasoline cars to become fast and reliable. In the case of aviation, it was the military, where price was no factor, and then BOAC, a national airline serving national interests with respect to the colonial administration.

All of these cases illustrate the kinds of processes taking place when transitions happen. In each of Geels's cases, however, that was not policy makers' primary intention. They were trying to achieve other things, and the transition was simply an unintended consequence. At the same time, however, the cases do reveal which of their actions proved to be effective at inducing change. Their actions were effective when they overcame a problem known as *lock-in*, which puts new technologies at a natural disadvantage simply because they are new. Their actions were effective when they managed the risks associated with individuals' and firms' moving over to new technologies. In standard economic theory, risk aversion plays a relatively minor role, but a great deal of recent experimental work reveals it to be one of the primary factors influencing people's choices. Finally, their actions were effective and politically sustainable when they appealed to the desires of different groups of stakeholders, each interested in different objectives. In the remainder of this chapter, I want to dive into each of these issues in some detail before turning to the question of how to get a transition to work in the energy sector.

Evolutionary theory and lock-in

The starting point for modern theories of change has to be Charles Darwin. In the *Origin of Species*, he proposed a simple algorithm that can get us all the way from cyanobacteria to humans, blue whales, and sequoia trees.[3] An organism has more than enough offspring. The offspring differ slightly from their parent and from each other. Because of these differences, some will be better at staying alive than others. Those that survive pass their traits on to become the new starting point, whereas those that don't survive, don't. That's it.

Since Darwin, others have pushed this algorithm into the social sphere. Richard Dawkins, in 1976, extended evolutionary theory into the realm of human ideas and ideologies.[4] He proposed the idea of a *meme* (to rhyme with gene) that represents key attributes of thought patterns. If the meme makes it more likely that the thought pattern will be passed on to other people, then it is likely to survive. Evolutionary theory metastasized to the discipline of economics fairly early. Karl Marx, a contemporary of Darwin, looked at how economic systems contained the seeds of their own destruction. Feudalism gave way to capitalism, and capitalism would give way to socialism.[5] But for Marx, the interesting thing was the economic system itself, along with its faults, and not the general processes leading to self-destruction. Thorsten Veblen, in 1898, coined the term *evolutionary economics*, but, somewhat like Marx, his concern was foreseeing where the economy was headed and not so much the process through which that change would happen.[6] Joseph Schumpeter came closer a few years later.[7] He postulated something of a punctuated equilibrium model: the economy could easily cruise along with some sort of steady-state allocation of resources, but, in reality, this never lasts because innovations come along and unleash a gale of destruction, with reallocative effects rippling through the economy. Various economists have proposed direct analogs to the elements in Darwin's algorithm: self-replication, the seeds of variation, and mechanisms that provide selection. They use these to explain how technologies become better adapted over time and, ultimately, spread.[8]

For our purposes here, however, I think the relevant question is not why technologies become better adapted. Showing that evolution happens with respect to technologies is not one single bit surprising. The more interesting questions, rather, concern when and why it fails to happen. Why, for example, are we not already all driving electric cars? They are better in lots of ways and on a trajectory to become better in all other ways. Why didn't that trajectory happen many years ago, and why are they so rare today? If we knew the answers to these questions, we might be able to come up with some interesting government policies to unleash the forces of change, if and where we think they ought to run wild.

The economist who started looking at this problem was Brian Arthur. A professor of economics at Stanford University, he was a leading figure in the establishment of the Santa Fe Institute for complexity studies. The theory of complexity is, in many ways, about identifying the conditions under which the evolution of a system, like the weather, cannot be predicted. Arthur's work in the late 1980s and early 1990s did the same thing for technological change.[9]

One of the critical elements of the equilibrium economic model that I presented back in Chapter 4 was the assumption of decreasing returns to scale. In the example of the furniture company already engaging in mass production, this showed up in two ways. First, I suggested that the more a factory wanted to scale up production, the further it would have to hunt for its raw materials, like wood or labor. This would drive up production costs. Second, I suggested that there was heterogeneity in terms of customers' desire to buy the product. The more people in the community who had already purchased a given piece of furniture, the less we would expect the next potential buyer to be willing to pay. In this way, the supply curve, moving left to right, pointed upward and the demand curve pointed downward. Where they crossed, you got a stable equilibrium.

Arthur questioned these two assumptions and proposed alternatives. First, as a company builds more of a product, the more it learns how to build it better and more efficiently. Scaling up production will accelerate this process. Essentially, the company is creating a new resource: knowledge. So there might be a temporary jump in production costs as it has to hunt further for resources, but the effects of the increase in the knowledge resource quickly overshadow this. And so there may be increasing returns to production, in that the more is produced, the less expensive it gets. The possibility of increasing returns means that the supply curve might actually point downward. Second, the value that consumers get out of a product may depend on how many other people already own a copy. In some cases, this pushes the value downward. If all my co-workers own black cars, and I want to find mine in the company parking lot, then I won't buy black. In other cases, this pushes the value upward. If all my neighbors own black cars, and I want to fit in, then I will buy black. In this latter case, the demand curve will point upward. Arthur documented a lot of cases where people really need to fit in. Fitting in means increasing returns in consumption and suggests a demand curve that might point up.

Figure 7.1 shows two graphs of the market, each with a supply curve and a demand curve. The left-hand graph is the same one from Chapter 4, with decreasing returns. Where the two curves cross lies the equilibrium, and it is a stable one. Consider, for example, if the market is at the equilibrium, but then a firm decides to increase production (a). The only way to sell all the extra merchandise is to lower the price below the marginal production cost at the higher level of production (b). The firm will lose money, and it will learn its lesson, going back to the equilibrium level of production (c), which will push the price back up (d). The same self-correcting thing would have happened if production had fallen below the equilibrium level: market forces would have pushed production back up. The right-hand graph shows the model based on increasing returns, and here things are different. The place where the two curves cross is an equilibrium in that all the goods

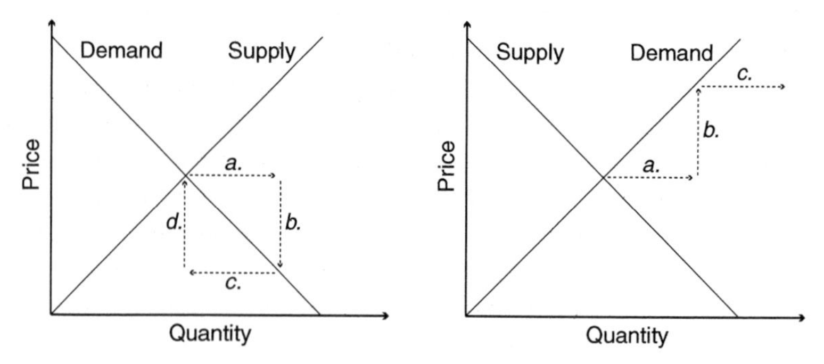

Figure 7.1. *Stable and unstable economic equilibria.*

produced will sell at a particular price. But it is unstable. If a firm increases production (a), it can raise the price it charges (b). It makes more money. Next time, it will produce even more (c). On other hand, if the firm reduces production, then the price would fall, and, most likely, the firm will go out of business. Production will fall even further.

The right-hand graph fits complexity theory in that it shows conditions under which a system is fundamentally unpredictable because it could spin out toward one direction or the other. What one can say is that there is some sort of threshold level of production running down the middle of the graph, below which lies bankruptcy and above which there exists a huge prospect for growth. If all of these curves were known to firms before they entered the market, then the strategy seems clear: start big, above that threshold. But the reality is that the curves' shape and position can only be inferred from experience. Not knowing where the threshold lies, the firm has every incentive to start as large as possible. But first they have to convince their creditors. The larger the firm starts off, but still below the magic size to succeed, the more the creditors have lost.

All this is fairly abstract, but Arthur explored it in the context of some far more concrete case studies. One thing he looked at was different formats for videotape (back when there was videotape). The VHS format was developed by the Dutch company Phillips, whereas the Betamax format was developed at about the same time by the Japanese company Sony. Most analysts agreed that the Betamax format was a little bit better in terms of reproduction quality and reliability. The curious thing, however, was that the VHS format took off, while the Betamax format never really established itself in the market. Arthur suggested that the case of videotape was like wanting to fit in with one's neighbors because people wanted to share their recorded television shows with their friends. If you had a VHS machine, you could only share with your VHS friends. For reasons having to do with corporate licensing, the VHS format started out with a little more of a market splash and gained a little extra popularity early on. In the eyes of the next generation of consumers, this increased the attractiveness of the VHS format, easily compensating for the slightly higher quality of the Betamax. From that point on, the trajectories of the two technologies diverged rapidly. Betamax died out fairly quickly, while VHS exploded in popularity until DVDs eventually came along to put it out of business as well.

Arthur described the phenomenon seen with VHS and Betamax as technological *lock-in*. The more ubiquitous a particular technology is, often the greater the attractiveness of sticking with it into the future. The idea of people trading videotapes of their favorite TV shows can be described as *network effects*. The same set of network effects explains why electric cars are slow to take off, even if, like Betamax, they are the better technology. There are a lot of gasoline and diesel cars on the road, which has given rise to a lot of gas stations. You could recharge an electric car in a matter of a couple minutes simply by swapping out the discharged battery for a fresh one, but this would take a network of such charging stations. The network doesn't exist because there aren't enough electric cars on the road to support it. But without the network, the demand for electric cars is necessarily less, perhaps too small to support a network of charging stations in the first place. The other factor holding back electric cars, as with all relatively new or niche technologies, is simply the relatively few number of units so far produced. As already described in Chapter 4, the existence of "learning effects" is fairly robust. The more of a particular product that is produced, the less expensive it becomes because engineers solve countless little problems associated with producing it effectively and efficiently. There have been millions of gasoline and diesel car engines produced, and they have become very cheap. Electric motors are also cheap. What are not yet cheap are the kinds of advanced batteries that can supply electrons to those motors for hours on end. Producing a few million of them will almost certainly lead them to become cheap, too. But, until that happens, the cars are expensive, and few will buy them. It is a chicken-and-egg problem.

Lock-in played a strong role in all of Geels's cases. It was the military that developed the jet engine for fighter planes. It was the BOAC Comet that brought these to commercial aviation. Similarly, it was British government that subsidized the construction of steam ships. In both of these cases, the intervention was aggressive in the sense of being enough to overcome a barrier to entry of the new technology, but it was also limited to a particular niche within which the technology was being promoted, usually for some other reason (e.g., national defense). In other words, the government promoted the new technology heavily in one particular band of application, rather than across the board. That band was wide enough to give the new technology room to grow and improve, although not so wide as to disrupt the entire economy with the new way of doing things. In the case of cars, government intervention was not to subsidize the new technology, but rather to address the problem of networks, and here it was equally aggressive in the form of road construction and the development of traffic laws.

For the case of climate policy, an important lesson emerges from all this theory. The energy system has both network and learning effects, and technological lock-in is the major problem to overcome. When there is lock-in, even if there are really good new technologies out there, the fact that they are not yet established makes them inherently less attractive and typically more expensive. In the world of VHS dominance, Sony could have discounted its Betamax players, and it probably would have made no difference. If Sony had given away a few million Betamax players, raising their popularity above some critical threshold, that might have worked to secure a position in the market. Then, perhaps, Sony could have offered the mild discount to build up market share even more. And then, it could have sold them at full price, and people would have bought them because, after all, they were a little better than VHS.

Most analyses of carbon pricing mechanisms suggest that they should start small and then ratchet up over time. The European Union, for example, has always planned to gradually tighten the number of allowances within the Emissions Trading System (ETS) to an extent that would push up the allowance price over time. There is a good reason for this. Going from no carbon price at all to a very high price would disrupt the economy in ways that would make the whole idea politically untenable.[10] It would be horribly inefficient.

But when you look at climate change and see it as a problem of technological lock-in, the critical insight to gain is that exactly the reverse is necessary. Fossil fuels and their associated technologies are not inherently better than their alternatives: the reason they are currently dominant is simply a result of history. We don't need a gradual build-up to a permanent policy regime. Whatever policy intervention a firm or a government wishes to take, that intervention needs to start big, overcoming a hurdle that is highest at the beginning, and then get smaller over time. Eventually, the hurdle can disappear altogether.

Behavioral economics and risk

Imagine yourself as the subject of a psychology experiment. When you walked in the door they gave you $40, which is yours to keep, and which you put in your pocket. And then the experiment is this: you get a choice between two very different kinds of gambles. One is to flip a coin and, if it comes up heads, you win another $20. The other is to answer a quiz question and, if you get the answer right, you win $10. The good news is that we will tell you the quiz question in advance: what is the capital city of Canada? You're not too bad in geography, and you know that the answer is Ottawa. Which would you choose? If you are like most people, you would choose the geography question. You assure yourself a 100 percent chance of walking out of the ten-minute experiment with $50 in your pocket.

But what if the experiment were a little different? When you walked in the door they gave you $60, which you put in your pocket. That $60 is yours for now, the experimenter tells you, but you have to make a choice that will influence whether you get to keep all of it or give some of it back. One option is to flip a coin; if you win, you get to keep your entire $60, but if you lose, you have to give up $20, give it back to the experimenter. The other choice is to answer a quiz question; winning means keeping your entire $60, whereas losing means giving up $10. The good news is that they tell you the question in advance: what is the capital city of the Turks and Caicos Islands? The bad news is that although you thought you were good at geography, you've never even heard of the Turks and Caicos Islands, much less of any city that might be its capital. What do you choose? If you are like most people, you would choose to flip the coin. At least you get a 50 percent chance of getting to keep your entire $60.

Of course, the two experiments are identical in terms of their possible financial implications: $50 with 100 percent probability or equal chances at $40 or $60. The difference is that, in the first version, you faced the prospect of gaining something new, whereas in the second, your prospects were framed in terms of giving up something you already had. For most people, this framing influences their approach to risk. Most people really dislike the prospect of losing anything they consider theirs and are willing to risk even larger losses in order to gain the possibility of losing nothing. Gamblers at the race-track or casino, for example, typically start off by taking relatively conservative bets; if they begin to lose money, then they start taking bigger and bigger risks in order to try to recoup all of their losses.[11]

Simple as it may seem, this finding, based on a set of experiments like the one I just described, has shaken the world of economics. The psychologists Daniel Kahneman and Amos Tversky published it in 1979,[12] and it has gone on to be one of the most frequently cited papers in the field of economics, some argue the most frequently cited.[13] Their work has blossomed into an entire subdiscipline known as *behavioral economics*, and, although Tversky met an untimely death from skin cancer before the value of his work was fully recognized, Kahneman won the 2002 Nobel Memorial Prize in Economic Science for the work they had done together. More recently, behavioral economics has itself given rise to another subdiscipline: *neuroeconomics*. Whereas behavioral economics studies the decisions that people make in particular situations, often making use of experiments like the one just described, neuroeconomics links these patterns to actual physical processes taking place inside our brains, with the experiments taking place in a medical laboratory.

What is the big deal of behavioral economics? The big deal is this: neo-classical economics, which is to say mainstream economics, rests on a foundation that is a set of assumptions or axioms about our inner desires and the rational way of achieving them. The axioms make complete sense in the abstract. They include, for example, the idea that more of something is better than less, or that the more of something you have, the less an additional unit of that thing is worth to you: having two bicycles is better than having one, but less than twice as good. Building on these axioms, using mathematical proofs, neo-classical economists have developed a wide-ranging set of insights into human behavior. Until behavioral economics came along, most economists believed that these insights were both normative and descriptive. That is, they offered guidance into how people ought to behave while also being a good guide to how people actually behave in practice. Why, after all, would someone behave in a way that did not best achieve his inner desires? If a person's inner desire is to take risks, then he will choose the option that gives him equal chances at $40 or $60. If his inner desire is to avoid risk, he will take the $50 for sure. Then Kahneman and Tversky came along and showed how people's behavior really depended on which experimental group they were randomly assigned to.

Now, for the past thirty-five years, behavioral economics has been showing how it is possible to substantially improve on neo-classical economics for a descriptive model of behavior. In other words, neo-classical economics can tell us how people ought to behave, but behavioral economics can tell us how they actually do behave. The difference between the two, which is often quite substantial, can be very important for the design of any public policy designed to influence people's actual behavior.

Some of the biggest differences show up in situations in which there are substantial risks and uncertainties. Behavioral economics demonstrates that people will often sacrifice substantial benefits in order to avoid risk, whereas in other cases they will pay money to take on risk, just as the experiment I described demonstrated.[14] It also matters whether there is some degree of personal attribution for the outcome.[15] People do like to win gambles based on the actions they take, but what they really hate is to take actions that directly lead to losses; if they are going to lose money, it had better be because of bad luck.[16] At the same time, people tend to underestimate large risks and overestimate small ones.[17] Combined, these effects can lead to very large degrees of risk aversion in many cases where events determining the outcome of a decision are beyond their control, a phenomenon identified as *status quo bias*. When people are given the choice between sticking with what they already have or trying something new, even a small likelihood of a negative outcome associated with the new option will often outweigh a large potential for gains, and they will end up sticking with what they already have[18] Neo-classical economics, by contrast, generally predicts much more risk-neutral behavior in individual instances. Consider that one of my goals, for example, is to increase my net worth to a particular level over the next twenty years (the point at which I will retire), and this will be determined by thousands of relatively small investment decisions. My best strategy would be to generally go for the highest average outcome from each of these, regardless of the level of risk. The reality is that I won't do this.

Neuro-economics has also shed some interesting insights into risk-taking behavior. Researchers at Cambridge University recently published findings linking people's risk-taking behavior to their blood levels of the hormone cortisol.[19] On the one hand, they found that greater levels of risk and uncertainty led, over the course of time, to higher levels of cortisol. On the other hand, people with high levels of cortisol were more likely to overestimate the risk of losses and be much more averse to taking on new risks. The implication is that people tend to have a particular appetite for risk and uncertainty, and the degree to which they have already satisfied it in a given time frame determines how hungry they continue to be, just like the sweetness of the salad dressing during dinner could influence whether you choose cheese or chocolate cake for dessert. People's willingness to make a particular risky investment may rise and fall according to the overall perceived riskiness of the market in which they are operating. In a bull market, when it's hard to lose, risk taking rises.

The issue of risk appeared in Geels's case studies in exactly this respect. In all three cases, a critical aspect of the transition was that the overall economic sector was expanding. Essentially, there was a bull market for shipping, for road transportation, and for aviation. The mood was of optimism, and the risks within each sector correspondingly low. Even then, some held back. The Americans did not jump into iron-clad steamship construction (preferring to stick with wood) until the British had demonstrated that steamships would work. The airplane manufacturer Boeing did not jump into jet production until BOAC had demonstrated its technical and market viability. Risk also shows up implicitly if you look at where the technological innovations took place. In all cases, it was in winner-take-all situations where the risks of innovating were lower than the risks of not innovating. One such situation was war, with the military developing the fighter jet. Another situation was colonial dominance, leading the British to develop the steamship and the commercial jet. Another was sports competition and the development of high-performance gasoline engines for auto racing.

Theories of risk-taking behavior are theories of change, precisely because change means sailing into new waters, where there is less experience and fewer data, and uncertainty is hence greater. Change is risky. This is no exception for climate policy. Climate policy demands, above all, the diffusion of new technologies for producing and transforming energy. The decisions of real people stand behind this shift. Real people need to put solar panels on their roofs, or build factories that make those solar panels, or loan money to the people building those factories. Neo-classical economics tells us to design policies that will increase the long-run average profitability associated with each of these decisions. It is relatively silent with respect to the range of outcomes from which that average is derived. Behavioral economics, by contrast, tells us that people can often be very sensitive to that range. If the purpose of policy making is to stimulate particular kinds of risk-taking investment behavior in the energy sector, then it needs to work to decrease the risks associated with those particular investments while also decreasing the overall level of risk and uncertainty in the market within which those investments take place.

Cultural theory and clumsy solutions

Mike Thompson is a guy in his 70s, of small stature and a big laugh. A few days every month, he shows up at my former institute, IIASA, outside of Vienna. The rest of the time he spends at home in Bath, England, or working with friends at Oxford. He is one of those brilliant people whom you would describe as "quite a character," a man full of surprises. Characteristically, while he reveals a slight limp while walking, the limp vanishes when he is roped up and climbing vertical rock. When he was younger, he was an officer in the British army, and one of his activities as a soldier was to climb the north route up Mount Everest. After he got out of the army, Thompson wrote a dissertation in anthropology. But he loved Nepal and the Nepalese people he had gotten to know while stationed there, and he returned there as a researcher. He began to talk to people about their views on all the development measures that the government was instituting to lift people out of poverty, with support of foreign donor agencies. They told him three things.

First, they told him, the measures weren't working very well. Although dams were being built to generate electricity and roads constructed to link the villages with the cities in the valleys, the benefits were usually not reaching those people who were really poor. They still spent their evenings in darkness, in villages still remote and in most cases only reached by foot.

Second, all of these development projects designed to benefit the people turned out to be designed by – and supported by – a very narrow segment of society, namely the elite levels of the government bureaucracy in Nepal and the countries providing it with aid. One of the reasons why the projects were not working well was because their designers appeared blind to a great deal of information. Mountain roads, for example, were built in areas where landslides and avalanches frequently passed through, destroying them for months until repair crews could completely reconstruct them. Dams were built to provide electricity to communities, but those communities still lacked the infrastructure to actually distribute and use that electricity, and, by many accounts, the people there didn't really want electricity, at least not as much as they wanted other things. Because the projects were being built anyway, in seeming ignorance of local knowledge and local desires, they were meeting with a great deal of opposition from local leaders of the very segments of society that they were intended to benefit. Because those leaders appeared powerless to stop the projects, their own standing in the community suffered.

Third, they told Thompson that, in the rare event a project did manage to get input from a wider swath of society, from different social and cultural groups, it was likely to be far more successful. One example was the construction of a cable car to transport milk from mountain pastures down into the valley, where the markets were. The bureaucrats in the capital city, Katmandu, didn't really want it as much as they wanted a road because, for them, a road symbolized progress, truly opening up the mountain to development to a degree cable car couldn't accomplish. Roads also were more expensive to build, which meant a greater flow of aid money through their offices. But the bureaucrats couldn't argue with the fact that a cable car, with strategically placed pylons sheltered from landslides, would be far more reliable. Village leaders on the mountain were concerned about their communities opening up too fast to the outside world. But they couldn't argue with the fact that a cable car to transport milk, and not people, would bring in highly desirable revenue. Local business leaders might have preferred a mountain road or railway that allowed easier access to a greater number of villages. But they couldn't argue with the fact that the cable car would at least be a step in the right direction. The "milk-way" was a solution that nobody really wanted, but a lot of people could still support, albeit for different reasons.[20]

Thompson's findings meshed nicely with a growing body of similar stories known collectively as *cultural theory*. According to cultural theorists, it is useful to categorize people and their political actions along two dimensions, or axes, seen in Figure 7.2. The vertical axis is the degree to which people believe that they are bound by existing social norms and practices, what cultural theorists call *grid* behavior. The grid is what constrains us. The horizontal axis is the degree to which people believe that their own actions ought to benefit the wider community, what cultural theorists call *group* behavior. Dividing people according to these two axes creates four distinct ways of looking at the world. People who see the connection between themselves and their community going both ways, high on both *grid* and *group*, tend to see things fitting into a social hierarchy. Those who see a freedom to chart one's one course as long as it is for the good of

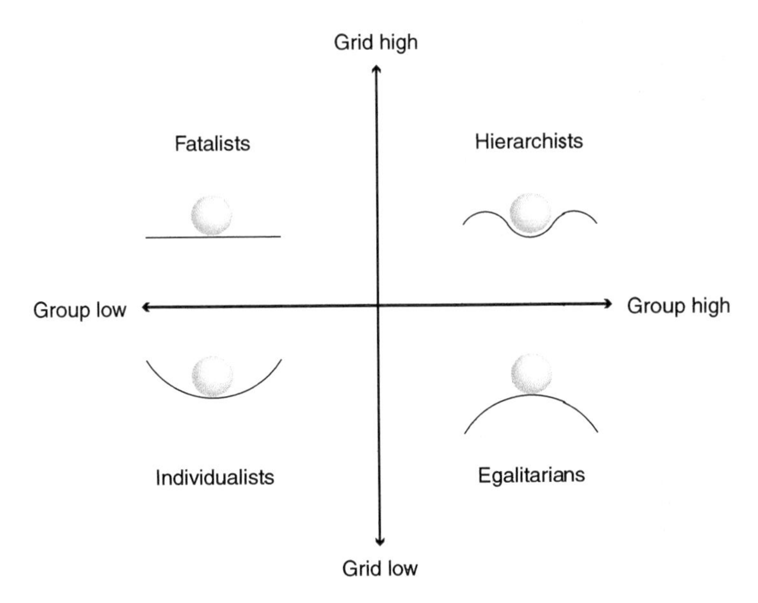

Figure 7.2. *Cultural theory.*

the community fit an egalitarian mold. Individualists see no social ties, whereas fatalists see their actions as constrained by, but of no value to, the community.[21]

Although there is nothing new in dividing people into different groups, cultural theory draws on two observations that are interesting. The first is that the characteristic beliefs of each group go well beyond the grid-group distinction and hence comprise a more complete worldview. One example is the view toward nature, illustrated by the little ball and lines on the figure. Egalitarians, the evidence suggests, see nature as balanced precariously with a high likelihood for disaster. For them, like the World3 modelers of *The Limits to Growth*, radical change is often the only solution to avoid collapse. If the little ball starts rolling, it will fall off. Individualists, by contrast, see nature as robust; it can be pushed this way and that by market forces, and yet these will always pull it back to an equilibrium close to today's, just as the little ball will always return to center. Market solutions are the ones for them. Hierarchists see the need for nonradical deliberate action to keep nature within a safe operating range, and, for this, government regulation is essential. Fatalists see nature as random. The ball is lying on a flat surface: maybe it rolls, maybe it doesn't.

The second observation of cultural theory, one that Thompson noticed in Nepal, is that policy solutions to complex problems work better when they represent a point of consensus or compromise between the different worldviews. The question of what transportation mode to build in the mountains was this kind of a complex problem because there were competing ideas from different worldviews about the proper degree of engagement between the villages and the valleys. The question of where to put a pylon for a cable car so that it wouldn't be swept away by a landslide was not a complicated problem. For that, you just needed a good engineer and geologist working together.

Thompson and his colleagues call those policy solutions that work – the ones that represent a compromise between different worldviews – *clumsy* solutions.[22] They are clumsy because they stumble along, with no strong basis support from any one group, and yet they nevertheless do manage to get the job done. They contrast these with what they call *elegant* solutions. Elegant solutions are perfect but only from the perspective of one particular worldview. From the perspective of the other worldviews, they aren't very good at all. A core observation made by cultural theorists and clumsy solutionists is that people simply don't change their worldviews, at least not in the context of a politicized decision-making process. Rather, they tend to dig in their heels. For elegant solutions to work, more often than not, the nonbelievers need to be converted. But this never happens. So the elegant solutions fail.

The idea of clumsy solutions also shows up quite prominently in Geels's case studies. In all of the cases, the technological transition occurred because of policy actions actually aimed at achieving some set of social goals, none of which was explicitly the elimination of the old technology. The transition to steamships benefited from the desire to maintain communication with colonial administrators and also to protect the British shipbuilding industry in the face of a shortage of raw materials: trees. In the case of cars, the government interventions were made in the interest of public safety and also the desire of cities to promote more rapid transportation. In the case of airplanes, the objectives were military and colonial in the first instance, and then, once the risks had been lessened, financial interests played an important role. The policies that had a large effect were those that satisfied different objectives of multiple stakeholders.

Climate change policies offer us contrasting cases of clumsy and elegant solutions. Indeed, the journalist Elizabeth Kolbert once documented both in a single *New Yorker* article.[23] The article describes a trip to Europe, where she visited two very different communities, each with its own approach to solving climate change. Her first stop was Samsø. Samsø is a Danish island that had managed to become a net exporter of energy based entirely on the strong winds with which the island was blessed. The island's residents were mostly farmers, and one thing that stood out to Kolbert was that they had put wind turbines on their land for different reasons. Some had put them there because existing national subsidies meant that building a wind turbine was a good way to make money. For others, the money was important, but it wasn't the primary motivation. Some cared about climate change and saw the building of a wind turbine as doing their part. Others wanted personal energy autonomy. Still others had strong community spirit and wanted to help achieve a goal of energy autonomy that island leaders had set. There were also plenty of farmers who hadn't put up wind turbines. Success didn't depend on universal participation.

Stop two on Kolbert's European trip was Zurich, Switzerland. There, a group of professors at my university, ETH Zurich,[24] had founded a movement called the *2,000-watt Society*. They reckoned that if every person on the planet were to consume energy at a rate of not more than 2,000 watts, about a third of what the typical European consumes and a sixth of the typical American, climate change would become a solvable problem. A number of Swiss institutions, like the society of architects, had started promoting the idea. But for the 2,000-watt Society to work, participation needed to go much further. And that is what the Swiss professors told Kolbert. Their job as scientists, her job as a journalist, was to educate people as to why theirs is a solution that would work and why it is essential.

Kolbert acknowledged that the 2,000-watt Society didn't seem to be taking over the world – not even Switzerland – the way windmills have taken over Samsø. Mike Thompson would say it has entirely to do with the need for universal buy-in. For lifestyle change to make a difference, there will need to be a critical mass shifting cultural norms away from materialism. Whenever somebody says that the first important step is educating people to get their buy-in, watch out for an elegant solution. The third guiding principle for effective climate policy is the idea of clumsiness. Effective policies don't need to make a lot of sense to a narrow group of people united around a common objective. Rather, they need to make a little sense to a wide group of people, often with very different objectives.

Reframing the central challenge

On September 1, 2014, the *New York Times* ran an article that quickly climbed to the top of their *Most Emailed* category.[25] It reported on a scientific study, the results of which had just been published in a leading medical journal, that for the first time employed a randomized control trial to examine a recurring question. Which type of diet appears healthier: one relatively high in carbohydrates and low in fat, or one relatively high in fat and low in carbohydrates?[26] The study had tracked the weight and cardiovascular risk factors of 148 healthy adults – men and women – over the course of twelve months, during which time equal numbers, decided randomly, followed each diet, and all of them had access to regular nutrition counseling. The result? The members of the group eating less carbohydrate and more fat came out on average 3.5 kg lighter than the members of the low-fat group. All of their cardiovascular risk indicators were better. Each difference was highly statistically significant.

The result shouldn't be surprising by now. For the past couple of years, there has been a growing belief that the culprit behind weight gain and obesity is eating a lot of sugar and simple starches; through a mechanism that involves spikes in the body's production of insulin, these lead to a lowering of the metabolism and the storing of excess energy in the form of fat cells. Fats are also high in calories, but they also have the effect of reducing appetite quickly and do not have the effects on metabolism that sugars and starches do.

The interesting story here is how conventional wisdom on diet has turned essentially 180 degrees in the past decade or so. Starting in the 1980s in the United States, public health official began really starting to worry about the growing problem of obesity. With greater and greater volume and clarity, the medical community told people that they needed to avoid eating fats, especially saturated fats, and get their energy instead from sources high in carbohydrates: rice, bread, pasta. These were clean, fat-free calories that would give people plenty of energy to stay active and not make them gain weight. Meat, butter, full-fat milk and yogurt: these would kill us. For thirty years, most people believed that eating fat meant getting fat. Supermarket shelves were dominated by items declaring themselves to be fat free. And yet obesity rates continued to rise.

Then, starting around 2000, the Atkins Diet started to gain popularity and turn things around. Centered around cutting carbohydrates and increasing fats and proteins, the Atkins Diet had actually been developed in 1972, based on scientific studies conducted in the 1950s, before the conventional wisdom about how fats would make you fat developed in the first place. But the diet went largely unnoticed until people started discovering that it worked, the conventional wisdom notwithstanding. And indeed what has worked in the context of people adopting the Atkins Diet appears to have been successful in public health interventions as well. In February 2014, the *Journal of the American Medical Association* reported that childhood obesity rates had fallen 43 percent over the eight-year period ending in 2012.[27] Moreover, almost all of the drop occurred in the final two years of the study.[28] The researchers attributed the decline to the recent effort to limit the intake of sugary beverages among young children combined with an increase in the rate of breastfeeding.

If it is interesting that our view of a healthy diet has changed, then it is also noteworthy that this is not the only recent case of conventional wisdom being turned on its head.

A similar story is about running. I started running about the time that running became popular, in the mid-1970s. Two other things started then as well. The modern running shoe was born in the form of the Nike Waffle Trainer, the first shoe to incorporate serious amounts of padding to reduce the shock to the foot of every stride on the pavement. The other thing was that running injury rates began to climb. As Christopher McDougall documents in his book *Born to Run*, the sports medicine community attributed the rise in injuries to the growing popularity of the sport and the fact that more people were taking it up, people who may have been less fit to begin with.[29] The doctors' prescription was better running shoes, with even more padding and with growing amounts of support to the arch of the foot. That is certainly what they told me when I had my first running injury at the age of fourteen. For the next fifteen years, I wore custom arch supports – orthotics – to reduce the natural rotation of my foot. Over that whole time, the pain never really went away, although it was manageable, and I attributed the success to the orthotics. Then one time when my orthotics wore out, I didn't replace them. And the pain became less. Later, I followed McDougall's advice and started running in shoes with less and less padding. Over time, the pain went away altogether. The evidence now suggests that the rise of running injuries resulted not from the sport's growing popularity but from the running shoes themselves. Padded, supportive running shoes felt nice, but they also caused a change in people's stride – causing them to strike the ground with the heel rather than the forefoot – and also made it impossible for the arch of the foot to act as the shock absorber that nature intended.

There are other examples that I won't take the time to explain, and all of them show the same pattern.[30] In all cases, the initial mistake came from having an overly simplistic model of the underlying problem and holding on to that model too tightly. We get fat from eating fat, so let's engineer the fat out of foods. Running injuries come from pounding the pavement, so we can cut down on them by putting more padding in our shoes. In all cases, the solution came once we recognized that the systems we were worried about also were a little more complicated than the initial model suggested, usually with some feedbacks built in. When we eat a traditional diet in a traditional manner, we give our bodies the signals they need to get us to stop eating before we are absolutely full. When we run more, our feet get stronger and do a better job of absorbing shocks. What is also the case in all of the examples is how deeply we were convinced that we were right, despite growing evidence to the contrary. The interventions we thought were right were having no effect, and so our response was to try them even harder – until somebody came along and suggested that the evidence might actually be telling us something.

The central thesis of this book is that the simple models guiding climate policy are inadequate – our reliance on them so far is the reason we have made no progress – and that there is a different model out there that can work. In the previous three chapters, I walked through the simple models and, I hope, demonstrated why they just aren't quite right. Now it is time to implement policies that are better.

To think clearly about the new policies, one needs to slightly reframe the core problem. Rather than seeing the core problem as being one of limiting emissions, limiting the impacts on a global commons, limiting the effects of economic growth, one can also see the core problem as being one of transformation.

Interestingly, the models for climate policy that haven't worked are all implicitly static in terms of their view of the world. Start with the model of the environmental externality. It is an add-on to the model of market equilibrium, indeed a special case where the market equilibrium does not result in an optimal allocation of resources. But the critical feature here is that there is an equilibrium at all. With a given set of production costs and consumer tastes, the production of goods will settle toward some steady state. With the externality, that steady state is an undesirable one, and yet the addition of a tax on pollution can put things right. The model of the tragedy of the commons, on which the current international climate regime is based, is really no different. Given a set of private costs and benefits for engaging in an activity that depletes a common-pool resource, actors will settle into an equilibrium behavior pattern that overconsumes the resource, making everyone worse off. A fixed legal framework can change this. The model of the limits to growth is not so much different, although the attention to growth might be deceptive. Society engages in activities that allow consumption to rise exponentially at some relatively fixed rate. Even though they grow in magnitude, however, neither consumption nor the production of goods and services needed to feed that consumption changes much in character. Indeed, Jackson and the Sustainable Development Commission basically reject the idea that they can change in any qualitatively important manner when he suggests that there has never been and likely will never be an absolute decoupling of material consumption and wealth.[31] What we need, then, is a fixed set of policy interventions that will counteract the forces causing growth and put the economy into a steady state, or even one where consumption begins to fall.

Within the old policy models built around limiting greenhouse gas emissions, people have seen a role for government to support new, climate-friendly technologies. But, ultimately, the purpose of such policies, as the advocates of the traditional approaches have consistently maintained, is to lessen the pain associated with the inevitable restrictions on emissions. In this sense, they label the promotion of new technologies "technology policy" or "industrial policy" and reserve the name "climate policy" for the restrictions or for those policies that would hopefully achieve the restrictions by creating a strong disincentive to exceed them, such as a carbon tax. Here is where thinking about climate change in terms of a true energy system transition is very different. If we get a real energy system transition, there won't be any need for such restrictions or for disincentives. The promotion of new technologies is not the preparatory work for climate policy. The promotion of new technologies *is* the climate policy itself. Can it work? Let's look.

8

Strategic technologies

I have mentioned the Intergovernmental Panel on Climate Change (IPCC) several times in this book. At the most general level, the IPCC reviews the science of climate change – thousands of scientific articles published every year – and synthesize it in a manner that makes sense for policy makers. But to synthesize information, it is necessary to organize it around some sort of framework and that in turn requires a judgment call. In its most recent assessment, the Working Group III report, which covers climate mitigation and mitigation policy, one of the critical questions was whether to frame the mitigation challenge primarily around the intellectual framework of the global commons (and hence, implicitly, externalities) or around that of the technological transition. There were debates among a select group of authors culminating at a week-long plenary meeting in Berlin in April 2014, where national representatives needed to approve the report's Summary for Policymakers, known as the SPM.

I was a member of the SPM writing team, and it should come as no surprise that I was part of the small faction within the SPM writing team suggesting that we place greater emphasis on the transition framing. Ultimately, the SPM draft that the authors asked policy makers to approve presented both framings. What surprised many of us was what happened next. The policy makers turned out to be far more receptive of the transitions framing than of the global commons problem framing. Indeed, several countries, as a precondition to their signing off, more or less forced us, the authors, to remove the references to the global commons. We ended up having to gut those sections of the SPM that spoke to the need for global treaty formation based on the concept of the global commons.

What's with that? The policy makers appeared unwilling to agree to language that suggested anything close to a legal and moral obligation to participate in a Kyoto-like treaty. It is one thing if a bunch of scientists and philosophers write about such legal and moral obligations, but by agreeing to such language in the SPM, the policy makers would implicitly be making such statements themselves. They didn't want to do that. Perhaps it would then make them look bad when, within the United Nations Framework Convention on Climate Change (UNFCCC) negotiating process, they took positions that were inconsistent with these types of obligations.

One response to these events would be to curse those damn policy makers and say that they just have to wake up to scientific facts and moral truths. To be honest, that is what most of us in the climate change science community have been doing for the past two or three decades. A very different response would be to listen to what the policy makers have been saying. What they were saying in the Berlin plenary meeting, I think, was that they were extremely uncomfortable with moving from a world in which nation states have the freedom to do what they want with their energy systems to one in which they don't have this freedom. They have good reasons for wanting to retain that sovereignty. For one thing, it means that policy decisions can be made at a level of government that is closer to voters.

The clumsy solution to the climate change problem, by which I mean a potentially successful solution to climate change, would be one that eliminates greenhouse gas (GHG) emissions while not forcing nation states to hand over sovereignty. That is what the transition approach to climate policy promises. But is it a real promise or an empty one? Cars quickly replaced horses, and jet planes quickly replaced propeller planes, but, in both cases, a prerequisite to their doing so was being good enough, of being capable of taking people where they needed to go. If alternative energy technologies are to quickly replace fossil fuels anytime soon, which is what needs to happen, then it is essential that they exist already and be good. Are they?

This chapter explores that question. To do so meaningfully, it is useful to examine not just whether the technologies needed to solve climate change exist, but whether they are or could soon be competitive. By this, I mean that the technology or system of technologies offers most people something that they prefer over what fossil fuels offer, without the government having to make a sustained effort to shape or influence those preferences. Only if such technologies do so will it be possible for them to push fossil fuels out of the picture, without the need for such things as a high carbon price supported by a global treaty. There are a lot of ways that one can describe competitiveness, but I am going to boil it down to three critical elements:

- *Cost.* Is energy from the new technology affordable? Ideally, it should be less expensive, in the time and place where it is used, than energy from fossil fuels.
- *Reliability.* Is the energy from the new technology there when you need it? This really covers two different issues, one of which is geopolitical and the other technical. The first issue concerns energy security and whether one needs to buy the energy from people one doesn't completely trust. Or, if one does trust them, is it because the United States, Europe, or China has propped up their government with military assistance? Could that government turn against you? The second issue, which turns out to be more important, is whether it is a form of energy that you can produce where and when you need it or store in sufficient quantity to be sure that there is enough inventory on hand. This is especially relevant for electricity. Relative to fossil fuels, electricity can be very difficult and expensive to store.
- *Potential capacity.* Can society produce enough of this energy, passing the tests of cost and reliability, to make a serious impact on fossil fuel use? At one level, potential capacity depends on technical and geographic factors, like whether there is enough land area to install all the solar panels one wants. At another level, potential capacity depends on social and legal factors, like whether a person wanting to raise a windmill can obtain the needed government permits, which in turn is a function of political acceptability.

To be most competitive, a new technology would offer benefits, compared to fossil fuels, on all three criteria. A technology could still do well, however, if it outcompetes on one criterion while not presenting any serious disadvantages on the other two.

I am going to go into some depth on several different technologies. To give this some structure, it is useful to follow the structure of the energy system itself in terms of where and how energy is produced and used. Figure 8.1 provides an overview of this in terms of the three main uses of primary energy in industrialized countries and, within each of those uses, where the energy both comes from and gets used more specifically. Consistent with the figure, I divide the rest of the chapter up into a consideration of electricity generation, transportation, and heating. Within each of these, I divide the discussion

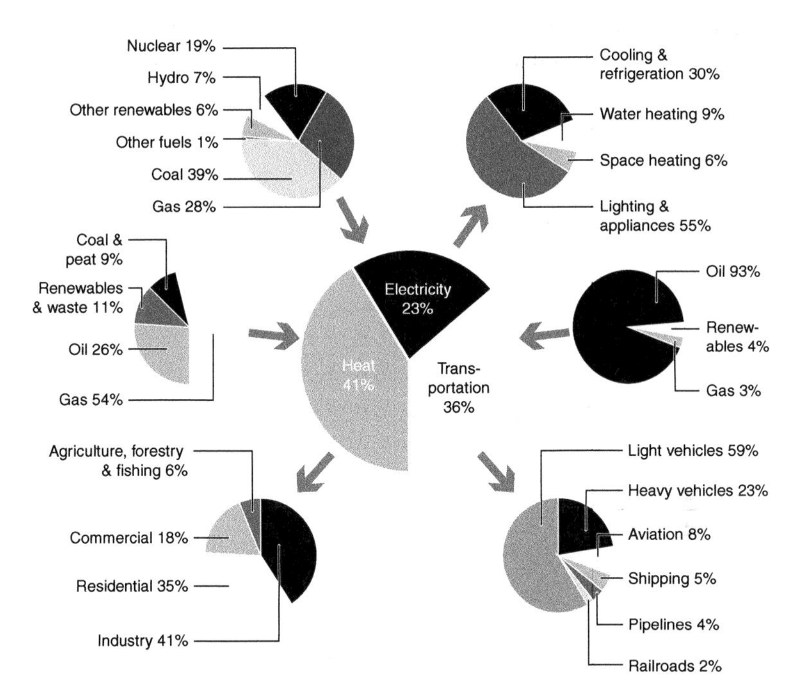

Figure 8.1. *Overview of the energy sector.*[1]

according to either inputs or outputs, depending on which makes more sense in the particular case.

Importantly, you may notice that this omits discussion of all the sources of GHGs that are outside the energy sector, most prominently changes in land use. These other sources, by and large, are not contingent on new technologies, but rather on changes in human behavior, which is why I feel comfortable omitting them here. One exception is the production of cement, which I will cover as well. Two chapters from now, I will cover the policy issues related to nonenergy sector emissions. But, for now, let's jump right into electricity, the smallest slice in the big energy pie.

Electricity supply

About a quarter of the energy that people use comes to them through a wire. Currently, most of that electricity is produced through the combustion of fossil fuels: coal, natural gas, and diesel fuel. Of these, diesel generators are the easiest and cheapest to set up but are by far the most expensive to operate and make up only tiny share of the market, somewhat more so in developing countries where access to investment capital is extremely limited. Most electricity comes from the combustion of coal or gas. With both of these two fuel sources, there are a number of different ways of constructing a power plant, with typically some tradeoff occurring between the cost of constructing the plant and the efficiency that it will achieve. The more hours of the day, on average, that a utility expects a power plant to be operating, the more it is worth investing in the more expensive and more efficient design.

The metric used to describe the costs of different technologies is *levelized cost of electricity* (LCOE). To calculate LCOE, one adds up all the costs associated with constructing, operating, and ultimately decommissioning the plant, discounting those costs to their net present value at the time when construction starts; one then divides the sum by the total amount of electricity that one expects to produce over the lifetime of the plant. The latter number is a function of three things, only the first of which is known for sure at the time of the plant's construction: the *maximum rated capacity* of the plant, measured in units like megawatts or kilowatts. The second is the *capacity factor*: the average fraction of its maximum rate capacity at which it actually operates. The third is the length of time that one expects the plant to be in service. To ensure that utilities don't provide investors with inflated profit statements, accounting rules dictate that they have to depreciate their investments over a certain period of time, and this period of time is based on a conservative estimate of the investment's lifespan. In reality, many plants live longer than the accounting rules suggest.[2]

If the utility can sell the power from the plant at an average price that is at or above the LCOE, then it will make a profit. In a market dominated by coal and gas, the LCOE values for these two technologies have the greatest influence on the market price and hence set the standard for the profitability of competing technologies. There are some complicating factors, as I will explain in the next chapter, but, by and large, LCOE is a good number by which to evaluate cost competitiveness. Analysts present LCOE either in terms of cents per kilowatt hours (kWh) (typically U.S. dollar cents, but sometimes euro cents) or dollars per megawatt hours (MWh). A megawatt is a thousand kilowatts, so 5 cents per kWh is the same as $50 per MWh. Typically, people use cents per kWh when speaking of the retail power market (where you and I buy our electricity) and dollars per MWh when speaking of the wholesale market (where power producers sell to power distributors). Here, I will use the former.

Back in Chapter 4, there was a graph (Figure 4.4) comparing the ranges of estimates for the LCOE of different energy sources. Now it is time to go into greater detail, examining how the numbers in that graph have been changing over time.

It is heavily debated what the LCOE is for coal and gas plants and what it will be in the future, but a common estimate has been about 5 cents per kWh. Historically, electricity from coal was less expensive than from gas, but that has changed in recent years as a result of fracking for natural gas in the United States, which has increased its supply and reduced its price. The U.S. Department of Energy recently projected the costs for both coal and gas to rise somewhat, especially for coal, suggesting that, for plants being planned now and brought online by 2019, the LCOE from coal would be roughly 8 cents per kWh and for gas roughly 6 cents.[3] I have no reason to question these estimates, but I also believe that they fail to capture a lot of uncertainties. I would suggest that the price of both coal and gas could be as low as 5 cents, but could also be higher, perhaps more than 10. All these numbers, by the way, are for a new plant that a company might consider building. Old coal and gas plants, ones that have already been fully depreciated, can run for significantly less money, as little as 2 cents per kWh, covering just their fuel costs and maintenance. That is like driving around in a twenty-year-old car on worn tires. At some point, you will need to replace it.

Let's move from cost to reliability. Coal and gas have both been seen as quite reliable, although for slightly different reasons. Coal is something that a lot of countries have, and so any issues associated with energy security are extremely minor. Gas, by contrast, is something that many countries need to import. Although the United States used to need to import gas, now it does not thanks to fracking. Europe west of Russia still does need to import most of its gas, and most of that comes from Russia via a set of pipelines. A lot of gas traded on world markets also comes from the Middle East, Qatar in particular. Neither Russia nor the Middle Eastern countries always see completely eye-to-eye with the West, and so there may be some reasons to view gas as slightly less reliable than coal for reasons of geopolitics. On the other hand, gas is more reliable in terms of flexibility to be turned on or off. Coal plants, in general, are most efficient when run at a steady pace, typically taking some time to warm up and cool down. Gas plants, by contrast, don't typically have this problem to the same degree. For this reason, the two technologies have historically done well together as complementary parts of a single power system: coal power plants stay on most of the time, providing base-load electricity, whereas gas plants get turned on and off to match fluctuations in load above that base.

In terms of potential capacity, gas has faced no problems in recent years, whereas coal is starting to come under fire in some countries. The reason for the constraint with respect to coal is the fact that it is an extremely dirty technology; in addition to CO_2 emissions, coal power plants give off a lot of other pollutants with serious local and regional consequences for the environment and for human health.[4] In countries with strong environmental laws, as well as strong empowerment of local interest groups within their permitting processes, it has often proved difficult for utilities to build new coal power plants. Given the debates currently ongoing about the environmental safety of fracking to extract shale gas, it may come to pass that the supply of gas, at least in the United States, becomes domestically constrained. One could speculate that this could limit the potential capacity of gas power, but that would be pure speculation.

There are three possible technological approaches to decarbonize the electric supply system. At least one of these technological approaches needs to offer the reliable promise of being able to outcompete coal and gas, ideally across all three dimensions. Exactly one of them does.

Nuclear power

It isn't nuclear power. Many people do pin substantial hopes for reducing CO_2 emissions on nuclear power, and the reasons for doing so appear superficially to be good. The most important of these is that nuclear power is well established. In many countries, such as the United States and Germany, nuclear power provides a substantial share of the total electricity that people use, which is in turn a large and growing share of the total energy budget. In my home country of Switzerland, it is about 40 percent of electricity. In neighboring France, it is more than 75 percent.[5] If France has done this, then other countries could too, the reasoning goes. Decarbonizing the electric power system with nuclear power seems really possible.

One thing that makes nuclear power really attractive, as a decarbonization option, is simply the fact that nuclear power plants produce a ridiculously huge amount of electricity. These days, even a modest nuclear power plant has a capacity of about 1 GW and a capacity factor – the proportion of its rated capacity that it actually generates on average – of more than 90 percent. A large windmill today has a capacity of 2 or 3 MW and, depending on geography, probably has a capacity factor lying between 20 and 40 percent, meaning that it would take about 1,000 of these things to equal one nuclear power plant. A fairly large residential rooftop solar power panel array would have a capacity of 4 kW and, depending on geography, probably has a capacity factor of below 20 percent, meaning that you would need more than a million of these things to equal one nuclear power plant. Take a country the size of Germany. Germany currently has seventeen operating nuclear reactors. It would take only sixty-five power plants, each at 1 GW capacity, to power the entire country. But each plant could also be bigger. The Fukushima Daiichi plant in Japan encompassed six reactors with a combined capacity of 4.7 GW, whereas the largest nuclear power plant in the world, also in Japan, has a capacity of 8.2 GW.[6] So maybe as few as fifteen or twenty relatively big nuclear power plants could cover all of Germany's demand.[7] Put this into perspective. Germany's largest car company is Volkswagen, and, in the past fifteen years, Volkswagen has built ten new factories.[8] Just one of these, in Chattanooga, Tennessee, spreads over 1,400 acres,[9] occupying more land than the 860 acres devoted to the Fukushima Daiichi complex.[10] So, if a single car company like Volkswagen can build all these factories, building enough nuclear power plants seems like it could be a very doable task.

I have nothing in particular against nuclear power, but despite these big plusses for nuclear power, I don't think it can begin to outcompete fossil fuels, at least not within the next ten to twenty years, on the basis of cost and potential capacity.

Famously, back when the U.S. government was first developing the prototypes for today's nuclear power plants, a credible authority predicted that nuclear power would eventually be "too cheap to meter."[11] Based in part on that belief, governments around the world turned to nuclear power, from the 1950s through the 1990s, as a major source of energy, promoting it with public funding for research, as well as direct and indirect subsidies to get private electric power companies to build such power plants. After all of this, however, nuclear power just isn't cost competitive and does not appear to be headed in that direction.

Arnulf Grübler, a researcher at the International Institute for Applied Systems Analysis and a professor at Yale University, studied the cost evolution of nuclear power in the context of the French build out.[12] He did what nobody else had done, which was to disentangle the cost projections for nuclear power from the actual costs incurred and, in the process, arrive at a realistic estimate of nuclear power's true costs. He found that considering the costs associated with the entire fleet of French nuclear power plants, the cost of electricity averages $0.045 per kWh. This is roughly comparable with the cost of electricity from coal or natural gas. What he also found, however, was that the later plants proved to be substantially more expensive than the earlier ones (by a factor of about four), thus putting the cost of power for the later power plants up close to $0.10/kWh, which is at the high end of the possible cost range for coal and gas. He speculated that the reason for the cost escalation was associated with the complexity of building increasingly larger power plants, combined to some extent with rising standards for safety. Other researchers have reached qualitatively similar results. In 2009, researchers at MIT suggested that the costs of new nuclear power plants have roughly doubled over the past decades and in 2009 stood at $0.084/kWh.[13] Most estimates that I see today are consistent with this latter figure. More importantly, there is nothing to suggest that the price is going to fall anytime soon. This means that nuclear power has no obvious cost advantages over fossil fuels and may be at a disadvantage.

The more difficult issue to quantify is public opinion and how this affects potential capacity. The question of what to do about the spent nuclear fuel from nuclear power plants has never been completely resolved, although it has often been a lesser concern to people than that of nuclear meltdown. Four years ago, many people were speaking of a "nuclear renaissance" and anticipating the countries around the world were going to engage in a new construction project, echoing what had happened in France in the 1980s and 1990s. But then came the accident at the Fukushima Daiichi power plant, following a tragic earthquake and tsunami, and opinion suddenly shifted – and with it the politics.[14] In the United States, enthusiasm for nuclear power dropped off sharply in 2011, with only 43 percent of Americans supporting the construction of new nuclear plants.[15] One of the major governments in this issue area, Germany, announced that it would accelerate the phase-out of its nuclear power fleet. Switzerland more or less did the same. It is simply difficult to imagine that public opinion would turn around quickly enough to lend strong support to nuclear power and make it easy to build a lot of new plants over the next decade or two.

Others are more optimistic. They point to fundamentally new nuclear technologies, such as pebble bed or thorium reactors, or a different approach to current reactor design, such as making them really small and modular, able to be mass-produced.[16] Such innovations could bring improvements to safety (or at least the perception of safety) as well as critical reductions in cost. If these views are right, that would add greater support to my basic policy argument in this book because it would mean that nuclear could conceivably outcompete fossil fuels. If they are wrong, which is what I suspect, then nuclear power will remain a technology that is less attractive than fossil fuels unless something like a carbon tax comes along to change that.

Carbon capture and storage

The second technological approach is carbon capture and storage (CCS). The idea behind CCS is that it is possible to separate out the CO_2 from the exhaust stream of fossil fuel-burning fires and then put that CO_2 someplace out of way, like underground. For many people, this is an attractive option, especially in the context of electricity production, because it means that we could still rely on coal and natural gas for a lot of our power, but we would take the additional measure of installing a CCS system on top of the existing system for producing electricity. CCS is also attractive when combined with the combustion of biomass to produce electricity or hydrogen gas. Doing so would have the net effect of removing CO_2 from the air: trees suck CO_2 out of the air, you burn them to produce electricity, and you bury the CO_2. That may prove important in case CO_2 emissions do not fall quickly enough to stabilize concentrations at an acceptable level. Actually, if society were to rely on CCS to decarbonize the energy system, some proportion of that CCS would have to be connected to burning of biomass. The reason is that CCS does not actually remove all of the CO_2 from the exhaust stream of a power plant, but rather more like 90 percent. Somehow, we would need to deal with these residual emissions.

In a sense, CCS is like nuclear insofar as it would enable the continuation of an electric power system that looks more or less like the one we have today. Nuclear power plants are common, but coal and natural gas power plants are even more so. If we could just add CCS on top of them, then we could have a nicely decarbonized power system, one built around a basic architecture that we know works. But as I am pessimistic about nuclear, I am also pessimistic about CCS, essentially for the same two reasons: cost and potential capacity.

Start with cost. Nuclear, as just argued, is currently more expensive than coal or gas, although some people suggest that this state of affairs could change. With CCS, the problem is intractable. The whole idea of CCS is that one adds it on top of a coal or gas power plant or perhaps a power plant burning biomass. This takes additional capital investment. Once that is in place, you need to use some of the energy that the power plant produces to drive the carbon capture process. This reduces the overall efficiency of the power plants, costing money again. Finally, you need to put money into the process of storing the CO_2. All together, these three costs make producing power with a coal or gas plant using CCS significantly more expensive than it would be without the CCS, a cost penalty of between 50 and 150 percent.[17] That may make CCS attractive in comparison to some other technologies. Coal with CCS may be less expensive, for example, than burning diesel fuel to generate electricity and perhaps less expensive than using photovoltaic (PV) panels, at least for now. But coal with CCS will always be more expensive than coal without CCS. The only way that CCS competes favorably is if there is some sort of a government policy, like a carbon tax, to make it relatively more attractive.

There are three other problems with CCS, and these can be subsumed under the issue of potential capacity. The first is that of safety, and, in this respect, it is somewhat similar to nuclear. The actual evidence that CCS is unsafe is thin, just as the actual numbers of people who can be shown to have been harmed from the Fukushima Daiichi accident would appear to be low. But the fears of CCS are significant. There have been cases of massive bubbles of CO_2 rising up from lakes and killing all the people and animals around. The safety issue hasn't really arisen because CCS hasn't really been built on a commercial scale, but I suspect that getting public approval for a carbon storage facility would be like getting approval for a nuclear waste facility. Which is not easy. This may limit the scale and speed with which CCS can be increased.

The second factor that could limit the potential capacity of CCS is the availability of places to store CO_2.[18] Early analysis from 2003 suggested that some regions of the world would have enough storage capacity, whereas other regions would not.[19] More than ten years later, the debate is still ongoing.[20] My interpretation is that we simply won't know until we try because that is the only way we are really going to find out what happens to compressed CO_2 when you put it underground in places other than oil wells. We may find that storage sites are plentiful, but we may also find that geological reservoirs that we thought would hold the CO_2 don't actually work. The most likely result of this is that it will somewhat push up of the cost of CCS because we will have to invest significant resources searching far and wide for underground locations and then connect those, via pipelines, to the places where the CO_2 is created.

The final problem with CCS is that it doesn't actually exist yet, and that raises doubts about its potential capacity. All the constituent pieces of CCS do exist and have been used for decades. The fact that people can survive in space is testament to the ability of engineers to scrub CO_2 out of the air. A pipeline to transport CO_2 to storage facilities would look very much like a natural gas pipeline. Burying CO_2 underground has been done, especially in the process of enhanced oil recovery, whereby it is pumped into oil wells to force more oil out. Experience suggests that the CO_2 stays there. But nobody has put the elements together into a reliable system that works at the scale that would be necessary to decarbonize the energy system. There is no good reason to think that it would not work, but there is every reason to think that it will take some trial and error to get things right. Processes like this typically take a decade or more.[21] I doubt very much that CCS is a technology we could confidently rely on, at a meaningful scale, before about 2030 at the earliest. This means that its potential capacity would be high in the long run, but not in the time within which we would need it. In other words, it will come too late to support the initial decarbonization of the energy system, although, in combination with biomass combustion, it could play an important role later on in drawing CO_2 out of the air.

So, neither of the first two technology options supports the central thesis of this book: they might present viable ways of decarbonizing the power system, but they don't offer an immediate chance to outcompete fossil fuels and hence decarbonize the power system without the need for continuous government intervention, something like a carbon tax. But the third technological option is different, and it is on this option that the thesis of this book rests. That option is renewable energy.

Renewable sources of power

All the energy we have available to use on the planet comes from one of three sources. The first is the kinetic energy of the Earth and Moon orbiting each other, which gives rise to the tides. The second is nuclear fission, which people have been making happen with nuclear power plants, but which happens on its own within the Earth's crust, the source of heat for geothermal energy. The third is nuclear fusion happening in the Sun, giving rise to solar radiation hitting our little third planet out.

Tidal energy is a relatively new and experimental technology, limited in terms of the total amount we can tap into – supplying only a tiny fraction of the total energy we need – and relatively expensive. Current costs are more than $0.20/kWh,[22] and I am skeptical that it will be any more than a niche player in the energy market, even in places where the tides are really high. Geothermal energy is more developed and, in a limited set of locations around the world, is already an energy source of choice. What matters for the cost of geothermal is how deep you have dig to find it, and so in those places where the Earth's crust is particular thin, it is cheap and abundant. The best example of these is Iceland, where it is the major source of the country's power. In other places, it is far less attractive simply because you have to dig so deeply to find enough heat, often employing other processes like fracking in order to increase its potential flow. I participated in a study to explore its application in South Africa, with the ultimate conclusion being that it is far from cost effective.[23] In other places, like Europe, the situation is mixed. Globally, geothermal could provide up to about a tenth of our total energy needs at affordable prices.[24]

Of the three, energy from the Sun is by far the most abundant in forms that we can use and has always been the backbone of our energy system. We harvest it in four ways, three of which involve natural processes that make it more concentrated and hence easier to tap into. The first concentration process is the growth of plant matter, either having taken place long ago and given rise to fossil fuels, or else happening right now, giving rise to biomass and biofuels. The second concentration process is the evaporation and subsequent condensation of water, giving rise to rain and snow high in the mountains and hence hydropower when that water flows downhill. The third concentration process is sunshine-driven atmospheric convection, which means wind. Average wind speeds are really high in some places and really low in others, and people typically make use of it where speeds are high. The fourth way to make use of the sun's energy is directly, collecting raw sunshine for heating or the generation of solar power.

Until recently, solar energy has been by far the most expensive of the four for the simple reason that it doesn't benefit from a natural (and hence free) process to do the work of concentrating it. We have had to build the concentrators ourselves, and that costs money. It offers, however, the promise of being the most abundant. The journal *Science* published an overview of the global economic potential of different renewable sources for electricity – the total amount that could be generated at reasonably affordable prices – comparing these to current demand. Hydroelectric could economically cover about 10 percent of current demand, biomass about 60 percent, wind 130 percent, and solar well over 500 percent – essentially unlimited.[25] If we are going to have a decarbonized energy system, it is on biomass, wind, and solar that we will have to depend. For reasons that I describe soon, it probably makes sense to devote what biomass we can grow as the feedstock for biofuels, to be used in those places where no other good alternative to fossil fuels exist, and not use it to generate electricity. For electricity, then, that leaves wind and solar. I take each in turn and consider whether each is or could soon be competitive with fossil fuels.

Wind energy

Wind energy has long been used to power sailboats and, more recently, to provide the mechanical energy to pump water and grind grain, with windmills dotting the landscape of the Netherlands and the U.S. Great Plains. The use of wind to generate electricity was developed in 1887, simultaneously in Glasgow, Scotland and Cleveland, Ohio. Throughout most of the twentieth century, relatively small windmills were used to generate electricity, often in off-grid locations. The development of modern wind energy got going on both sides of the Atlantic in the 1980s, as a result of public policies to stimulate renewable energy. The early leader was California, where three large wind farms were developed in mountain passes, supported by subsidies from the U.S. government. When these subsidies ended, leadership shifted to Denmark, and, by 2000, wind was supplying more than 12 percent of Denmark's electricity. That year, wind power supplied 30 terawatt hours (TWh) of power globally. Since then, wind has taken off in more places, spreading across Europe (notably to Germany, Spain, Portugal, the UK, and Ireland), the United States (notably in west Texas and the Great Plains), and in China and India. In 2012, wind supplied more than 530 TWh, with installed capacity continuing to grow at 10 percent per year.[26] Globally, electricity production is more than 23,000 TWh, and so wind production accounts for less than 2.5 percent of that. In some countries, however, it supplies much more. Top of the list is Denmark, where wind provides 33 percent of the country's electricity.[27] In the United States, wind supplies about 4.5 percent of total power demand.[28]

As wind has grown as a sector, the costs have fallen for three main reasons. First, windmills have gotten bigger, and this means that they reach higher in the sky, where the winds are stronger. Second, technological improvements have led windmills to become more reliable. The mechanical forces inside a windmill are immense, leading to breakdowns that take the windmill offline and require calling in the repair crew. Over the years, engineers have figured out how to build them so that they break less frequently. Finally, manufacturing costs have fallen as more and more components are mass-produced. The one countervailing factor driving costs up has been the price of raw materials, and, for a time – culminating in 2007 – this led wind power to rise in cost and make people wonder if the cost reductions were over. Since then, things have returned to normal, and costs have continued to fall.

The most recent credible estimate for the cost of wind power was put together by Ryan Wiser and Mark Bolinger at the Lawrence Berkeley Laboratory, which is run by the U.S. Department of Energy. They analyzed projects built in the United States in 2013, most of which were located in relatively windy onshore sites in the interior of the country. The found the average price for selling this wind on the market to be $0.025 (2.5 cents). This caught some journalists' attention, and some reported that the LCOE for these windmills is that low. In fact it isn't. The owners of the windmills also benefit from subsidies and tax credits that give them at least another 1.5 cents per kWh. But still: prices potentially starting around 4 cents. That is consistent with recent estimates for the LCOE of wind power in Europe, which is typically under 6 cents at a good site.[29] The European sites are simply not as windy as the North American ones, and some simple calculations suggest that they might produce a third less power because of this. But if the costs of coal and gas are somewhere between 5 and 10 cents, then it is safe to say that onshore wind has already won this race.

There is also offshore wind, as anybody who has flown into Copenhagen has seen. Compared to onshore wind, offshore wind is more expensive, with levelized costs of typically between 10 and 20 cents per kWh. The main reason for this is the much greater installation and repair costs: everything is more expensive when you have to do it at sea rather than on dry land. Its costs, too, have been falling, however, and one can easily imagine them reaching something like 6 or 7 cents within the next ten years. At the same time, the potential for offshore wind to be installed in deeper and deeper waters has been increasing. The latest development, now becoming commercially viable, is the use of floating platforms. This allows offshore wind parks to be located in very deep water, anchored to the bottom with cables, rather than having to build them up from a foundation on the sea floor. For places with a narrow continental shelf, such as Japan, this makes offshore wind a real possibility.

If wind, especially onshore wind, is already so cheap, then why aren't we already using it for everything? One answer could be that we are trying to: it is only a recent phenomenon that its costs have fallen to below those of coal and gas, and, since then, it has been expanding almost as fast as the equipment supply chain will allow, most recently at about 10 percent per year. The other reason is that, as I will discuss with solar, it is intermittent. Wind is not an inherently reliable source of power.

Recently, my family and I went on vacation in western Norway, renting a cabin on a lake just a few kilometers inland from one of the country's spectacular fjords. The lake itself was a natural body of water, but by now both the two streams into it and the one stream out had all been dammed. Water both entered and exited our little lake through a network of high-pressure pipes running through turbines on the way. When we arrived there, we discovered that the power station on the inlet pipe was only about fifty meters away along the lakeshore, and it seemed not to be running, which suited us just fine. The second day there, we woke up to clear skies and light winds from the southwest, but big waves hitting the rocks in front of our cabin. Why the waves? The power plant had turned on during the night, and water was gushing out, creating the waves not far from our swimming and fishing spot. I had a hunch why it had turned on, and I went online to double check. Sure enough: a high-pressure system, meaning the absence of wind, had developed to the southeast. Over Denmark.

The reason that Denmark can get 33 percent of its electricity from wind is that is tapped into the much larger German and Scandinavian grids. When the wind is blowing, Denmark produces more wind power than the country itself can use, and a lot of that extra power gets exported, especially to the north. Norway and Sweden themselves get most of their domestic electricity from hydropower, and so when the power starts flowing from Denmark, they shut off some of their dams. When the wind in Denmark stops, the flow goes the other way, and all the dams turn on. All of this illustrates the point that large shares of a single renewable power source work well only when there is some other feasible source of power, able to be turned on or off with the flip of a switch. If Denmark weren't connected to its neighbors, or if all its neighbors had wind power shares and also a flat landscape like Denmark, problems would arise. This problem of intermittency also exists for solar power and is probably, right now, the biggest factor constraining the two technologies' growth. As long as each technology constitutes only a small share of the overall power system, reliability does not become a problem. When the technologies grow to become bigger players, as they have in Denmark and Germany, then reliability does become a problem, and they are suddenly much less attractive. There are ways of dealing with the reliability issue, but, by and large, these ways are common to both wind and solar, and so I will discuss them after I have already described solar itself.

With wind, then, there is one issue left, and that is its potential capacity. A large wind turbine, today, has a peak capacity of about 3 MW, and a capacity factor of below 40 percent. To replace a 1 GW coal plant, then, one would need an awful lot of these windmills. Once, during an especially boring seminar, I did a back-of-the-envelope calculation, asking myself how many windmills one would need to build in order to power all the cars on the road in the United States, if all those cars were electric. The answer was a numbingly large answer, which also happened to be roughly equal to my estimate of the number of gas stations currently serving the vehicle fleet. Somebody built all those gas stations, which tend to be in places with relatively high real estate prices. So I suppose that somebody could build that many windmills, too.

Instead of the actual numbers of windmills needed, which, like everything else in the energy sector is huge, two other issues are more relevant. The first is how much land and sea area is available, with high enough average wind speeds to make sense, on which one could plant windmills. Globally, as I have already suggested, the numbers suggest that there is more than enough space available to build windmills to satisfy all of our current electricity needs, which is about a third of our total current energy needs. Regionally, however, there are serious constraints. One study for Great Britain, for example, concluded that the combination of onshore and shallow water offshore could not come close to covering the country's electricity needs. Even there, though, wind could offer enough capacity to make a substantial difference, even if on its own it couldn't completely replace fossil fuels. The far more important issue with wind, however, is public acceptance. In more and more places, in countries like Germany, France, the United States, or Brazil, there are cases of people deciding they don't like living next to windmills or having the windmills infringe on a landscape that they value.[30] In some cases, such as in the Czech Republic, public opposition has put an effective halt to development.

For now, I want to flag the issue of public acceptance of wind power as being an important one. A lack of public acceptance for wind could make it a noncompetitive technology, every bit as much as with nuclear. The evidence suggests, however, that public acceptance for wind is often contingent on factors that policies can influence, such as whether stakeholders were consulted in the early stages of project development, and, in this sense, it appears to be somewhat different from the acceptance issue with nuclear power. I am saving that discussion for the next chapter, in which I describe the policy instruments we need more generally. Let us say that the potential capacity of wind is good and will not stand in the way of its competitiveness, if and only if policies are put into place that deal effectively with the public acceptance issue.

Solar power

Solar power encompasses two different groups of technologies. The first group is photovoltaic (PV), which relies on the fact that when light hits certain kinds of materials, it causes electrons to do things that create the electric charge that is the basis for generating electricity. There are several different types of PV, grouped into those that use crystalline silicon and those that use other elements deposited in a thin film on some sort of backing material. Of the two, crystalline silicon generates more electricity per square meter facing the Sun and so is the preferred technology where space is limited, as on rooftops. *Thin film*, the newer technology, can be more easily mass produced and now offers a lower LCOE; it is the preferred technology where space is not a constraint, as in large arrays built on low-value land.

The second group of technologies is concentrated solar power (CSP). CSP relies on mirrors with a large surface area to reflect sunlight onto a much smaller area, thus concentrating the energy to create high temperatures – just like those boys who weren't my friends used to fry ants with a magnifying glass, back before there were video games. The high temperatures are used to create steam, and this powers a turbine, just like a coal or gas plant. There are four different CSP configurations.

The most common arranges the mirrors in long rows, with a parabolic cross-section focusing the sunlight on a tube running down the row, in the focal point of the parabolic troughs. The trough has to rotate on one axis to continually focus the reflected light onto the tube. The next most common configuring utilizes a central tower with a heat-receiving element at the tip. Arranged around the tower, in all directions, are hundreds of flat mirrors, each tilted separately to shine sunlight onto the heat receiver. These have to be rotated along multiple axes to keep reflected sunlight exactly focused on the tower's tip. The two other technologies utilize a *Fresnel mirror array* and a *Sterling engine*, respectively, and I won't go into them; they are currently marginal and offer no clear benefits relative to the other two. Three other aspects of CSP, common to both parabolic troughs and central towers, deserve note. First, CSP plants require a cooling system, and traditionally, cooling systems use a lot of water. Second, CSP plants can incorporate a thermal storage system. This allows the operator to introduce a time lag – currently up to about sixteen hours but theoretically much longer – between when the concentrated sunlight hits the receiving unit to generate heat and when that heat is sent to the turbine to generate electricity. For reasons I will soon discuss, this could be important. Third, CSP plants, at least compared to PV arrays, require constant attention to cleanliness. A dirty PV panel will generate a little less power. A dirty mirror will scatter the light and generate almost no power at all.

In terms of their use applications, PV and CSP differ in several critical respects. The amount of electricity that a PV panel generates depends on the amount of total sunlight hitting it. This means that it generates no electricity at night, a fair amount when skies are cloudy, and a lot when skies are clear, thus providing a slight benefit to building it in places with generally sunny skies. Slightly compensating for this, the efficiency of PV panels falls off a bit (a percentage point or two) with higher temperatures. With CSP, by contrast, the power output depends on the total *direct* sunlight hitting it, which means no clouds. For this reason, CSP is far more economical when it is built in arid landscapes, where cloud cover is rare. PV panels are modular, and there are no major benefits to building them bigger rather than smaller: the LCOE from a residential rooftop array isn't that much different from that of a commercial array producing 1,000 times more power. CSP, by contrast, offers significant economies of scale. Indeed, putting this together with the need for direct sunlight, the only kind of CSP facilities that really make sense are those built in the desert, far away from people, with each plant covering a large area of land and with a small but full-time staff to attend to their operations. PV works well in these environments, but it works almost as well when integrated into the fabric of our communities: on top of roofs and on spare bits of land.

Let's jump to the three criteria: cost, reliability, and potential capacity. In 2007, when I first started studying solar power, most people were predicting the LCOE of CSP to fall to the point of being competitive, whereas for PV it would take a while longer. Seven years later, the picture is quite different; indeed, to some extent reversed. Let's take each technology in turn, starting with CSP because it is the simpler story.

Back in 2010, my colleagues and I published one of the first papers foreseeing how soon CSP could reach a point of competitiveness.[31] We estimated the LCOE of CSP then, in 2009, at about 15 cents per kWh. Some industry experts told us we were too high, and the true number was closer to 10 cents, while others told us it was closer to 20 cents. There were several new plants being developed in Spain and Southern California and talk of many more in North Africa. Based on this information, we assumed that CSP could easily grow in total capacity by 25 percent per year, matching what PV was doing at the time, and keep up this growth rate until it got to the size that PV was at the time. We were thinking about "Desertec" in particular – the idea of producing solar power in the North African desert and transmitting part of that electricity to Europe – and so we took into account the costs of power lines from North Africa up to Europe, adding between 1 and 2 cents to the LCOE. We also assumed a price for coal of 5 cents per kWh, which may be too low. Given reasonable estimates for the learning rate – the speed at which LCOE falls as a function of growth in installed capacity – we estimated that the LCOE could fall to 5 cents per kWh – and hence be competitive with coal – by as soon as 2020 or as late as 2030, depending on a few key assumptions. One assumption was where exactly the plants would be built: most of Egypt gets a lot more direct sunshine than Tunisia or Morocco, and this gives rise to a higher capacity factor and lower costs. The other was the interest rate. Projects that people see as risky have to pay investors much higher rates of return on equity and take on debt with a substantial risk premium. Indeed, as we examined in a second, related paper, a crucial aspect to achieving a quick drop in CSP costs was the implementation of a policy instrument to reduce investor risk.[32] If policy makers could do this, and we had reason to believe that they could, then the 2020 date for hitting the 5 cent LCOE threshold didn't seem at all crazy. Several of our contacts in the CSP industry agreed with us then. If the true cost of coal were higher, say 8 cents, then the 2020 date for CSP competitiveness seemed even to be a good estimate.

That was, however, before the Arab Spring and before the European debt crisis. The Arab Spring made new investments in most North African countries appear riskier than ever and, with the exception of a CSP plant in Morocco, shut off new CSP development in North Africa.[33] The euro debt crisis resulted in Spain putting an abrupt halt to its support for CSP. As a result of both factors, the only place where CSP has continued to move forward has been California. Back in 2009, when we were writing our first paper on CSP costs, we expected the LCOE to be well below 10 cents by the time I am writing his now, mid-2014. Instead, current costs appear to be closer to 12 cents. That is a lot slower progress. I would say that 2020 is out of the question for hitting the 5 cent mark, but there is reason to believe that 2030 is still feasible, with the right policy intervention.[34] If the cost of coal and gas is higher, say 8 cents, then a 2025 date for cost competitiveness seems possible.

Although the news on the cost front has not been particularly good for CSP, it has positively exceeded all expectations for PV. Back in 2009, the LCOE of PV was more than 40 cents, and cost reductions had stalled because demand for new PV modules – fueled by generous subsidies in many European countries – had exceeded supply. Then things began to change. First, demand cooled off somewhat; several countries, most notably Spain and the Czech Republic, drastically reduced their levels of support, to a large extent in response to the financial and European debt crisis. Second, a whole lot of new production capacity came on line. In the area of thin film, the market leader was a U.S.-based company called First Solar, and they simply replicated their successful factory floor design in a number of locations around the world, thus bringing down average costs with each new duplicate factory they brought online. In the area of crystalline silicon, the Chinese jumped into the market, supported by high levels of state support, and were able to undercut the prices of suppliers from places like the United States and Germany.[35] There is some disagreement among analysts, but the evidence suggests that the cost reductions that have coincided with the Chinese market entry reflect genuine improvements in both product design and manufacturing efficiency.[36]

Even as the costs of PV change monthly, they also vary a great deal by location. Partly, this is a reflection of differences in sunlight, which affects the capacity factor. A PV system installed in Los Angeles will have a capacity factor that is about 70 percent higher than one built in Munich, Germany.[37] Surprisingly, a greater source of variance is the difference in costs other than the PV module itself. The price of PV modules has fallen dramatically, from more than $5 per watt of installed capacity back in 2006, to below $1 per watt by the end of 2011, and, most recently (August 2014), falling to just under 35 cents.[38] But to get a PV system up and running, there are other costs involved, and these are divided into "hard" and "soft" costs. Hard costs include the mounting frames for the PV modules, as well as the power inverter to go from DC to AC. Soft costs include whatever the company actually mounting a PV system on your rooftop has to spend money on, including paying its workers and advertising on Google. Combined, the hard costs plus soft costs reflect the state of development of the local PV market. In the United States, they typically add up to between $3 and $4 per watt.[39] In Germany, they average just about $1.[40] As a result, the LCOE for PV ends up being about 20 cents per kWh for a rooftop system in Los Angeles, but about 11 cents in Munich.[41]

The question of what constitutes competitiveness for PV is slightly different from that for CSP. For CSP, it makes sense to compare it to the LCOE for coal and gas, as I have done. For PV, things are a little more complicated, at least in the case of residential roof-mounted systems. From the perspective of the individual homeowner, you have to take into account the fact that a PV system lets you purchase less power from the grid. The costs of running the grid are not trivial, which is one reason why residential power prices can range from about 10 cents per kWh in the United States to well over 20 cents in much of Europe. So, by one measure, PV is competitive when its cost falls below this number, which in some European countries it already has. Unfortunately, the PV system will generate the most electricity during the middle of the day, when residential power consumption is typically low,[42] and nothing at all after dark. For most systems, the majority of the power generated will not be used in one's own home, but rather be sold to the grid. For this share of the power, the relevant point of comparison is wholesale power market price, determined in most places by the LCOE of coal and gas.

To see how this would work out in practice, I did a cost calculation for my own home, which is located near Zurich, Switzerland, and has a rooftop oriented to the south-southwest. I made convenient use of an online calculation tool put together by the Swiss solar industries association for this purpose.[43] The Swiss policy regime currently in place is this: the government would provide me with a direct subsidy, paying for 30 percent of the total cost of the installed system,[44] and would also guarantee that I can sell all the power that I do not use in my own home to the Swiss grid at the wholesale power price of about 7 cents per kWh. Given my family's current consumption patterns, and taking into account the residential power price of about 14 cents per kWh, I calculated a payback period on the PV system of just about twenty years. That is exactly how long we can safely count on the PV panels lasting, and it means that PV already makes sense for us. But it would not make sense without the 30 percent subsidy. What would change that? The thing that immediately comes to mind would be to wait another couple of years, by which time the cost of installing a rooftop PV system could well have fallen by another 30 percent. So that gives an indication of PV's cost competitiveness for this application: in Switzerland, at least, it really is just around the corner. The current boom in many U.S. states suggests that, there, it has already arrived.

What about large commercial PV systems designed to feed all of their electricity into the grid? In general, the point of comparison is the LCOE for typical coal and gas plants, as with CSP. There is one important exception, however, and that is something seen in California over the past few years. There, the factor determining peak power consumption is the use of air conditioning, and that hits its maximum value in the early afternoon during summer months, corresponding to when PV produces its greatest output. What does that PV compete against? It turns out that it competes against gas power plants built for the sole purpose of satisfying this relatively narrow time-band of peak demand. Correspondingly, these gas power plants have a low capacity factor, which means they have a high LCOE. Their LCOE sets the wholesale power market price during these hours, and it is a market price that PV (and also CSP) has been able to beat. So, for the last few years, grid-scale PV has been competitive in California, and power companies have responded by building a lot of it. Once there is enough of it to close this summer peak gap, which should happen quite soon, conditions in California will return to normal, and PV will become uncompetitive. But this, too, is almost certain to be a temporary phenomenon. There is every reason to believe that grid-scale PV will make it down to the 7 or 8 cent range in the next five to ten years. In a place like Germany, that would require total system costs of under $1 per watt. A module price of $0.25 per watt, which seems fairly imminent, would leave $0.75 for the nonmodule costs. The evidence from Germany suggests that this, too, is likely, given the pace of cost reductions seen so far. So, grid-scale PV, like rooftop PV, is rapidly becoming competitive on the basis of cost.

The next issue is reliability, and, as with wind, this is where the real problems with solar power arise because of its inherent intermittency. And, as with wind, I am going to push much of that discussion aside for a few pages and cover it under a section dealing with the issue of electricity load management and storage. First, I need briefly to cover the issue of potential capacity.

Solar power is the one source of energy for which potential capacity vastly exceeds any reasonable estimate of humanity's future demand for energy. At a global scale, it is simply not a problem. At national scales, it is in most cases not a problem either. The only exception to this is those countries with very high population densities and very high energy demand, places such as England or Germany. In these places, there may well not be enough available space, taking into account both rooftops and land not being used productively for other purposes.[45] In these countries, then, it would be necessary to import solar power from other countries and regions where the available capacity does exceed supply. One example of this is Desertec, the idea to generate solar power in the Sahara Desert for export to Europe.

When considering the generation of large amounts of CSP power in the desert, one issue that does arise is the limit to the amount of water that could be used for cooling and mirror washing. My research group has analyzed this, looking at CSP built in the Middle East and North Africa (MENA) region. I can summarize our findings in three ways. First, we found that if CSP plants were to use wet cooling, which is the least expensive way of cooling, then indeed water availability would pose a substantial constraint.[46] For MENA countries to supply themselves and Europe with 50 percent of both regions' total demand, they would end up using over a quarter of their sustainable water supply. I think this would probably be seen as unacceptable. Second, if CSP plants were to use dry cooling, it would only push up costs a bit, but push water consumption down to levels that probably would be acceptable. With dry cooling, they could supply 50 percent of MENA and European power demand while using about 2.5 percent of their available water, pretty much all of this for mirror cleaning. Doing so would raise the cost of CSP by about 1 cent per kWh, compared to building it with wet cooling. We estimated that this would delay the point of cost competitiveness by about two years. Third, exporting electricity generated in dry-cooled CSP plants would use substantially less water than exporting the same amount of energy in the form of oil.[47]

The other issue to take into account when considering potential capacity is public acceptance. Unlike with wind power, public opposition has generally not been a problem for solar, at least not for PV.[48] For CSP, there have been issues of public acceptance in the United States, mainly as a result of concerns over water use; there is good reason to believe that use of dry cooling would be a reasonable solution. Nobody really knows whether public acceptance of CSP would be an issue in other places, such as the Sahara. A few months ago, members of my research group started examining this; so far, we have completed a household survey near a CSP plant being built in Morocco and are planning a second survey for Jordan. The Moroccan results suggest that there are no objections, at least not right now.[49]

So, let's summarize solar. The cost estimates are not quite as good as for wind, but still very good. There is every reason to believe that rooftop PV will become competitive, absent any subsidies, within the next couple of years. It seems very likely that grid-scale PV will become cost competitive within the next five to ten years, and that CSP, assuming some level of government support to enhance capacity, will become cost competitive within the next ten to fifteen years. As with wind, the issue of reliability is of little concern as long as it represents a small share of the overall market but becomes a major problem for PV as it continues to grow. The potential capacity of solar is generally not a problem.

Load management and stationary electricity storage

The preceding two sections have each demonstrated that the major factor standing in the way of both wind and solar outcompeting fossil fuels, across the board, is the issue of their reliability. A perfectly reliable source of electricity is one that you can turn on and off when you need it. Gas provides this. Coal and especially nuclear do not, because it is better to leave them running at a constant level. But wind and solar are far worse, each in their own special way. Wind shuts off when high pressure sets in. That is bad in the summer because high pressure means sunny skies and additional air conditioning demand. It is also bad in the winter because high pressure means cold nighttime temperatures and additional heating demand. Both PV and CSP typically provide less total power in the winter than in the summer. That's fine in the United States, where peak demand is in the summer because of air conditioning, but problematic in Europe, where peak demand is in the winter because of heating and lighting. PV always turns off at night, and the only good thing about that is that demand tends to drop off then as well.

As long as wind and solar represent a small share of the overall power system, their intermittency creates a bit of a headache for the person managing the power grid but doesn't have any real effect on its overall reliability. That begins to change once their combined share of overall power production approaches the 20 percent mark. In both European and North American power markets, this is an issue already starting to be faced. It will only grow worse. The good news is that there is a range of options available to manage the problem, in various stages of technological development. Eventually, if and when wind and solar become our dominant sources of electricity, we may need to rely on all of them. Between now and then, we can phase them in gradually, according to their availability, ease, and cost. The options fall into three broad strategies, and I discuss each in turn.

Smart grids

The first strategies can be described as the *smart grid* approach. The basic idea behind a smart grid is having a grid that not only transmits electricity, but also information about that electricity's price, combined with an increasing number of options to shift power demand from times when electricity is expensive to times when it is cheap. The easiest of the latter are household appliances, such as freezers and washing machines, which can be set to automatically turn on and off. More complicated solutions involve devices that use electricity in commercial and industrial applications, as well as a potential future fleet of electric cars. With the latter, analysts have studied the possibility of getting electric cars to charge when power is plentiful and cheap and even to give some of that power back to the grid when power is in highest demand and supply is low.

This idea of a smart grid is also associated with another technological improvement to the grid, namely the feature such that power generated along the distribution grid can be automatically stepped up in voltage to match that of the transmission grid, something that the current grid architecture typically does not allow. This would allow power generated in a distributed manner, such as by many residential PV installations, to be used in an entirely different community, such as where a steel mill might be, thus providing some level of smoothing of supply and demand. Indeed, it is impossible to imagine a power system relying on large shares of distributed generation, such as PV, unless this were possible.

Smart grid technologies are currently available, and there is evidence that they could make some impact on intermittency, albeit not a huge one. A study by the International Energy Agency considered the potential effects of using smart grids to draw power from a fleet of electric vehicles, so-called vehicle-to-grid technology.[50] The study found that by the time electric cars had achieved full market penetration, this application of smart grids could address roughly one-third of the intermittency caused by fluctuations in wind power. It is hard to estimate the effects of load shifting in other ways, partly because economists don't yet understand how much consumers and firms would be willing to shift their electricity demand in response to price signals, even if the technology were there to make it relatively painless. It seems reasonable to assume that such load shifting, as with that associated with vehicle-to-grid, could account for some share of the intermittency from wind and solar, but far from all of it.

Super grids

The second approach to dealing with intermittency uses a confusingly similar term, namely *super grid*. What a super grid represents is the idea that electric power would be shared over much wider distances than it is today. This would involve a technical aspect, most importantly the construction of an enhanced network of long-distance – and outside of North America international – transmission lines. These power lines don't come for free, but their cost is nevertheless modest compared to the generation of the power itself; transporting electricity 1,000 km with a high-voltage direct-current (HVDC) line adds roughly 1 cent per kWh to the LCOE.[51] It would also involve an institutional aspect, namely the integration of power markets, to make fluctuating trade in such power relatively seamless. I already have described how power currently moves north and south between Denmark and Norway. Imagine the same thing happening between Egypt and Germany, or even between the United States and Brazil. These are things that just don't happen now. For various reasons, super grids could do a lot to address the problem of intermittency and, potentially, could even address it completely.

One way that a super grid can address intermittency is by balancing production of power from a single type of source, say wind or solar, over large enough distances to take advantage of disparate weather conditions. In northern Europe, for example, the wind blows strongest in the winter months, whereas in North Africa it is strongest in the summer. Putting the two resources together would make one large system that is better balanced than either of the two systems separately. The same effect would take place if solar energy were to be transmitted across hemispheres, between north and south. Even at shorter distances, the effects can be important. A study of a super grid connecting offshore wind resources over a 350-mile distance on the U.S. Atlantic seaboard, for example, found a significant smoothing effect. Although the output from individual windmills frequently jumped back and forth between full rated capacity and no output at all, the combined system always produced power somewhere, and most often in the range of 40–80 percent of full capacity.[52] Another study of wind examined the effects of building a super grid extending from Iceland in the north, to Morocco in the South, and Russia in the east.[53] It found that by taking advantage of this wide area and installing just enough extra capacity in key locations, it would be possible to completely satisfy European power demand, all the time, at an average LCOE of just over 8 cents per kWh. That's pretty good, especially when you take into account that the cost assumptions on which that study was based are already out of date. In other words, the costs would be well below 8 cents.

In my own research group, we examined this same issue not for wind, but for CSP. We considered the case of networks of CSP plants connected to demand centers via a super grid. We looked at four different such networks in different regions of the world: North America, South Africa, the Sahara region, and India.[54] CSP, of course, offers the additional benefit of a few hours of thermal storage. This is enough to allow power plants to operate through the night after a sunny day, but it does not allow them to keep operating when the weather turns cloudy for a prolonged period. Even in these deserts, our weather data showed, all CSP plants would have to turn off sometimes. When you put the plants together, however, the situation improves. Indeed in two of the regions – South Africa and the Sahara – we found the improvement to be so substantial as to potentially solve the whole intermittency problem. In those two regions, we found, you could operate a coordinated network of CSP plants in such a manner as to guarantee providing at least 90 percent of peak demand without adding *any* additional amount to the overall system cost. That means that if you had another resource, such as hydropower, which could occasionally cover up to 10 percent of peak demand, then the entire system would function reliably. If that resource were not available, then you could get to 100 percent load following capacity of the CSP system by slightly oversizing it in key locations, adding up to about 5 cents per kWh to the overall LCOE of the entire system. The results for North America and India were not as good. In these two locations, you could build a CSP network that could guarantee to supply at least 50 percent of peak capacity without adding additional cost due to system oversizing. Beyond about 70 percent guaranteed capacity, the costs started to escalate sharply. Why the difference? Both the North American and Indian desert regions suffer from relatively widespread cloudiness during monsoonal seasons. But even in these two regions, our results suggested that CSP connected to a super grid could overcome a large share of the intermittency challenge.

A somewhat different way that a super grid can address intermittency is by connecting different regionally specific sources of renewable power, just like the connection between Denmark's wind and Norway's hydro. Indeed, several studies, in both Europe and the United States, have suggested that by matching different renewable resources together it is possible to eliminate virtually all of the intermittency issue and get total system supply to match total system demand nearly all the time, much as we did for our study of coordinated CSP networks. The best of the European studies, completed by a consortium of high-profile consulting firms and commissioned by the European Climate Foundation, showed it to be possible for renewable sources to supply more than 80 percent of power demand without a need for additional electricity storage.[55] Analysts at one of Germany's leading energy research institutes, Fraunhofer ISI in Karlsruhe, reached a similar result.[56] Analysts at the National Renewable Energy Laboratory concluded essentially the same thing for North America.[57] Importantly, none of these studies considered the possibility of integrating large amounts of CSP, as I described in the last paragraph. Thus, these are two alternative pathways for a super grid to deal with intermittency. Certainly, it would make sense to combine the two approaches, thus achieving complete reliability with even less need for coordination. Nobody has looked at that yet.

The smart grid and the super grid share two features. First, both would be inexpensive to implement relative to the overall cost of the electricity itself, given the current state of technology. Second, both would require substantial legal and institutional changes. I describe these in the next chapter, where I discuss the needed public policies. Most of these changes will not come overnight, and, indeed, some of these may turn out to be unattractive; many people are skeptical, for example, of the Desertec concept. The third and final approach is the exact opposite: easy from an institutional and policy perspective, but challenging from a technological and cost perspective. That approach is grid-scale electricity storage.

Power storage technologies

There are essentially three ways to store electricity: mechanically, chemically, and electrochemically. Mechanical storage currently takes two forms, one of which is already in widespread use: pumped hydro. In a pumped hydro facility, electricity is used to move water uphill, to a mountain or hilltop reservoir. Later, the water can flow back down again, generating electricity just as in a normal hydroelectric unit. Switzerland presents a good example of how pumped hydro is currently used. In neighboring France, the fleet of nuclear power plants operates round the clock, supplying somewhat more power than the French actually need during the night and less than they need during peak load periods. During the night, the Swiss pumped hydro companies buy the excess power at a price close to the average LCOE of the French nuclear fleet, which, as I described before is about 5 cents per kWh. During peak load periods, they let the water flow back down, recovering up to 80 percent of the energy needed to pump it, and sell that power back to the French. At those times, the market price is set by peaking gas plants with a low capacity factor and is up above 10 cents. Even taking into account the capital costs of the pumped hydro and the 20 percent of the energy lost in the whole process, the Swiss manage to make some money.[58] The LCOE for every kWh of electricity stored with pumped hydro is between 3 and 4 cents. A message to take away from this is that pumped hydro does come at a significant cost, but it is not prohibitive, and it works.

The second mechanical storage option is compressed air energy storage (CAES). CAES is a much newer technology[59] and does the same with air as pumped hydro does with water: pumping the air into a closed chamber, building up the pressure, and then generating electricity when it is released. CAES is complicated somewhat by the fact that compressing air makes it hotter, and unless this heat is utilized (rather than dissipated), then the efficiency of the whole system goes way down. In theory, through effective heat exchangers, the efficiency of the entire system can approach 100 percent. In practice, right now, efficiencies are less than those of pumped hydro. The capital costs of

CAES are also comparable in cases where there are naturally occurring high-volume chambers, such as underground salt domes. As with mountains for pumped storage, these are unevenly distributed around the world. The cost becomes much higher if one has to build the compression chamber.

Of chemical storage means, the most common involves the chemical separation of water into hydrogen and oxygen using the process of electrolysis. A cutting-edge chemical storage technique is to combine the hydrogen gained through electrolysis with CO_2 to produce methane (CH_4) or, with some additional effort, the longer chain hydrocarbons that constitute the various liquid fuels such as gasoline or diesel fuel. When the end product is hydrogen, then running the reaction in reverse generates electricity. The efficiency of such hydrogen storage is somewhat less than that of either pumped hydro or CAES, currently around 30–40 percent, and is foreseen to rise to 50 percent.[60] This could make it an attractive technology where neither pumped hydro nor underground CAES is an option.

Increasingly, however, people are discussing the primary benefits of such "power-to-gas" and "power-to-liquids" technologies not in terms of their use as a means of storing electricity, but rather as a method of producing fuels that in turn can be burned in the transportation sector. This doesn't directly help the issue of grid reliability, but it can do so indirectly. The idea is that it would make economic sense to build enough wind (and potentially solar) power stations to be able to cover peak electricity demand even when the output from the power stations is relatively low. That would mean that, a lot of times, the output from the power stations would greatly exceed power demand. During these times, one would use the excess electricity to generate fuels for use in the transportation sector. For such a business model to succeed, the overall costs of producing some ratio of electricity and fuels would have to be less than obtaining the same mix of electricity and fuels from fossil sources. Right now, that is not the case, and I don't know of any analysis suggesting when it could be the case. But there is reason to believe that if the cost of wind power continues to decline, then it could start to be competitive, perhaps within a decade or two. In such a case, this could become an important source of jet fuel, a problematic area that I will describe later in this chapter. It could also prove useful as a source of hydrogen for a fleet of long-distance trucks and ships running on fuel-cell technology.

The final approach to electricity storage is electrochemical, which most people know of as batteries. Batteries involve a reversible chemical reaction that doesn't create new molecules, but rather new *ions* – atoms or molecules with different numbers of electrons. Unlike the other storage options just described, batteries can offer very high efficiencies, up to and even in excess of 90 percent. On the other hand, the batteries themselves are much more expensive and typically have a limited lifespan. There are many different battery technologies involving different types of ions, and, for each of them, calculating the extent to which they add to the LCOE of the underlying power source involves a consideration of their capital costs, the number of years they will be in service, and the frequency with which they are used. Which battery technology is preferred depends strongly on the particular use application; for a homeowner seeking to store PV-generated power, one technology may be the best, whereas for the power company seeking grid-scale storage, another may be preferable.[61] But whichever the technology, batteries right now are expensive. A recent study compared four of the leading technologies. In the case of a utility company wanting to balance intermittency, sodium sulfur batteries currently offer the least-cost alternative and yet even these come at a cost of roughly $0.24 per kWh of electricity stored, plus or minus a few cents depending on use conditions. This suggests that batteries, at today's prices, are an extremely unattractive way of storing a lot of electricity.

Will the prices of stationary batteries come down? The answer almost certainly will be yes, but there is a tremendous amount of uncertainty as to how much and, more importantly, how fast. Until fairly recently, the need for grid-scale storage was simply not that high, and pumped hydro could satisfy all demand at a price that batteries could simply not touch. As a result, there was little research going on and virtually no capital investment. That has changed dramatically, and now battery research is at the top of many countries' energy research agendas. There are technologies now in development that promise much lower costs, as little as 4 or 5 cents per kWh of electricity stored. Cost reductions of this magnitude do not seem crazy when one considers that technological developments have led to the cost of PV modules falling by similar fractions over the past five years. But still, it is an open question whether these promised cost reductions are real and, if they are, how soon they will be realized. Over the past six months, I have had two students investigating this very question, interviewing the leading experts. They tell me they just can't find a consistent answer.

When calculating the costs of electricity storage, it is important to bear in mind that only a fraction of the total power produced and consumed would need to go through a storage system, whether it be mechanical, chemical, or electrochemical. It is for this very reason that it makes sense to pursue all of the options for load management, grids, and electricity storage at the same time and not count on a single silver-bullet solution. Analysis suggests that a power system relying almost entirely on wind and solar – and better yet integrating some share of CSP – covering a wide but still politically negotiable area, with some smart grid features allowing for demand-side load management, would only need to use storage to time-shift a small share of the total power generated.[62] Let's say that share is 10 percent. That is, by today's standards, a huge amount of storage. But if the LCOE of electricity storage is 10 cents, which does not seem to require unreasonable advances in battery technology over the next decade, then the effect would be to raise the average LCOE by only 1 cent per kWh. Everything we have seen suggests that this is a cost penalty equivalent to the cost reductions seen in wind and solar every year or two. In other words, employing all these means together to make wind and solar competitive on the basis of reliability would not substantially push back the point of their becoming competitive on the basis of cost. There is very good reason to believe this will happen in the next five years in some applications, perhaps ten years in most of the rest, if the policy interventions are the right ones.

Light-duty vehicles

It's time to move from electricity to transportation. Transportation today relies almost exclusively on oil, refined into gasoline, diesel fuel, and jet fuel. There are two very different approaches that one can take to replace that oil. One is to replace the engines that burn the oil with electric motors and to supply those motors with electricity that comes from decarbonized sources. The second is to leave the engines more or less alone, but to fuel them with substitutes for gasoline, diesel fuel, and jet fuel that derives from plant matter; that is, biofuels. I am going to start with the first of these options, for the type of transportation that currently uses the most energy: driving around with cars, pickup trucks, and motorcycles, collectively known as light-duty vehicles. And for this I will start with a simple question: does driving an electric car save money?

Since things are changing fast in this market, I will base my answer on what the Internet tells me today, for where I live. The electric car that looks good to me is the Renault Zoe, an all-electric five-seat four-door hatchback, with a 22 kWh battery giving it an estimated range of 210 km under optimal driving conditions. In Switzerland, a new Zoe currently costs CHF 22,900 (CHF is the symbol for Swiss Francs, worth just a little more than US$1), plus another CHF 1,300 for the charging unit to mount in the garage.[63] Additionally, the buyer leases the battery, with the monthly fee determined by how much you drive the car. I would to drive it 20,000 km a year, resulting in a monthly battery fee of CHF 125. Driving it that much, I would spend another CHF 25 on electricity, given a residential price of CHF 0.14 per kWh. Assuming that I drove the car for five years, the total cost of owning it would be CHF 32,238, using a 5 percent discount rate for the monthly costs. That works out to CHF 537 per month, or, in a more familiar currency, $586. This is plus maintenance and insurance and minus whatever I could sell the car for at the end.

I compared this to the closest conventional car I could find, which was the Renault Clio, a similarly sized four-door five-seat hatchback. I chose an automatic transmission (to match the fact that the electric car is also automatic) and the cheapest models available in both gasoline and diesel. Both are a little bit slower to accelerate than the electric version, but, like the Zoe, have a top speed well over the legal speed limit. I have no idea how they compare on resale value or maintenance costs compared to the Zoe, with one exception: the Clio models need regular oil changes, whereas the Zoe does not. Adding up all the costs in the same way and converting to dollars, the gasoline version came out to $585 per month and the diesel $542.

So, the answer is no: driving an electric car does not save money. Compared to the gasoline model, the electric car costs $1 more per month to drive. Compared to the diesel, the electric car costs $44 more. It is important to note that Switzerland offers no subsidies or other incentives for electric cars. So these cost differences are real differences, purely driven by market prices.

What would it take for the Zoe to compete on price? Compared to the gasoline version, the answer is essentially nothing: to me, $1 is the same as $0, and the two cars cost the same. Compared to the diesel, it would require the battery lease to drop from CHF 125 per month to CHF 72 per month, a reduction of 42 percent. Does this seem possible? Yes. Electric vehicle battery prices dropped 40 percent between 2010 and 2012,[64] and analysts expect this trend to continue for the next decade.[65]

If true, that would mean that the point of cost competitiveness, at least in Switzerland, is just about two years away, four to be on the safe side. In the United States, where electricity is a little bit cheaper, gasoline a lot cheaper, and diesel cars much less common, it will probably be the same, perhaps a year or two longer. After that, electric cars will start to save money, and the cost savings associated with driving an electric car will grow and grow.

Having tackled cost, let's turn to reliability. The issue for a car is whether it will get you where you need to go. With a range of 210 km, the electric car would easily get me to work and back every day, and handle most of my family's weekend trips to the mountains. But not all. And certainly not the family vacation. Two hundred and ten kilometers gives me what the experts call "range anxiety." In fact, I need to count on the Zoe going somewhat less than 210 km, the number calculated according to an optimistic set of assumptions, like not needing to turn on the heat or the air conditioning. Renault suggests that I should really count on more like 150 km.

Let's take my family's recent trip to Norway, where we stayed next to the hydroelectric plant, as the worst-case scenario, since it was the longest drive that I have done in the last decade. To get there, we drove 1,000 km from our home in Switzerland to Kiel, Germany, then boarded the overnight ferry to Oslo, and then from Oslo drove another 450 km to the cabin we had rented near the west coast. On the Zurich–Kiel leg, we stopped three times, for reasons that involved eating, filling up with diesel fuel, and using the bathroom. These were spaced fairly evenly, but not quite; I think the farthest that we drove between stops was just under 300 km. In Norway, the roads were a lot slower, and the distance between stops correspondingly less, since bodily functions don't depend on driving speed.

The drive from Zurich to Kiel suggests to me that for the Renault Zoe not to make a hellish day like that even worse, three things would have to change. First, the range would have to go up from 210 km to at least 300 km. That means a battery that is 50 percent bigger and, at today's prices, likely 50 percent more expensive. A few paragraphs ago, I suggested that battery costs are falling by about 20 percent per year, so this extra 50 percent of battery pushes the point of cost competiveness out another two and a half years, from late 2016 to somewhere in 2019. A complementary approach to extending range, something that BMW is investing in heavily, is making electric vehicles lighter, primarily by substituting carbon fiber frame elements for steel. Second, I would need a place to charge the battery about every 50 km. On the highway, that corresponds to 25 minutes of driving, which seems to be the absolute limit that a child can hold it after announcing a need to pee. Third, I would like to be able to charge the battery in the time it takes to use the toilet at a highway rest stop, roughly five minutes. I suppose that I could live with ten to fifteen minutes.

Are these latter two conditions reasonable? I think so. The 50 km rule means that every existing full service rest stop on the highway – the kind that currently offers a gas station and a place to eat – would need to have the equipment to charge the car battery. If I want to satisfy the five-minute rule, right now the only way to do so is through battery swapping. Battery swapping means driving up to a mechanical unit that reaches under the car, takes the old battery out, and puts a new battery in. The Renault Zoe doesn't come with a swappable battery, but other electric cars, like the Tesla Model S, do, and more are planned for the future. In the future, it may also be possible to charge the car directly in as little as ten or potentially even five minutes. That would require improving grid connections so that these charging stations can deliver a lot of power quickly, as well as a car battery that can accept it quickly without overheating. The Chief Technology Officer at Tesla thinks that ten minutes is likely within the next few years and five a real possibility.[66] Of course, either battery swapping or high power charging would require a major investment in infrastructure at each of these highway rest stops. That is a subject for policy. In the meantime, there is an opportunity for plug-in hybrid cars – electric cars with an additional gasoline engine to take over if need be – to occupy a place in the market.

What about potential capacity? Could we build enough electric cars to eventually replace all the gasoline and diesel ones, not to mention the growing demand for cars in places like India and China? The only serious concern in this respect has to do with the availability of lithium, the element that current state-of-the-art batteries require. The balance of the evidence suggests that there is more than enough lithium available to enable a transition to electric vehicles, at least until the year 2100.[67] Long before then, one can safely assume, there would be other battery technologies using other elements that could take over. The other issue with respect to potential capacity is whether we could generate enough renewable energy to charge all of those batteries; with the potential capacity for PV and CSP an order of magnitude greater, the answer is clearly yes.

I started this section on transportation with a discussion of electric cars for two reasons. First, cars are the dominant user of fossil fuels within the transportation sector. One can argue for behavior change, such that everybody currently driving a car switches to bicycles and public transportation. This would be fantastic in a lot of ways. But for all the reasons that I argued in Chapter 6, I doubt whether this will happen. The fact is that cars currently offer a lot of people a form of local and regional transportation that is unparalleled in terms of convenience. Second, switching from gasoline and diesel fuel to electricity is by far the most practical among the sustainable options to make driving carbon free.

Some people suggest that the future is hydrogen, whereas others suggest that hydrogen will always be in the future.[68] I side with the latter view when it comes to cars and light trucks, whereas I am on the fence when it comes to long-distance trucks and ships. Compressed hydrogen offers an energy density that is much higher than that of batteries and thereby permits a greater driving range, and yet in two other respects it has serious drawbacks. The first is that fuel cells are, currently, prohibitively expensive. Several car companies have announced plans to launch fuel cell vehicles, and even the most modest of these would cost close to $70,000. Even if this problem is temporary, the issue of efficiency is not. When electricity is used to make hydrogen, and then the hydrogen is used to make a car move forward, more than 60 percent of the original energy is lost, six times more than with batteries.[69] Finally, a transportation system based on hydrogen would need a whole new network of pipelines before hydrogen cars would be practical at all. Granted, by the time everyone is driving electric cars we will need to upgrade many parts of the power grid, but these upgrades can go hand-in-hand with a rising market share.

Biofuels for transportation

The other option is biofuels, and, for a while a few years ago, they seemed like the answer. The big positive aspect of biofuels is that they are already competitive with respect to cost. As the IPCC has reported, production of first-generation biofuels, which rely on either sugars/starches to produce ethanol or lipids to produce diesel and jet fuel substitutes, can already happen less expensively than fossil fuels.[70] Indeed, partly for this reason, Brazil has managed to switch over to biofuels almost completely for its transportation sector. Second-generation biofuels, which use enzymatic processes to convert other types of plant material (such as wood) into a usable liquid form, are not there yet but likely will be soon.

One big problem with biofuels, however, has to do with potential capacity. With other issue areas, I have suggested that the barriers to capacity are both technical and social, with the relevant issue on the social side being whether the technology, at large scale, would encounter political objection.

Scaling up biofuels probably would encounter political objection and for good reason. Until now, the use of arable land to grow the feedstocks for their production has been highly problematic: it has led to deforestation, a substantial release of soil carbon, competition with agriculture, and rising food prices. The European Union, for example, initially developed a renewable energy policy that promoted biofuels heavily. It later retracted the biofuels part based on scientific evidence that biofuel production was causing environmental and social harm. There are reams of scientific articles documenting this, and other reams of articles analyzing the possible policy solutions.

Let's assume for a moment that all of these policies work. Do we actually have enough land available to grow the biofuels feedstocks needed to substitute for fossil fuels in the transportation sector?[71] Here, the experts give different opinions. To understand why, I supply some numbers. The International Energy Agency (IEA) in Paris compiled a review showing biofuels to vary widely from estimates of 33 EJ (exajoules) per year of primary energy production to more than 1,000 EJ per year,[72] with the scenarios differing according to assumptions about land availability and yields per hectare of land.[73] One of the important assumptions in the higher end estimates is that land currently not being used for farming would be both productive and available for biomass production; others argue that the reason this land is currently not devoted to agriculture is that it is simply too infertile and would not be productive. Of these studies reviewed, however, one of the more frequently cited, produced by scientists at the International Institute for Applied Systems Analysis (IIASA), suggested lower and upper uncertainty bounds of 350 and 450 EJ per year.[74] Let's take that as a starting point, given as well that it falls in the middle of the total IEA study range and recognizing that even this estimate may be too high in practice. It also corresponds to the upper end of the range for technical potential suggested by the IPCC.[75]

It turns out that it takes a fair amount of energy to grow the biofuel feedstock and ultimately convert the biomass into liquid fuels. Where does this energy come from? Currently, it comes primarily from fossil fuels. If we were in a world relying primarily on biofuels for our primary energy, especially for transportation, then it would come from other biofuels or the combustion of biomass. That is the world we are investigating, and, in this case, the ratio of total primary energy required to the energy in the biofuel itself is likely in the range of 3 to 4, using a life-cycle analysis.[76] Combining the two sets of numbers and using the necessary conversion factors suggests a global potential to produce biofuels of between 2.5 and 6.4 trillion liters annually. By contrast, if we take the lowest estimate of biomass potential from the IEA study, the number is about one-tenth of this or 250 billion liters.

How does this compare to demand? Currently, the world uses about 2.3 trillion liters each year in the form of gasoline and diesel fuel for motor vehicles, and that number is growing at about 1.5 percent per year, taking into account the growth in vehicle use on the one hand and the improvements in automotive efficiency on the other.[77] If that growth were to continue to 2050, which of course is a hugely simplified assumption, then it would imply demand of just over 4 trillion liters. Four trillion liters lies right in the middle of the range derived from the IIASA estimate of global potential, which in turn lies in the middle of the range taking the full set of studies that the IEA summarized. So there you go: it might be possible, but we simply don't know if biofuels would suffice to replace fossil fuels for all of our driving needs.

But it is also worth considering one additional piece of data. In 2010, the world also consumed 301 billion liters of jet fuel.[78] That number is growing much faster, close to 5 percent per year. If you take this growth out to 2050, it suggests a possible annual jet fuel demand of 2.1 trillion liters by then.[79] If we believe the IIASA estimates as our confidence range, this suggests that the world could meet all of its 2050 aviation needs with biofuels with certainty and could potentially meet its needs for both aviation and road transport. Again, the IIASA range falls in the middle of all the estimates. This means that we could have some degree of confidence in satisfying the world's thirst for jet fuel, at least through 2050, only if we aren't already burning all of this in our cars. If I were a policy maker, I would want to discourage rather than encourage the use of biofuels for road transportation and instead go the route of electrification, at the same time as I would promote the use of renewables to generate that electricity. Both pieces of this solution make sense.

It is true that planes have been developed that use hydrogen or run on batteries, and, indeed, there is a Swiss-made prototype that is planned to fly around the world using only the energy generated by PV cells on the tops of its wings, the Solar Impulse. But none of these technologies is anywhere close to being commercially viable simply because the energy density of batteries and compressed hydrogen are so much less than jet fuel.[80] "Drop-in" biofuels, by contrast, could be, and with little needed to change in terms of the planes themselves. To be sure, there are technical challenges. A group of airlines working in conjunction with nongovernmental organizations like the Worldwide Fund for Nature (WWF) has been experimenting with mixing biofuels into their fuel mix on some commercial flights, gradually obtaining the data needed to make sure that performance and safety are not compromised. In 2012, the Canadian government sponsored the first jet flights testing the use of 100 percent biofuels, with results that prove promising.[81] To most analysts, the primary challenge associated with switching aviation over to biofuels concerns the question of how to get enough biofuels in the first place and not whether they make economic or technical sense.[82] So this is why the policy flag needs to stay up. Aviation is a tough nut to crack. Over the coming years, policies, to my mind, will have to address two issues. First, primarily by addressing issues of land-use decisions, but also the basic technologies used in biofuels production, policies will have to make it possible to substantially scale up our sustainable production of biofuels. Second, policies will have to scale up the pace of research and development (R&D) to find other ways to keep planes in the sky. Perhaps this will involve different kinds of airplanes. It could also involve radically different ways of producing the fuels themselves, relying on inorganic processes rather than plants.

Beyond cars and planes

To complete the issue of transportation, it is worth mentioning the other areas where we use energy. Heavy-duty vehicles - trucks and buses - account for a not insignificant share of road transport in terms of total energy use. There isn't a huge amount of attention paid in the media to electric heavy-duty vehicles, and for this reason it is tempting to be more pessimistic about the potential for a rapid changeover to electric drive trains in these vehicles compared to cars. At the same time, however, there are a few reasons why electrification could go quite smoothly.

Many trucks are utilized exclusively for short- to medium-distance deliveries in urban locations. In this application, the issue of range is less problematic because charging could be done during the loading and unloading phases. At the same time, the advantages of an electric drive train are greater in the frequent stop-and-go driving conditions that these vehicles encounter. One manufacturer of such vehicles suggests that because electric vehicles essentially shut down when stopped, rather than idling, overall wear and tear is so much less that maintenance costs can be cut by half.[83] A recent study in Switzerland, focusing on a locally produced eighteen-ton electric truck with a range of up to 300 km, suggested that the break-even point is 47,000 km per year or 180 km per day, five days a week; less than this, and the per kilometer savings do not add up to the increase in purchase price, whereas above this they do.[84] That isn't a crazy number. If the truck is averaging 60 km/h, it implies that it is being driven for at least three hours per day. One of the things that may make the truck market simpler to cope with than the car market is the fact that delivery trucks are often purchased for a specific route, thus allowing battery size to be tailored to the distance traveled. On the other hand, for long-distance highway trucking, the benefits of electrification are lower because these trucks tend to operate at a constant speed, thus reducing the benefits of the electric drive train and regenerative braking.[85] There is thus reason to believe that the technology that could replace fossil fuels for trucks is hydrogen fuel cells. It is difficult to say when this could prove to be competitive, but a time frame of 15–20 years does not seem to be unreasonable.

With buses, the distinction between short- and long-distance routes is perhaps even more important. For the former, typically in urban mass transit systems, buses have long made use of overhead trolley lines, which has the advantage of obviating the need for a battery but has the twin disadvantages of requiring the lines and making the routes inflexible. More recently, a number of cities have developed transit buses utilizing lithium ion batteries, which offer sufficient range for their end-to-end routes. An even more recent development is to use, instead of batteries, capacitors. In practical terms, capacitors differ from batteries in three main respects. First, they offer much less total storage capacity, meaning that they offer very little in terms of vehicle range (only a few kilometers). Second, they are much less expensive than batteries. Third, they can be recharged almost instantly. Urban transit systems in Asia are currently experimenting with such a system, where the capacitors are recharged at each bus stop.[86] Long-distance buses ought to be able to be electric, although there has been very little development in this direction to date.

Rounding out transportation needs are shipping, rail transport, and the use of pipelines. All of these are in fact easier to electrify than road transport, and, indeed, I can gloss over them very quickly. With shipping, weight is of lower concern, thus making batteries less of a problem, and indeed there has already been a push toward electrification for reasons purely economic and having nothing to do with climate change.[87] So far, that push has been absent in the area of long-distance shipping due to the very high investment costs associated with batteries. As these costs decline, it is possible to imagine change here, too. This may also be an area, in addition to aviation, where biofuels make sense. Rail transport is an area where the benefits of electrification have been around for so long that the sector largely is already electrified, a trend that is continuing. Pipelines require pumping, and a mix of electricity, gas, and oil powers today's pumps. To a large extent, the advantages of the latter two are especially pronounced when the thing being pumped is itself gas or oil since it is a fuel source that is right there.[88] To the extent that we phase out the use of oil and gas as fuel sources for electricity, heating, and transportation, so too will we have less need to transport them by pipeline, and that means that the remaining pipelines (for things like water) will make more economic sense when powered by electricity.

Space and process heating

It is easy to lose sight of the fact that the single most important energy service – the thing we rely on modern energy for – is keeping warm. And whether or not losing sight of this is a good thing, the fact that so much of the world's energy goes to fulfill this function is a bit of a blessing for the energy sector. The reason is that heating, at least most heating, is a place where decarbonizing and saving money most clearly go hand in hand. It is worth separating heating out into two basic categories, however: space heating and process heating. The former is used for commercial buildings (18 percent of energy demand for heat in countries belonging to the Organization for Economic Cooperation and Development [OECD]) and residential buildings (35 percent), whereas the latter is used for agriculture and forestry (6 percent) and industry (41 percent).[89] The decarbonization solutions are different across these two applications.

For space heating, it is clear that each of three different approaches is already cost effective. The first and most important for new building construction is improved efficiency through careful attention to building design. In the residential area, the best example of this is the so-called *passive building*. Loosely defined, a passive building is one that reduces heat exchange with the outside to such an extent that it is comfortable to occupy, year round, without any need for active heating or cooling. Even in the middle of winter, the body heat generated by the people inside it, as well as the heat from cooking and the use of electrical appliances, are enough to keep it warm. In summer, the passive house keeps hot air out during the day and takes advantage of cool temperatures at night.

To achieve this, the passive building relies on three features. First, the walls, windows, and roof are insulated very well to reduce radiative heat loss. Second, windows take advantage of sunlight to warm elements of the house with a high thermal mass but can be shaded in summer, when heat is not needed. Third, the flow of air into and out of the house is carefully controlled. When all the windows and doors are closed, passive buildings are completely airtight, and all ventilation occurs through a specially designed system incorporating a heat exchanger. This transfers the heat in the stale air being pumped out of the building over to the fresh air being pumped in. When cooling is needed, the heat flow goes in the other direction. Even with these three features, passive buildings allow a great deal of design leeway, such that they can look and feel virtually identical to a nonpassive building of any style. They are somewhat more expensive to build, but this added cost is typically made up for within a few years of cost savings. Most passive buildings have been houses, but the principles work just as well for commercial uses such as office buildings. My university, for example, has a fairly new passive building housing offices and scientific laboratories.

The second cost-effective approach to space heating, already described in the passive building design, is to use electric pumps to leverage existing reservoirs of heat and coolness. The most basic of these systems is the heat pump: outside air is drawn into the heating system and compressed, which increases its temperature. This hot air is then passed through a heat exchanger, giving off its energy to the air inside the building before being expanded and released back to the outside, colder than it had been in the first place. Heat pumps can be incredibly efficient, with the energy content of the heat that they transfer being several times that of the electricity needed to power them. That efficiency can be even higher if they make use of a heat reservoir that is warmer than the outside air. This can include the inside of a solar collector or warmer temperatures underground. In all cases, they make use of a small amount of electricity for compressing and pumping to achieve a large amount of space heating. The electricity, of course, can be generated renewably. Although the use of heat pumps, solar collectors, and geothermal systems is most cost effective in the case of new construction, it is also feasible to retrofit existing building stock. Sometimes this is easier than others. My wife and I looked into retrofitting our historic house in Austria with a heat pump; to make it work, we would have needed to replace our wall radiators with floor heating, which can operate with lower temperature water. This in turn would have meant tearing up our old parquet floors, something we simply did not want to do.

The third cost-effective approach is district heating, something that has become wildly popular in northern European countries and is gradually spreading southward. In a district heating system, a power plant is located in the community, typically burning some mixture of garbage and other waste products, including waste biomass. It generates electricity and also uses high-temperature exhaust gases to create steam, which is then pumped through a network of underground pipes throughout the city or town. The buildings along that network tap into these pipes for their heating needs. Building owners need to pay for the steam that they use, but this is typically much less than they would have to pay to run their own separate gas-fired heating systems. It is relatively easy to connect existing building stock, like our historic house in Austria, to a district heating system. Indeed, just as we sold that house to move to Switzerland, our community was beginning to install the pipes to connect our neighborhood to a district heating system.

Of course, there are other approaches. For half my life, including in my current home of Switzerland, I have heated with wood. The old-fashioned way takes a lot of work and isn't for everybody. The modern way is with wood pellets, typically pressed together from sawmill waste. In either case, the issues of potential capacity are the same as with biofuels. Given the wide variety of options, however, it is clear that for space heating there are technologies that already offer benefits across the board – cost, reliability, and potential capacity – comparable to the use of natural gas, oil, and coal.

Process heat is a little trickier, but the combination of biomass burning, heat pumps, and solar thermal energy offers a set of solutions that can be competitive across almost all applications. Of the three, biomass and biogas offer a cost-competitive alternative for process heat applications requiring high temperatures (generally, above 400°C), which is roughly half of the process heat total. In countries where supply chains are well developed, biomass already has made substantial inroads into many applications. In the Netherlands, for example, biomass and biowaste products account for more than 80 percent of the fuel used in cement production, whereas in Latin America, biomass burning now accounts for more than 30 percent of industrial process heat.[90] Across applications, biomass and biogas generally offer cost savings, provided the necessary infrastructure is in place.[91] Reliability is not a concern, but potential capacity is, and, as with biofuels for transportation, a red flag needs to go up: policies need to be developed and implemented to help make the supply of large amounts of biomass and biofuel feedstocks more sustainable.

In some high-temperature applications, such as steel production, electricity is already being widely adopted based on its across-the-board competitiveness. For lower temperature applications, the other half of process heat requirements, heat pumps and solar can be competitive, as they are for space heating.[92]

Cement production

The final industrial source of CO_2 that requires attention is the production of cement, more specifically known as *portland cement*, the binding element of concrete.[93] Concrete, in turn, is increasingly the world's favorite building material. Portland cement production generates almost 5 percent of global CO_2 emissions and is growing rapidly.[94] The emissions come roughly 50/50 from the use of energy – mainly to create process heat but also for the grinding of the original stone – and from the chemical reaction by which cement is created. That reaction takes place during the hardening process and releases 950 kg of CO_2 for every ton of portland cement ultimately found within the concrete.

Cement production offer two small pieces of good news, and one big piece of bad news. The first good news item is that there are technical options for reducing CO_2 emissions that are already competitive and may become increasingly so. There are cost-effective options to improve the efficiency of energy use during the entire production process. It is increasingly attractive to use waste products and biomass to generate the heat needed to melt the core ingredient of portland cement, known as *clinker*. As I already mentioned, the Netherlands is an example of country where already 80 percent of the needed heat is produced from fuels that are not fossil based.[95] It is also possible to mix other materials, such as fly ash, into the clinker, without substantially lowering its quality. This would then lower the total amount of CO_2 released during the hardening phase.

The second piece of good news is that there is an alternative technology that has been developed over the past twenty years that could substitute for portland cement as the binding material in concrete, either entirely or in a hybrid of the two. It is known as *geopolymer cement*, and it relies on an entirely different set of materials and chemical reactions that do not release CO_2 in the hardening phase. Although research into geopolymer cement is still young, there are indications that its structural qualities are similar to that of portland cement, and, in some respects, its durability may be slightly greater. The main materials needed for its production are various types of clay soil, which are just as widely available as the limestone needed for portland cement. The U.S. Federal Highway Administration suggested that geopolymer cement could be a "game changer."[96]

The bad news is that there is no reason to believe that geopolymer cement will ever outcompete portland cement on its own merits. For each of the energy technologies I have discussed, the new technology can be, in some critical respect, superior to the technology relying on fossil fuels. Not so with cement. Geopolymer cement could conceivably be just as good and just as affordable as portland cement, but there is no indication that it is in any important respect better.

Transforming the industry would take a major push from policy makers in two respects. First, they would have to support the gathering of data, over a long time period, on the structural integrity of concrete made from geopolymer cement. It is fine to lay a sidewalk using geopolymer concrete, but you wouldn't want to build a bridge or a skyscraper out of a material, the long-term durability of which was still unknown. Second, they would have to convince industry to switch over to an entirely new production process, perhaps locating their factories in different places to be close to the raw materials. This would represent a huge initial expense, even if the long-term costs of geopolymer cement are no higher than portland cement. For both reasons, even with substantial policy intervention and a fair amount of optimism, it will be several decades before geopolymer cement could occupy the role that portland cement now has.

A two-part strategy

In the past few pages, I have gone into some depth on a number of different technologies that collectively could transform our energy system in the next forty to fifty years. In every case, there is a huge amount more to think about and take into consideration. With just about every technology, for example, there are concerns related to the environment and pollution. Perhaps more importantly, there may be significant barriers for each of the technologies in a developing-country context. Driving an electric car across Germany is one thing. Driving it across Kenya is another. Frankly, for the drive across Kenya, it is going to be a long time before I would want anything other than a diesel engine.

But I have kept the discussion fairly short and limited it primarily to a developed-country context in order not to lose sight of the basic question: are there technologies out there that could, now or starting in the next few years, begin to outcompete fossil fuels under the right geographic, market, and regulatory conditions? Across the entire energy industry, the answer is yes. In the area of electricity production, we are entering a time when wind is the most economical way of generating power. This could soon be followed by solar in a growing number of applications. Both offer a potential capacity that is large enough to pose no major issues in most regions of the world, although in highly populated places like Western Europe it may be necessary to begin importing power generated by windmills or solar power plants located elsewhere. Wind and solar create major challenges with respect to reliability, but these can be solved through a combination of several different strategies. The strategies will not come for free and, as a result, could push back the point of cost competitiveness of a wind or solar power system by a few years. But still, we are looking at competitiveness within the decade. In the area of transportation, we are entering a time when electric power trains for everything but aviation make increasing sense and, within five years, could outcompete gasoline and diesel fuel on the basis of costs. As long as the electricity itself is from renewables, then this is a valid pathway toward decarbonization. To get electricity equal to gasoline and diesel fuel in terms of reliability will take a concerted effort but seems well within the range of possibility, starting in the same time frame. In the area of heat production for buildings and industry, a variety of technologies can lead to the cost-effective phase-out of fossil fuels. Indeed, many of these are already gaining rapidly in popularity. The only dark spot is cement and concrete production, where a substitute would be required for portland cement, given that the latter releases huge quantities of CO_2 during its hardening process. A substitute exists, and yet it would be unrealistic to expect it to begin to push aside portland cement absent a major and sustained policy intervention.

A single chapter is far too little space to go into complete depth on all of the technologies that could play a role in solving climate change. The set of technologies is growing quickly and certainly, by the time you read this, there will be new technologies to think about; I am sure that I have missed one or two, potentially more. One thing I have intentionally not done in this chapter is to look out into the distant future, more than about a decade away. A lot of technologies currently under development could be game changers for the energy sector, and most analysts agree that, in the long run, there are several different possible technological paths. Maybe hydrogen fusion will succeed. Windmills can be tethered to the ground and sent aloft to float in the high winds of the stratosphere. Solar panels can be placed in geostationary Earth orbit and beam their energy down around the clock. All these are exciting possibilities, and some of them may play an important role in the energy system 100 years from now, two or three investment cycles away. But we are not talking about 100 years. We are talking about ten years for a technology to become established on a growth path and then follow that growth path for another thirty or forty years beyond that. For that, there are far fewer technologies available. Yet the handful that we have, fortunately, is not only enough, but even leaves some room for a few of them to fail.

For every one of the technologies that I have described in this chapter, there exist counterarguments, reasons why one should not be as optimistic as I have been here that they can actually take us where we need to go. Such doubts are well-founded; each and every one of the technologies I have covered in this chapter has something going against it, making it challenging to scale up. That is why, at least now, an energy system transition to these new technologies probably won't happen purely as a result of market forces. But there is reason to believe that every one of the challenges we face with each technology is one that we can overcome with some concerted strategic action.

Concerted strategic action is synonymous with public policy, and that is where we turn next. I consider the policy challenges and solutions in two separate chapters, and these correspond to the two separate elements of the technological pathway that appear to make the most sense. The first element of that pathway is to get all of our electricity from renewable, carbon-neutral sources. The second element of that pathway is to switch as much of our energy consumption onto electricity as possible. Where electrification is not possible, we will need to use biofuels, and for this we need to consider the issues of agriculture and land use as well.

9

Energiewende in the German power sector

From 2006 to 2013, I lived and worked near Vienna, Austria. For most of that time, starting in 2008, I taught a class each spring semester at the University of Bayreuth, in northern Bavaria. I would typically travel there and back on a German high-speed train, sinking myself into lecture preparation on the way up and relaxing in the dining car on the way back. I wouldn't pay much attention to the landscape flying by outside because I had seen it so often. But every now and then, I would take a look. The interesting thing is that I always could immediately tell whether or not the train had passed the border between the two countries. It wasn't because of any particular change in the landscape or vegetation, in the architecture of the buildings, in the cars on the road, or the way the villages were laid out. All these things were essentially the same, whether I was in Austria or in Germany. The difference was on the rooftops. On the German side of the border, it seemed like every second house, barn, or commercial building had a set of solar photovoltaic (PV) panels on top. In some places, every building in sight had a PV installation. Not infrequently, the train would whip past a farmer's field that was now covered with PV panels as well, mounted on metal frames about a meter above the ground. In Austria, they were absent.

That was southern Germany. In northern Germany, the situation was the same, except that what you see isn't so much PV panels on the roofs, but windmills in the fields. Just about everywhere, gently turning in the wind that blows moderately across the flatlands. Crossing the border from the Netherlands in the west, or Poland in the east, the contrast of windmills in the sky is just as strong as it is with PV panels in the south. Denmark, to the north, has policies for wind that are not too different from Germany's. And the Danes, too, have lots of windmills. Policies make a difference.

The fact that Germany stands out in this way is a little bit strange. The northern part of the country is reasonably windy, but not particularly so, not like the U.S. Midwest or the European Atlantic coast. Southern Germany has moderately sunny skies, but nothing in comparison to Italy or Spain to the south, not to mention the U.S. Southwest. Moreover, the country is heavily invested in coal. During World War II, the country's war machine was kept running on liquid fuels created from the black rock. In the postwar period, the industrial heartland of the country and the source of its economic miracle was the Ruhr Valley, where coal mines dot the landscape, driving power generation turbines and steel mills. And then there is the German car industry. Mercedes, BMW, Audi, Porsche, VW, Opel. The politics of speed limits in Germany – and the fact that large parts of the Autobahn have none – comes down to the socially constructed truth that German cars are so good, so agile, so reliable, that they are inherently safe at any speed.

But Germany also has a strong environmental and naturalist movement, born out of a romantic connection that the German people feel to their forests. It starts with their children's books, on the one hand a magical fantasy land full of elves and dwarves in the woods, and, on the other hand, a practical primer in landscape ecology and forest food webs. A fantastic fact about Germany is that many villages have an official mycologist on call. If you have been out picking wild mushrooms in the forest and aren't exactly sure what kind they are, he or she is there to identify them and tell you whether they are safe to eat. Just about anywhere in the country, there is a network of forest trails close at hand, connecting you with those mushrooms and much more. One time, flying from the United States to Vienna, I had a four-hour layover at Frankfurt International Airport. It was a warm summer's day, so I stepped outside the terminal building, spotted a walking path sign, and within five minutes found myself on a well-maintained trail through the forest. I ended up at a beach on a little lake, and I had everything I needed to enjoy a refreshing swim before my connecting flight. Try that in a large airport in just about any other country, and I guarantee that you will be disappointed. The only other airport where I have managed a swim in a natural body of water was Wellington, New Zealand, and even there I had to dash across a four-lane highway and navigate my way through a maze of shopping mall parking lots.

Somehow, in Germany, these two sources of national identity – industry on the one hand and romantic environmentalism on the other – came together and coalesced around the idea of renewable energy. Political leaders decided that German industry would lead the world in saving the planet, setting targets such as the reduction in greenhouse gas emissions by 40 percent by 2020. They followed up on the setting of such targets by implementing a set of laws, regulations, and other government programs that have been amazingly effective. Several countries out there get huge amounts of their electrical power from falling water: Brazil almost 75 percent, my home of Switzerland more than 50 percent. Germany is not among them and, because of its geography, cannot be among them. But Germany now leads the large countries of the world in the proportion of its power that comes from other renewable sources, 25 percent and still growing. As the *New York Times* described it, this has transformed the country's landscape, both literally and more figuratively, in terms of its industry.[1] And yet, recently, the German government has started to reconsider whether the policies it implemented, those that have made it the global leader, were so wise after all. The entire case offers some lessons about the policies that can actually promote the technologies described in the previous chapter.

The birth of the German feed-in tariff

During the 1970, environmental and peace activists in several West German states began to run for political office. In each state, they identified themselves as members of a group of candidates with common principles. In 1980, the various groups came together to form a national political party, The Greens (*Die Grünen*), standing for four ideological pillars: social justice, ecological wisdom, grassroots democracy, and nonviolence. In 1983, they passed the 5 percent national voting threshold required to take seats in the lower house of parliament largely on the basis of their opposition to deployment of American and NATO cruise missiles. Their political strength increased following the Chernobyl nuclear disaster in 1986, as well as the attention that the issue of acid rain had gained in the popular media; in 1987, they increased their share of the vote to 8.3 percent. They struggled a bit during German reunification, but emerged in the 1994 federal election with a strong 7.3 percent of the vote, the third most popular party after the Christian Democrats and the Social Democrats. In 1998, they entered into a coalition with the Social Democrats under Gerhard Schröder, forming a government together that lasted through the 2002 election and until 2006. It was in 2006 that Angela Merkel became

Chancellor, leading the Christian Democrats in coalition with a string of other parties.

The Green Party played a pivotal role for renewable energy on two separate occasions. The first was in 1990 and took the form of collaboration with the Christian Democrats, which held the German Chancellorship as the main partner in a governing coalition.[2] A Green parliamentarian from Bavaria, Wolfgang Daniels, worked together with a Christian Democrat also from Bavaria, Matthias Engelsberger, to come up with a law that would provide a fair price to small-scale producers of renewable energy. For Engelsberger, it was about tradition: small-scale hydropower had been the early lifeblood of the Bavarian electric power system and yet now was not able to negotiate a satisfactory price with the electricity distribution company.[3] Together, the two men proposed a law that would guarantee a fair price, set to be roughly 80 percent of the residential price for electricity, payable from the wholesale purchasers of electricity to the owners of small-scale electricity generators, including not only dams but also windmills and solar PV panels.

In addition to protecting hydropower, an important justification for the premium was rooted in the economic model of externalities and was meant to compensate for the fact that renewable energy sources did not generate local and regional air pollution, including acid rain, then the big political issue. The proposed law gained the support of the Christian Democrats nationally and went into effect on January 1, 1991. The price was substantial enough to provide a stimulus to invest in windmills, and, indeed, between 1990 and 1999, Germany went from having 55 MW of installed wind capacity (equivalent to about twelve large modern wind turbines) to having 4,435 MW. This capacity pushed Germany into first place in wind power, passing Denmark and the United States. The price set by the German law was not high enough, however, to stimulate investment in more expensive renewable technologies, including solar PV.

Germany's second iteration came a decade later and this time grew out of a collaboration between the Green Hans-Josef Fell and the Social Democrat Herman Scheer at a time when the two parties together formed the governing coalition. Fell, consistent with Green Party principles, had a strong motivation to protect the environment, to phase out nuclear energy, and to promote social justice; he wanted a piece of legislation that empowered ordinary Germans to make a difference. Scheer, meanwhile, as a Social Democrat and the party of the German workers, was very interested in "Made in Germany." He wanted a piece of legislation that would, on the one hand, create new jobs in German industry and, on the other hand, ensure that Germany became even more energy independent.

Where Fell and Scheer could find common ground was in changing the existing law to make sure that it would provide a boost for PV. On the one hand, the potential global capacity for solar power was so large that one could imagine its truly powering the world, thus allowing for a complete phase-out from both nuclear and fossil fuels. On the other hand, PV was both technically complicated (perfect for German engineers to become leaders on) and a wide open market, the latter fact in contrast to wind, where Danish producers had already established a position of market dominance. Their big idea was to change the existing law in two ways. The first was to move from a price support mechanism based on the benefits society derived from renewable energy – the avoided pollution – to one based on the costs of production. "PV cost more than wind?" they asked. Fine, pay more for PV. It might not be the economically efficient thing to do, but it would get the solar panels built. The second change was to lower the risks for investors by moving from a fixed percentage model – a fixed percentage of the residential power price – to a flat tariff, now know as a *feed-in tariff* (FIT). What's the difference? The difference is that the price of power might fluctuate in the coming years, introducing some uncertainty into the revenues that power producers would collect. A flat tariff, by contrast, would not. The government would set the tariff just high enough to cover the costs of production plus a little profit. That profit would be virtually guaranteed; the only uncertainty lay in the weather, but there was enough data to show that annual fluctuations were small enough to be manageable. This high level of certainty would be exactly what ordinary German homeowners would need in order to borrow money from the bank to finance solar panels on their roofs.

The FIT in practice

So, here is how the FIT works. Government regulators decide on a set of technologies to support. For each technology, they identify what the current costs of producing power are in terms of the expected *levelized cost of electricity* (LCOE) for a project built right now, and update these cost estimates on an annual basis. This has meant distinguishing between different configurations for the same technology, such as differently sized installations of PV; whether the PV panels are on a rooftop or the ground; or the difference between offshore wind, wind built onshore along the coast, and wind built inland. They then add a premium to each of these costs to give a fair profit – about 10 percent – to the person or company making the investment. Together, the LCOE estimate and the profit premium become the set tariff paid for energy fed into the grid from a particular project. If you build a project, you sign a contract with the power transmission company (known as the TSO, for *transmission service operator*). You will sell to the TSO whatever power you don't use yourself, and they are obligated to buy it from you, even if that means that they have to improve the power line connection to the grid. The price is locked in to today's FIT over the entire expected lifetime of the project, typically set at twenty years. If you wait and the build the project next year instead of now, the cost of the project will probably be less (since the price of windmills and PV panels continues to fall), but so too will be the FIT you lock yourself into.

Meanwhile, regulations dictate how the TSO recaps the relatively high cost of FIT electricity from the power consumers. A set of local power companies around Germany purchase electricity from conventional producers (e.g., coal, gas, and nuclear) on a wholesale market that establishes the price for power, which varies according to the time of day and other factors (which I describe in detail in the next section). The power companies are also obliged to purchase power covered by the FIT at that same wholesale power price. They then turn around and sell the power to residential and business customers, charging a retail price per kWh that incorporates the average wholesale price, the costs of power transmission, and some degree of profit. On top of this, the government sets a surcharge attributable to the FIT electricity, which customers see as an additional element on their electric bill, along with other things like a sales tax. That surcharge then gets passed directly on to the TSO. Every year, the government readjusts the surcharge to ensure that it covers the costs that the TSO incurs associated with buying electricity according to the FIT. Just to put all of these numbers into perspective, in 2014, the average wholesale power price is about 4 cents per kWh, the FIT surcharge about 6 cents, and the final retail price, including the FIT surcharge, taxes, and everything else, about 29 cents.

The law that Fell and Scheer drafted, incorporating the FIT, went into effect in 2000. It was highly controversial at the time, precisely because it bucked the trend of economic thinking. It made no promises to get an economically efficient mix of renewables built but rather several different technologies built, even the more expensive ones. Moreover, where conventional economic thinking suggested that industry alone could judge the relative costs of the different technologies, the new law required government regulators to determine these on a continuing basis, in order to set tariffs that were fair. Finally, it was difficult to anticipate the effects it would have and the overall financial burden on consumers.

In all these respects, a useful point of contrast is the *renewable portfolio standard* (RPS) that was being developed in many states of the United States, as well as in several other countries under different names (such as *tradable green certificates*). The RPS is an instrument also intended to boost investment in renewable energy, but it was designed to do so in an economically efficient manner, although not in a manner that would necessarily empower the little guy to get involved. The idea behind the RPS is that retail power companies buy the power that they eventually sell from a portfolio of producers. State regulators would require, starting at some particular time in the future, each company's portfolio to include at least a certain percentage of power from renewables, such as 10 percent. Typically, the RPS law would allow for trading of renewable certificates among the different retail power companies. If one company sold a percentage of renewable power well above the RPS, then it could sell certificates associated with that excess to another power company operating in the same state.

By putting an RPS in place, the state would essentially be creating a separate wholesale power market for renewables, one guaranteed to be at least a certain fraction of the size of the larger conventional power market. Knowing that this second market exists, power supply companies would have an incentive to produce renewable power and perhaps to try to enter into long-term power purchase agreements with the retail power company. But the RPS leaves as an open question exactly what price power will be traded for on that secondary market and, indeed, whether any particular renewable power producer will be able to sell at all: as soon as the aggregate supply of renewable power begins to exceed the set size of this secondary market, the high-cost suppliers would likely be closed out of the market. In this way, the RPS provides an especially strong incentive for project developers to restrict their investments to the least-cost technologies. Early analysis of the RPS mechanism identified the potential for greater economic efficiency – through this incentive for low-cost production – as a definite advantage of the RPS over other renewable support laws, like FITs, that leaned more in the direction of a subsidy.[4] The RPS has been very successful, especially in promoting wind power in the windiest states of the United States.

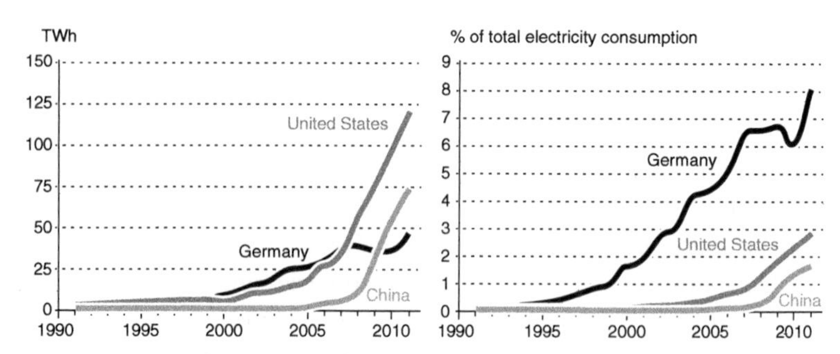

Figure 9.1. *Wind power production in Germany, the United States, and China.*[5]

But while the RPS was collecting praise among economists, it has been the FIT that has gone on to conquer the world of renewable energy and PV in particular. The FIT established Germany as the global investment leader, achieving this investment at the least cost to consumers and driving down the costs of PV (that had been an extremely expensive technology) to the point of being competitive. It has also spawned a number of imitators, especially in other European countries, although none of these has managed to perform as well as the German version.

Let's look at some pictures that document the German success story. Figure 9.1 tells the story of wind power production looking at Germany, the United States, and China.[6] The left-hand graph shows the total amount of power generated by wind in Terawatt hours (TWh), whereas the right-hand graph divides this number by the total power consumption in each country to yield a percentage of the countries' electricity that comes from domestic wind production. The law that Germany had in place for renewable energy between 1991 and 1999 led to a fair amount of growth and, by 1999, had put Germany into first place for total wind generation, overtaking both the United States and Denmark. Then, with the FIT, production began to take off even faster until 2007. At that point, new investment essentially stalled in Germany, for a few reasons. First, most of the good onshore sites had already been developed, leaving room for growth only in offshore wind or in replacing existing onshore windmills with larger ones. Second, an explosion in commodities prices, which preceded the global financial crisis, led to a temporary increase in the price of windmills; the regulators in Germany were slow to respond, and, for a couple of years, the LCOE for new wind exceeded the FIT, thus making new investment uneconomical. At the same time, there were several years in Germany with exceptionally light

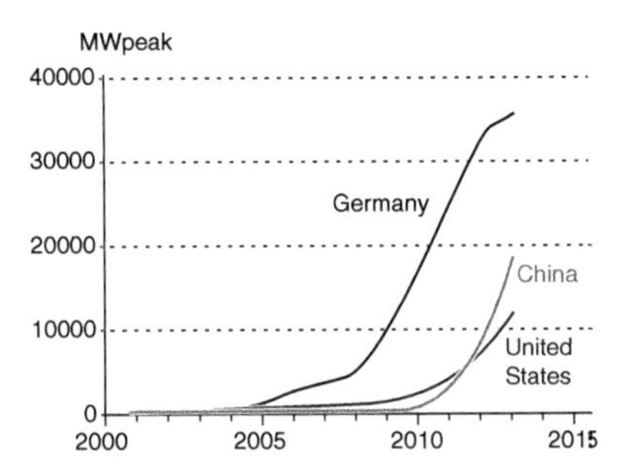

Figure 9.2. *Photovoltaic installed capacity in Germany, the United States, and China.*[8]

wind, meaning that production from the existing fleet declined. After three years of declining production, both the wind speeds and the investment climate began to pick up again in 2011.

The United States during this time had a patchwork of laws in place: RPSs in many states and a tax credit for new investment at the federal level. Investment was slower to grow, despite the fact that the United States is a much larger country with much winder conditions across its heartland than just about anywhere in Germany.[7] Things in the United States did begin to take off starting in 2007, exactly as the German market cooled down. There is a causal connection: the market had been overheated, with the prices and waiting times for new wind turbines rising over several years. The slowdown in the German market, along with a global recession, took the pressure off, and prices began to fall dramatically. In the United States, wind started to become a viable investment, even without the RPSs, and it has boomed. It has also boomed in China, which instituted a FIT system modeled after the German one. In fact, since 2009, China's growth has kept pace with that of the United States. But even as both the United States and China have surpassed Germany in total wind production, neither comes close when looking at wind production relative to the overall size of the power market, which the right-hand graph shows.

Figure 9.2 tells a similar story for solar PV, looking just at total capacity. In 2000, the year the German FIT came into effect, the country with the most installed capacity of PV was actually Japan, with 330 MW, followed by the United States. And then things began to change rapidly. By 2005, Germany had passed Japan and was only starting to gain momentum. By 2007, it was up to almost 4,000 MW; by 2010 more than 17,000 MW; and, in 2012, close to 33,000 MW.

As Figure 9.2 shows, it left the United States far behind; only in 2011, as PV was starting to become cost competitive in the California daytime market, did PV begin to take off, helped by a 30 percent federal tax credit and residential financing programs like the PACE program that started in Berkeley. The latter involves the city loaning the money for PV systems to homeowners and businesses; the payments on the loans are connected to the real estate taxes on the underlying property and continue with that property even if it is sold.[9] Meanwhile, starting around 2010, the Chinese market began to explode, driven by a FIT copied after the German one and augmented by the fact that the Chinese government was providing heavy subsidies to PV module manufacturing companies.

The total capacity of PV installed in Germany now stands at 37 GW, able to hit a peak production of about 24 GW on the sunniest day of the year, compared to the country's peak demand of 70 GW.[10] Of course, the capacity factor averaged over the whole year is quite low relative to other sources of power; in 2013, the PV fleet generated a total of 30 TWh of electricity, 5.3 percent of the total power consumed.[11] For the United States and China, the share of total power consumed is so small as to not warrant producing a separate graph: in 2012, the United States produced 0.25 percent of its power from PV and China 0.14 percent.

FITs have not only stimulated more investment than other instruments, such as RPSs, but have done so in a manner that has been more efficient, driving down costs more quickly. A number of scientists have studied this issue, and the reasons all come down to the relative certainty that the FITs create.[12] In Germany, investing in a PV system or windmill under a FIT became essentially a no-risk proposition.[13] Under an RPS, a project developer would have to put together a business plan based on the expected price at which the power could be sold for and accounting for the risks of the market becoming saturated with renewable power supply in excess of the RPS. The bank or other investor lending the money would have to look at all this and make an estimate of the risks involved. Usually, there would be some sort of risk premium attached, and usually the type of entity that could get financing would be a large company with plenty of other assets to act as collateral. Not so with the FIT. Back in 2007, I had some friends, farmers in Bavaria, who were considering putting PV panels on their barn. Doing so took two phone calls, two signatures, and almost no money down. Their local bank had already worked out a financing package at a risk-free interest rate before they ever walked in the door. The fact is that the risk-free nature of the FIT cut transaction costs to their bare minimum. It also led to a competitive industry within Germany to take PV modules – initially these were primarily domestic in origin, but eventually came increasingly from China and other Asian suppliers – and turn them into whole rooftop systems involving all sorts of collateral components like mounting brackets and power transformers. That competition drove these costs down substantially.[14] There is evidence that the combination of instruments employed in the Untied States do not do as good a job at handling uncertainty as the German FIT, thus resulting in much more modest cost reductions for installed systems.[15]

There is also evidence that FITs have led to faster cost reductions in the basic technologies for essentially the same reasons having to do with their certainty. Cost reductions come because factories scale up production, with both economies of scale and continual innovation on the factory floor. It also happens because start-ups enter the market with new advances in technology. When it comes to the former, the fact that FITs have led to the fastest growth in investment alone suggests that they drive cost reductions. When it comes to the latter, surveys of people engaged in financing start-ups, typically venture capitalists, show a clear preference for the greater degree of market security within a particular technology niche that technology-specific instruments like the FIT create.[16]

What China has accomplished in a short time, seen on Figures 9.1 and 9.2, is another testament to the success of the FIT because this has been their primary policy instrument as well. Other countries could show similar stories, including Spain, Italy, and the Czech Republic. Each of these countries instituted its own version of the FIT and saw a quick jump in investment for wind and PV. In each case, as well, there were criticisms launched because the governments generally set the tariff too high, which led to an overheated market and insufficient competitive pressure on the companies contracted to actually put wind and PV systems together.[17]

In addition to being the country that has best copied the FIT structure, China has also been the country that has most stymied the vision for a German PV industry. In 2000, Japan was the market leader for PV module manufacturing, producing modules with a combined capacity of 129 MW, followed by the United States with 75 MW, and then Germany with 23 MW. Production in all three countries scaled up substantially, primarily to supply modules to Germany. By 2011, Japan produced 2,753 MW of modules, Germany 2,331 MW, and the United States 1,056 MW. But by that time China had already been pushing hard, with heavy state-subsidized loans for the PV manufacturing industry. By 2012, China was producing over 21,000 MW, displacing production elsewhere. In 2012, the German PV industry imploded, with factories closing and production falling by 40 percent.[18]

The demise of the German FIT for PV

That year of implosion, 2012, also marked the beginning of the end of the German FIT program, at least in the highly successful form it had taken for so long. It was becoming clear that even though the FIT was leading to huge amounts of German investment in new PV modules, China had captured the market on manufacturing. In this sense, one of the two original justifications for the law – the boon that it would provide for German industry – ceased being believed. At the level of the household considering investing in PV panels, this didn't matter, and new installations continued to surge ahead through 2012. At the political level, however, the FIT could no longer represent the clumsy solution it had started out as – a meeting point between those wanting to help the environment and those wanting to stimulate the economy.

By then, the consequences of the unprecedented growth in PV were showing up on everybody's electricity bill, in the form of a FIT surcharge that was of noticeable size, costing the average household a few tens of euros every month. And, with this, the political chorus opposed to the FIT – representing two different sets of industries that the FIT was hurting – began to sing louder and with greater effect. In 2013, regulators took the first step by greatly reducing the tariffs for PV, to a point where, for the first time, it was unclear whether investors could actually make money. As soon as this happened, new investment slowed, as can be seen in the final year of Figure 9.2. Then, in April 2014, they revised the law in many additional ways, essentially dismantling its main attributes. They placed a cap on the total annual volume of new investment eligible to fall under the FIT. They provided additional exemptions for industry groups so that they would not have to pay the FIT surcharge. To compensate, they made it mandatory for owners of new PV systems to pay the surcharge on the electricity that they consumed directly, energy that had not entered the grid. Finally, they announced that, by 2017, they would get rid of the purchase obligation for power companies. This introduces a level of uncertainty similar to that seen with the RPS.

If you look at the German government webpage where they explain these reforms, you will find text that boldly highlights how they have taken the FIT and improved it by making it more efficient.[19] I find such a statement to be ironic. The whole idea behind the FIT was the intention to design a policy instrument not with the goal of efficiency in mind, but rather with the goal of inclusive social transformation. Surprisingly, the steps needed to make the FIT inclusive, which required dealing effectively with risk, are what made it so successful and ultimately resulted in its being the most cost-effective way to promote PV.

But what is done is done, and it is useful to understand why. The industrial coalition that led the campaign in 2013 and 2014 to "reform" the FIT represented two very different sets of companies. The first was the electric utilities, which in Germany means four large companies, each with its own regional base. For them, the boom in PV and wind led to a decline in the wholesale price of electricity and was ruining the business model that allowed their coal, gas, and nuclear plants to make a profit. PV was doubly bad because it was not something that they would invest in themselves (unlike wind, where they had become somewhat active in developing new projects). The second group of companies represented the manufacturing firms that used a lot of electricity. They saw their electricity bills rise steeply in recent years, with some of that rise directly attributable to the FIT. They were even more worried, however, about what was to come. Their fear was driven by what had happened to the first group of companies, the electric utilities. As the utilities' business model fell apart, they had stopped investing in conventional generating equipment. Eventually, the manufacturing firms foresaw, there would develop a scarcity of reliable electricity, the kind that a factory needed to run through the graveyard shift at 3:00 am. With that scarcity, their energy bills could really explode and, with it, their competitive position.

This industrial coalition made the argument that the FIT was costing German consumers a great deal, and more was to come. The first part of their argument was, arguably, correct, but the second part is harder to defend. When the FIT went into effect in 2000, the tariff for roof-mounted PV was more than 60 euro cents per kWh. By 2008, it had dropped to about 45 cents, by 2011 to 28 cents, and by mid-2013 it dropped below 15 cents.[20] Whatever tariff was in effect at the time a PV system went online would remain fixed for that system for twenty years, and that means that German consumers have been paying 45 cents per kWh to the owner of a system brought online in 2008 and will continue to pay this until 2028. That is about 40 cents above the wholesale conventional power price, which is a little less than 5 cents. By contrast, a system brought online today, within a FIT of 13 cents, would mean a price of only about 8 cents above the wholesale power price. Of course, 13 cents is low, barely profitable in most of Germany today, but it is still resulting in some investment and, were it to remain at 13 cents, investment would begin to pick up as the price of PV modules falls even further. Comparing the two premiums over the wholesale market price – 40 cents to 8 cents – means that a system built today would lead to only 20 percent of the surcharge as the one built six years ago. By the time the FIT could drop to 9 cents – which is foreseeable in the next few years – the cost differential would in turn fall again by half.

Germans are paying a lot of money for PV because a lot of PV was built during the years that PV was still expensive; that is what made it become cheap! By 2013, close to 19 billion euros was going to the owners of PV and other FIT power providers each year. That averages to about 230 euros for each of Germany's 80 million residents. At the same time, residential consumers have seen their electricity price rise from about 16 euro cents per kWh in 2002 to over 29 cents by the end of 2013.[21] It is easy to place exclusive blame on the FIT, but that would be dishonest. About half of the total price rise has something to do with the FIT, whereas the other half is attributable to other causes associated with the retail electricity market. Consumers see on their power bill a surcharge attributable to the FIT program that, for 2014, was set at 6.26 euro cents per kWh. In fact, only about half of this latter number is directly attributable to the cost of the total fed-in power and only 1.4 cents of this to PV power, while the rest covers various transactions costs that, to a large extent, result from some miscalculations in the past, including setting the surcharge too low.[22]

So let's do the math. Imagine that, in fact, future PV would lead to a surcharge of 20 percent of that of past PV simply because PV is now so much less expensive relative to fossil fuels. That means that you could double the total PV currently in the system, yielding a PV generating capacity roughly equal to Germany's total peak power demand, and add only about another 0.3 cents per kWh to the FIT surcharge.[23] In other words, the direct costs to consumers of continuing the FIT would have been miniscule. It also implies that if other countries were to institute a FIT, setting it at rates comparable to those in Germany, the costs to them would be similarly small.

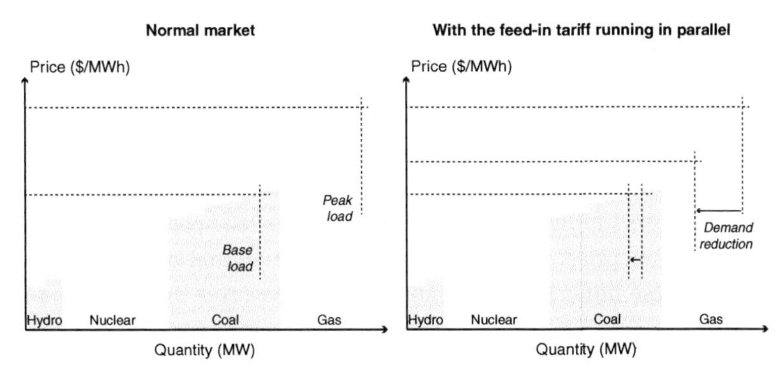

Figure 9.3. *The wholesale power spot market and the merit order.*

The FIT and incentives in the power market

But, leaving aside the low costs for consumers, in other respects the fears of industry are justified. The core of the problem stems from the structure of the underlying wholesale power market in Germany, and so it is worth taking a moment to explain that market. It hinges on contracts to supply power at different hours of the day, selling this into the high-voltage transmission grid, where it can be shipped around the country and ultimately distributed through a low-voltage grid to retail customers.

This wholesale spot market has a clearing mechanism known as the *merit order*. Figure 9.3 illustrates, in a stylized manner, how the merit order works. Let's start with the left-hand graph, showing the normal market on its own. All the conventional power producers submit bids to sell their capacity at a particular price, with the price they specify typically derived from their marginal operating costs. The market operator then ranks these bids from lowest price to highest price. The gray boxes in the left-hand graph illustrate this and reflect the fact that nuclear plants typically submit the lowest price bids and gas the highest. Even though the LCOE for nuclear and gas are fairly similar, the ratio of fixed to marginal costs are quite different: in comparison to the other plants, gas plants have very low fixed costs and quite high operating costs. Meanwhile, the level of demand in the market varies according to how many computers, hair driers, and light bulbs are turned on, ranging from a low point in the middle of the night (base load) to a high point in the afternoon or early evening hours (peak load). Again, the left-hand graph illustrates these two demand levels.

The market operator, meanwhile, buys all the power in the merit order to the left of the level of consumer demand, starting with the cheapest offers and moving to progressively more expensive offers until demand is satisfied. This implies that prices are lower during base load periods and higher during times of peak demand. It also implies that the fuel mix tilts toward gas during peak demand periods. Importantly, all the power plants supplying power during the peak load period receive the same higher price, and this exceeds their marginal operating costs during these times by an amount sufficient to cover the plants' fixed costs.

But now consider what happens with the FIT. The FIT requires the market operator to buy all the PV and wind power that is available, outside of the normal merit order market. The costs of this PV feed-in power get passed on to the consumer in the form of the FIT surcharge. Meanwhile, the merit order on the wholesale market remains largely unchanged. The only difference is that the demand level that this market has to satisfy is now lower, to the extent that power covered by the FIT is already supplying some fraction of consumers' needs. In the right-hand graph, one can see what this does to the market price and the fuel mix.

Base load typically happens at night, when the windmills may be turning but the PV panels are in the dark. The drop in demand hitting the spot market is relatively slight; in this case, it might have little or no effect on the wholesale power market price. Peak demand typically comes during the day, when both wind and PV are active. This has two effects. First, the utilization of gas plants drops off substantially, but with no change in coal plant operation during peak demand periods and only a modest effect during off-peak periods. Second, as a direct effect of the lower utilization of gas plants, spot market prices fall considerably during these peak load periods.

The former of the two effects has been a reason for criticizing the FIT on environmental grounds. After all, if you are going to build all that wind and solar, wouldn't you want to have the biggest possible effect on CO_2 emissions? In Germany, wind and solar have primarily displaced gas rather than coal, and, for this reason, their effects – in terms of CO_2 emissions – have been only about half as big as they could have been simply because the CO_2 emissions kWh are much lower for the gas than for the coal plants. A carbon price instrument would not have had this outcome: it would have penalized coal the most on account of its very high CO_2 emissions and caused it to fall out first. But one also needs to see this as a transient problem. If the FIT were to continue and wind and PV were to continue to grow (something that, I will soon argue, can only happen if policy makers also address the issues having to do with their intermittency), then eventually they will push aside all the gas and then move on to coal, pushing this aside progressively as well.

The latter of the two effects seen in the wholesale power market is the drop in prices for wholesale power. In Germany, right now, the wholesale power price averages about 3 or 4 euro cents per kWh. The fact that the market design, in the presence of the FIT, leads to lower power prices has benefited consumers, mitigating some of the added cost showing up on their bill in the form of the FIT surcharge. But the market price of 3 or 4 cents is less than the LCOE for any of the major technologies. The only way that it works is because of the fact that the four main companies generating electricity have a lot of old coal and nuclear power plants that they have long since depreciated and can operate simply for the cost of fuel and continuing maintenance. In today's market, they could not build any new power plants and hope to make money with them; indeed, for the past few years since PV started to explode, nobody has been building anything new. These companies are now in a position where they are losing money and will continue to lose money as long as the situation remains in its current configuration.[24]

There are even times when the power price has fallen below zero. That's right: the owners of power plants must pay people money to take their power. This has happened on a few very windy nights, when wind production is very high and demand is very low. All the natural gas power plants have been turned off, as have the coal power plants that can be turned down easily. But there has still been too much power in the system. Turning down the remaining coal plants and the nuclear plants creates problems when they have to increase their generation again, and so the power companies have chosen to leave them running. But the only way to get rid of that power has been to pay someone to take it.

And it is not just the Germans who are facing such problems: the level of German PV and wind has had ripple effects throughout Europe. A few months ago, my students and I visited a hydroelectric plant operator in Switzerland who showed us around the facility and gave a brief presentation on the economics of Swiss hydro. I mentioned in the previous chapter that part of the Swiss hydro business model is to arbitrage the price of power throughout the day with their pump storage fleet, buying power from French nuclear power plants at night when prices are low and exporting the power back to France during the day when prices are higher. The person presenting to us suggested that the boom in German PV has cut into that business substantially. On sunny days, the prices of peak power on the German wholesale market have fallen, and that means that the owners of coal and other power plants have started to export increasingly more power to neighboring countries, thus depressing daytime power prices there, too. As a result, the opportunities for price arbitrage are much less than they used to be.

Across the whole Alpine region, power companies in France, Switzerland, Italy, and Austria have plans for new pumped hydro facilities. Many of them are putting such plans on ice because the current supply of German PV has suppressed daytime wholesale power prices to the point where the pumped hydro plants can't turn a profit. That, unfortunately, is the last thing one would want. If we are to have a power system relying on a lot of intermittent capacity, such as wind and solar, then we will also need a lot of storage, like pumped hydro. That takes time to build. It would be great if the hydro companies were building it now, but right now they have every incentive not to do so.

Moving forward with and beyond the FIT

To see where Germany needs to move forward and whether other countries should and should not copy them, it is useful to separate out three different problems associated with the *Energiewende* in the power system. Each problem suggests the need for different public policies.

The first problem is getting renewable generating capacity built at all, when the LCOE for renewables exceeds the cost of power from conventional sources. We need this capacity to be built, not necessarily because it will make an immediate difference in CO_2 emissions, but rather because it is a necessary ingredient of falling technology costs. When executed well, as in Germany it largely has been, the FIT has proved to be the best policy instrument yet devised. It provides certainty to project developers, reduces transaction costs, and mobilizes all sorts of people to get in the game of investing in PV and wind. In short, the FIT gets a lot built quickly at what appears to be the lowest possible price to consumers. The result of all this has been falling prices, and, indeed, for onshore wind, guaranteeing a tariff above market rates may no longer be needed since prices have fallen so far.

There are some possible innovations to the FIT that policy makers could consider. One of these is the introduction of bidding on the part of project developers, something that Germany is introducing; others include FITs that vary more precisely according to location or time of day. Some of these may prove to be useful changes, although the evidence is still lacking as to their results. Whatever precise shape it takes, however, a few more years of FIT-covered PV would be a good thing, as would a FIT or similar support for offshore wind and concentrated solar power (CSP). But let me be clear: such FITs should not be considered in any way permanent. The purpose is not to use the FIT to finance the construction of enough renewable power to displace large amounts of fossil fuels. Rather, it is to finance enough renewable power to get the price to fall to the point of being competitive. That is the point at which deployment can really start to accelerate and begin to make a dent on emissions.

The second problem is getting the wholesale power market to function efficiently when there is a lot of renewable electricity in the system. The main problem that Germany faces, and that can be attributable to their build of wind and solar, is the drop in prices on the wholesale market. This is hurting companies selling conventional power, and they are understandably upset. Importantly, however, it doesn't really matter whether that renewable electricity is covered by a FIT or not: the results for producers of high-carbon electricity would be the same. There is sound economic analysis to show that a carbon market, employing an auction mechanism to allocate permits, would have a similar result if that carbon market were to be substantial enough to stimulate wind or PV development.[25] Simply put, when there is a lot of electricity in the system with very low marginal costs, the merit order market clearing mechanism just doesn't work very well, at least not in a way that allows utilities to earn money.[26] This is a bad thing because these companies are needed to maintain reserve capacity in the system, something they now have little incentive to do. It is also a bad thing because they are politically powerful in most countries. If they aren't happy with change, change isn't going to happen.

A lot of good economists are studying this problem right now and are very far from consensus on the solution. The elephant in the room is the extent to which one believes that it is important to have a market in place at all, by which I mean a mechanism that sets the price through a competitive process, as the merit order does. The fact is that there is a simple solution to the problem of disappearing profits, but it is one that a lot of people don't like: get rid of the merit order and have government regulators set prices for all the technologies used in the system. In essence, this would be akin to extending elements of the FIT to all technologies. There is some evidence that this may be happening already; recently, politicians in the United Kingdom decided on setting a fixed tariff for nuclear power in order to get a new plant built. Moreover, having tariffs set by government regulators is how things used to work when some countries had state-owned utilities companies, and other countries tolerated monopolies in power production but regulated those monopolies intensely to make sure they acted in the public interest.

Then people had the idea, starting in the 1980s, to liberalize the market, opening it up for competition. Liberalization has turned what used to be a fairly boring industry into a much more interesting one. In theory, it should lead to falling consumer prices for electricity, although in practice it often has had the opposite effect either because of poor implementation or because of insufficient competition within the industry.[27] We could go back to how things were before the wave of liberalization, and it wouldn't be a disaster. At the same time, it might be a shame to abandon liberalization altogether, and it may also be politically impossible. There are a variety of more incremental approaches that policy makers can take that would retain competition, including making fixed payments to firms for maintaining capacity or setting price floors during peak load periods. It makes sense to continue to search and try out options, recognizing the need for some sort of change to the market structure. We are in new territory here, and the terrain is full of potential pitfalls. The emphasis here needs to be on learning and being willing to correct mistakes quickly.

The third problem is a technical one looming large on the horizon: how to ensure that there is enough dispatchable capacity in the system, power that can be modulated to cover consumer demand every hour of the day, every day of the year. Germany is already facing one side of this problem, visible when the power price goes negative because there is too much power available. If that eventually causes gas and coal plants to be decommissioned, which is a longer term possibility, then it will face the other side: not having enough capacity when it is needed, with the resulting threat of brown-outs. Simply building more PV and wind won't fix this, but will only make the problem worse. What Germany needs, and what other countries will need soon after them, is a set of technical solutions outlined in the previous chapter, which enhance the overall reliability of the power system. The next wave of policies are those that make sure that the power grid is up to the job, as well as promoting the installation of energy storage capacity.

What Germany can learn from Texas

Given where Germany now stands, with a huge and still growing capacity of wind and PV, the most urgent need is to get ahead of the curve on power lines, building them before they are needed rather than after. Power lines would enable a country like Germany to export power more easily when there is excess wind and to import power when there is not enough. In Germany, however, the construction of new power lines is languishing, facing the twin problems of inadequate financing and insufficient public support.

Right now, the best example of a place that is building power lines with a sufficient sense of urgency is Texas. Texas constitutes an autonomous power market of its own within the United States. In 2005, state regulators in Texas already foresaw the huge potential that wind could offer the state and, with that potential, the huge need to get enough power lines built. Rather than leave this for market forces to solve, they mandated the construction of 6,000 km of new power lines, at an aggregate cost of $7 billion. The grid operator, a private company, has been responsible for getting the lines built, but it did so with the assurance that the costs would be borne by consumers, with a state-mandated surcharge on their monthly power bill not unlike the surcharge that German ratepayers see for the FIT. Texans, on average, see a $6 surcharge each month for the new grid. Given the low cost of wind, however, many believe that this has already been compensated by lower prices for the power itself, as more and more low-cost wind parks have come online, spread far enough apart geographically to ensure a fairly high degree of aggregate reliability.[28]

In a gigantic and sparsely populated state like Texas, building 6,000 km of new power lines can happen fairly easily once state regulators decide that they want to do so. In more densely populated regions, it would be far more challenging, at two levels. First, with so many more people around and many more competing land uses, it would face potentially higher public opposition. This is something observed around the United States as well as throughout Europe. In Europe, moreover, new grid developments would often have to cross national borders, thus requiring a greater level of international cooperation than has existed in the past. In the previous chapter, I flagged the need to deal with the issue of local support or opposition to wind power if we are to avoid running into a capacity constraint. The same problem faces the construction of power lines needed for renewables more generally.

It is certainly a mistake to say that all people objecting to new windmills or power lines are wrong or should have no influence on these things being built. There are places, perhaps of high natural beauty or particular economic value, where they should not be built. People opposing power lines have concerns that are legitimate, associated with the visual landscape, noise, and the health effects of electric fields. But a decade of research has shown us that, above and beyond these issues, other factors stand in the way of getting the needed infrastructure built quickly. First, many people react negatively, very negatively, to proposed infrastructure not because they dislike the infrastructure per se, but rather because they feel disempowered, disrespected, or taken advantage of by the institutional process by which that infrastructure is built. For example, there is strong evidence that when utilities planning a piece of infrastructure like a power line involve local residents in the early stages of identifying the need for the project, they will be far more likely to support it, in contrast to cases where a company tells them that "experts" have established the need.[29] Likewise, there is evidence that public acceptance of a project is higher when the project developer is a locally owned company rather than an outside giant.[30] There is room for a significantly greater amount of transparency and "procedural justice" in project planning. Another issue related to procedure is the time it takes to reach the resolution of a permitting dispute. In Europe, as in parts of the United States such as California, it can often take a decade or more for a planned power line project to ultimately be approved. Contributing to this is simply the slow pace at which regulatory agencies and the courts operate because speed has historically not been a priority. But speed is becoming a priority. The European Union is considering legislation that would set a time limit for such permitting decisions of under four years.

In addition to better process per se, there is also room for better and more complete information within that process. The research of my own group, in Europe, shows that people are more supporting of new power lines when there is better information available about the overall environmental effects and in particular when it can be demonstrated that the power lines will have the effect of preventing harm to the environment. This is not trivial: for many people, a power line represents a vivid example of industrialization, a swath of metal cutting across a green landscape. How could this be good for the environment? It is important to identify the environmental consequences, not only in terms of global climate change, but also with respect to the local environment. The best example of this that I know is a computer model that has been developed at the University of California, Berkeley, called the SWITCH model.[31] This computer model predicts the local environmental effects, including those on water resources, air quality, and land availability, associated with different regional energy scenarios. Because the model has a high resolution with respect to both time and space, it can account for issues of local intermittency in renewable energy production as well as the effects of power transmission to balance these effects. So far, the model has primarily been used to investigate options in the western United States. It shows that the energy scenarios relying on high shares of renewable power have the effect of lessening the negative local environmental consequences of electric power production. Furthermore, the model would allow one to investigate how the local and regional environmental effects would differ and whether the overall infrastructure needs would be lower if a given power transmission line were in place. I believe that models like this could play a critical role in providing needed information in the context of power line permitting processes.

The second factor, particularly relevant in Europe, is the lack of harmonization across different political jurisdictions. Spain, France, and Germany may all have different procedures for determining rights-of-way as well as divergent political interests, and so to build a power line from the first country across the second to reach the third is extremely difficult. A recent study asked leading stakeholders for their opinion and found that they believed the problem could not be solved given the existing regulatory framework: there is a need for a thorough overhaul of the permitting processes across Europe in response to the need for a much more international power grid.[32] There most likely needs to be a shift in the planning authority from individual European countries to a regional authority, most likely an arm of the European Union. The current decentralized planning authority makes it virtually impossible to plan a power line that crosses multiple national borders when it is unclear which countries or states will benefit the most. Indeed, the benefits are likely to be disproportionate even if the aggregate benefits to humanity are huge.

Across the United States and Europe, wind and PV are going to continue to grow, which is a good thing. But, just as Texas got ahead of the curve with power lines, the rest of the United States and Europe needs to do the same thing. At the same time, more power lines alone won't solve all of our problems. Regulators on both continents also need to plan ahead for wind and PV capacity with top-down attention to several other elements of the reliability puzzle: CSP, smart grids, and storage capacity.

As described in the previous chapter, CSP is a good thing because it can be dispatchable. The policy instrument of choice would be the FIT, for all the reasons described, but it would need to be structured to provide an incentive for coordinated use of storage capacity. Like wind and unlike PV, CSP typically comes from large installations and hence gets built by large energy companies rather than private households. For this reason, the need to guarantee an absolutely predictable rate of return is not quite so crucial. Hence, the FIT could be structured more in the form of a premium above the wholesale power price, thus providing project developers with an incentive to build and operate their plants to follow load and supply power into the system when prices are highest.

It is easy to imagine a support scheme for CSP working well in the United States, and indeed this is where CSP may make the most immediate sense. What makes Europe a harder nut to crack for CSP is the fact countries like Germany would need to institute a FIT for power generated in other jurisdictions, including North African countries like Morocco, Tunisia, or Egypt. As long as CSP is a relatively expensive technology, this makes it politically unattractive because it would mean a significant power surcharge in one political jurisdiction to finance construction (and create jobs and profits) in another. The lower the price of CSP becomes, the easier it becomes politically to build even more of it. There is every reason to believe, however, that the need for CSP as a source of dispatchable electricity will grow over time.

One of the drawbacks with CSP is that it, too, requires new transmission capacity because the CSP plants themselves are located far from the urban areas where the power is likely to be consumed. The biggest hurdle to integrating a lot of CSP may in fact be the pace at which power lines can be approved, even if countries do their best to streamline the approval process. Partly for this reason, reliability can and should also come from other avenues as well: the smart grid and storage.

In the case of the smart grid, the main barrier that policy needs to overcome is not obtaining rights-of-way but rather the high initial investment cost associated with upgrading the existing grid to include data transfer as well, combined with the regulatory changes needed to make use of real-time information in the setting of electricity prices. There may not be any alternative to a strong governmental hand, mandating investment in new infrastructure and providing the system to pay for it, just as Texas did with its new power lines.

In the case of storage, and perhaps most crucially batteries, there is a huge amount of room to expand the range of basic technologic options and hence for large government R&D budgets supporting a variety of emerging technologies.[33] Many countries, such as the United States and Germany, are indeed providing this and hopefully will not stop. But there is also a need to start getting storage scaled-up out in the market. Germany started this in 2013 by subsidizing the installation of storage capacity: it set aside 25 million euros to subsidize new battery systems, with the exact format of the subsidy varying according to the size of the system and whether it is coupled to an existing PV installation.[34] The data are not yet available, but analysts have been predicting that the German budget of 25 million euros will be used up quickly.[35] There is room to provide a lot more funding, in Germany and soon elsewhere, as well as to experiment with different types of subsidy packages, in particular when it comes to joint investments in PV and battery storage together. Japan is currently positioned as the world leader with respect to many key battery technologies, and there is evidence that it is adopting such a strategy.

Financing continued growth

In moving forward, the issue of finance needs to guide some of our policy choices. In total volume of finance, the needs required to transform the current energy system into something new are not that much greater than they would be if we stuck with our present system architecture because, in either case, we would be replacing infrastructure when it becomes old and outdated. The most recent Intergovernmental Panel on Climate Change (IPCC) report indicates that total global investment in the energy sector currently stands at about $1.2 trillion per year. In terms of geography, roughly half of this is in countries belonging to the Organization for Economic Cooperation and Development (OECD), and, in terms of activity, about half is connected with the extraction of fossil fuels, while most of the other half is associated with electricity generation and distribution.[36] Looking into the future, one can compare the investment needs under a business-as-usual scenario (no climate policy) with those under a strong mitigation scenario. The IPCC found that within the electricity sector, the mitigation scenario implied an increase in global investments of between $20 and $200 billion, with a median estimate of about $125 billion per year. Within the fossil fuel extraction sector, the mitigation scenario implied a change in investment ranging from no change to a savings of more than $350 billion, with a median estimate of about $60 billion less per year. With these two factors roughly balancing each other, the report's authors suggest that climate mitigation implies a large shift in investment but no major change in the total volume of investment.[37]

The fact is that as renewable technologies become competitive, their overall total costs will not be higher, and, across the sector as a whole, the total financing costs will not be much different. Where the finance challenge lies is not in coming up with huge new amounts of new finance, but rather with its redirection into new forms of capital investment. A large share of the existing finance comes from energy companies themselves, out of their profits and stocks of cash reserves. If these companies stick entirely with the fossil fuel business, then the money they would have invested in new energy infrastructure will be of little use to the new energy system, and the companies and people actually making investments will have to seek alternative sources. On the other hand, if the existing energy companies do transition into renewables themselves, then the need to find new sources of funds will be less. In fact, what we are seeing is something between the two extremes. In Germany, for example, roughly 10 percent of the investment in renewable energy sources has been coming from large, established energy companies.[38] For the United Kingdom, the anecdotal evidence suggests that the number is much higher. One reason that the German number may be so low, in absolute terms and relative to the United Kingdom, is that such a large share of the recent German investment has been in rooftop PV, whereas in the UK it has been in wind, including offshore wind; wind is simply something that large companies invest in whereas rooftop PV is not.

The German FITs that drove the PV boom have managed to reduce investor risk to the point where raising sufficient finance is not an issue. In Germany, you did not need to be a large company to be able to borrow money at attractive terms for a set of PV panels. That has been fine so far, but may not be the case in a few years, when the subsidies for PV can and should disappear. In parts of the United States, this is already starting to happen, and it is there that alternative financing models are developing. One is the PACE program that I described earlier, whereby a municipal government provides the financing for new PV installations and attaches the loan payments to the property tax liability. A very different model has come from the private sector. The company SolarCity has emerged as the leading installer of PV panels, and they vertically integrate the financing of the system with the installation. Homeowners can either lease a system from SolarCity or else borrow the money for its purchase; in either case, there is no initial cost to them, and payments are made on the system once that system is up and running and saving the homeowners money by reducing purchases of power from the grid. In both the PACE and SolarCity models, the important thing is that both the high upfront cost of the system and the risk of its not working out are borne by an entity that is larger than a single household and better able to raise financing at attractive rates and to spread risk across a portfolio of investments.

The lessons learned

I believe that the actions that Germany has taken over the past twenty-five years have transformed the climate change problem. Before 1990, climate change was like overfishing; technology was the source of the problem, and only the most imaginative dreamer would see it as the possible solution. Then the Germans took early leadership on wind power and pushed it during the 1990s to the point where it could begin to take off in other markets as well, most prominently in the United States. Immediately thereafter, Germans started doing the same for PV. As a result, PV is set to boom elsewhere, having entered the zone of competitiveness in various applications and in various places. We are entering a period of a couple decades during which all new investment in electric generation will have to switch over to decarbonized sources. We are doing so with two winning technologies. We are in a good position.

In the past two years, the German government has taken a series of actions to make the FIT less effective and to rein in the growth of renewables. I think this is unfortunate, but it is also understandable, a result of several different political arguments. One such argument is that the rest of the power system needs to catch up by putting in place market reforms, installing needed transmission capacity, and beginning to scale up the pace of investment on power storage. The fact is that the FIT led to faster growth in renewables than most people had anticipated, and, in some ways, that growth was simply too much too fast. Another argument is that what ordinary Germans are now paying for having taken the leading role on wind and solar, although not astronomical, is also not trivial. They chose to push both technologies while they were still expensive. They hoped to get a major new export industry out of doing so, but this did not really prove to be the case for PV, although for wind it did. Rather than complain about the tide turning one way or another, I would rather simply be thankful for what Germany has done.

But I also think that we can learn a lot about what effective climate policy can look like. More or less, the Germans turned their backs on the accepted wisdom concerning climate policy as it existed back in 1990 and as it existed even more strongly in 2000. They turned their backs on the idea of one all-encompassing integrated policy solution, a system of linked carbon markets within a global framework of national emissions targets. They turned their back on the idea that this would really get them the kinds of changes in their energy system that they wanted.

They wanted several things from the energy system. They wanted an energy system that would continue to have some hydropower and support a growing amount of wind power, and this was reflected in the first policy framework put in place in 1990, which guaranteed premium prices paid to renewable generators. Next, they wanted an energy system that included a lot of PV, and this they achieved through the FIT laws introduced in 2000. There were two big reasons for this attention to PV. First, PV offered the possibility of smaller scale, more local power production, and systems in which individual households and communities could produce their own energy. Second, PV offered the hope of developing a new export industry based on German technical innovation. Most recently, Germans have recognized that they want their energy system to be more affordable and to deliver renewable energy without sacrificing reliability. To achieve this, they have scaled back on the FIT for PV in order to slow deployment and introduced new policies to begin to boost innovation and deployment of power storage technologies; the early indications are that these are working.[39] What Germany has not yet done is to really address some of the legal and regulatory reforms that would allow power markets to work more effectively when most of the power produced is renewable and that would accelerate the pace with which power lines could be planned. It also hasn't addressed the potential need for CSP imported from the Mediterranean region, including North Africa. Germany started to develop policy approaches to this about five years ago involving frameworks for financing and importing renewable power from other countries, but then political events unfolded in such a way as to put such ideas on ice.

The big lesson to learn from this is that there are a number of separate problems to be faced in the process of restructuring a power system, and the different challenges require targeted policy solutions. The German story is not the story a single policy instrument, the FIT, and how it alone has transformed the German power sector. Rather, it is a story of constant evolution and change, with policy instruments tailored to the challenge at hand. The problem that the FIT responded to was that of boosting investment in PV, and it worked because it created a profitable and risk-free investment climate. But Germany also shows us how the FIT can create new problems – such as the need to enhance storage and transmission capacity – for which the FIT on its own is insufficient. Other instruments, primarily geared around supporting new investment rather than penalizing it, have risen to the challenge. The German story offers us a picture of messes being created and then cleaned up. Walk around any construction site in the world, and you will encounter a certain amount of messiness. It's how big things happen.

Moving forward, the main policy needs for the *Energiewende*, not just in Germany but everywhere, are going to be related to such mess, the second-order challenges that come with change. Fortunately, there is a good set of options out there: subsidy and financing options for grid infrastructure and storage capacity; planning and communication practices to avoid polarization at the community level; market reforms to cope with changing cost structures. Policy makers in Germany, elsewhere in Europe, and in parts of the United States are already experimenting with them. Moving from isolated experimentation to widespread practice will be our climate policy challenge of the next two decades.

10

Policies beyond power

Germany has led the way with policies to transform the electric power system from coal and gas to renewables. Year by year, as a result of those policies, the power sector in Germany is changing, and the country is well on track to have a completely decarbonized sector by 2050. Where they have led, other countries have started to follow. These have included the United States and China, as well as a number of other countries in Europe.

A large share of the financial cost is now behind us, to a large extent because of Germany's initiative. Indeed, there are now credible estimates that the remaining actions will come at no net financial cost.[1] Photovoltaics (PV) does still need to become a little less expensive to be able to compete in increasingly more places, and offshore wind and concentrated solar power (CSP) do still need generous financial support, at least for a few years. But the past twenty years of growth in onshore wind and PV have left both technologies in an extremely attractive position. For several years now, global investment in these two technologies has surpassed investment in fossil fuel generation.

A decarbonized electric power system will play an increasingly important role in a decarbonized society more generally as electricity begins to play a larger role as the energy carrier for both transportation and heating. It is also potentially the hardest piece of the puzzle to solve because of the difficulty associated with storing electricity and hence making the system reliable. But it is, still, just one piece of the puzzle. We still need to make a changeover from gasoline and diesel cars and trucks to electric ones. We still need to transform the process of heating and those industries where CO_2 production is intense. Finally, we need to scale up the production of biofuels at exactly the same time as we slow down and then halt the processes of tropical deforestation. These are all large challenges. And they will take public policies that are different from the ones that Germany has used for electricity.

One of the pieces of good news is that these challenges are distinct and separate. It may strike you as odd that I would list that as a piece of good news. After all, one of the supposed advantage of a carbon price instrument is that it can address all of these separate problems at the same time. And yet that is part of what makes a carbon price instrument so difficult. The stakes are so large, and so many different sectors of the economy are affected that, on the one hand, it is extremely risky to move too quickly, and, on the other hand, the political forces lined up behind the status quo are large. If you break a big problem into lots of little ones that can be addressed independently and in parallel, however, then the potential for political gridlock is simply much less. So let's look at what the separate building blocks are.

In this chapter, I cover policies outside the area of decarbonizing the electric power system. I start with ground transportation, where the goal of policies needs to be to promote electrification. From there, I move on to space heating; this requires policies to ensure that new buildings are efficient and electricity based, as well as coordinated programs to retrofit existing construction. After that is the need to mandate the use of biofuels as the basis for high-temperature process heat in industry, as well as to make geopolymer cement a viable proposition for the future. I conclude with the interconnected issues of agriculture, deforestation, and biomass production. In this case, the core challenge is to end the incentives for deforestation, such that an increase in the production of biofuel feedstocks does not prove to be counterproductive.

Electric cars

Two chapters ago, I described a trip that my family took to the fjords of western Norway. On the way back, we decided to take two days to drive to Oslo, where we would board our ferry to Germany, spending the night en route at Gjendesheim Lodge, at the edge of the Jotunheimen National Park. Years before, my wife and I had stayed there during a cross-country skiing trip; we remembered it as being stunning in terms of its location, and so we wanted to show it to our kids. It didn't let us down. In the winter, years before, it had sat in the middle of a frozen landscape, covered with a thick layer of snow, right out of one's childhood images of the North Pole and Santa's workshop. Seeing it in the summer, we discovered that it lay on the banks of a rapids-filled river bursting with trout, and low birch trees and blueberry bushes all around. In winter, it was inaccessible, the access road covered in snow and the only approach was by skis or the weekly four-wheel-drive bus of the Norwegian outdoors club. In summer, it had a dirt parking lot filled with cars, including our own with two sea kayaks on top. Except that there were a few cars not sitting out in the parking lot, but right up against the building. These were the electric ones, plugged into a charging station next to the front door of the lodge.

It wasn't an unusual sight. Norway, it turns out, is the country that has seen the largest market penetration rate of electric cars, from small cars to large. The electric drive Nissan Leaf is the best selling car overall, whereas the Tesla Model S is the best selling luxury car. Electric cars in total already command 12 percent of new car sales, a number that is growing.[2] To some, this is surprising, given that Norway is a major oil exporter. But Norway taxes its gasoline and diesel fuel as much as anywhere else in Europe, and the country doesn't seem to have any particular affinity for big cars.

What Norway does have is a lot of separate little government incentives to buy and drive electric cars, and, put together, these have made a huge difference. First, and most importantly in terms of financial incentives, electric cars are exempt from the new car sales tax. In Norway, as in many European countries, this is linked to the weight of the car and can be as high as 100 percent of the pretax value of the vehicle. Second, and altogether different, drivers of electric cars are entitled to use the bus lanes. These are similar to the high-occupancy vehicle lanes seen in the United States, which a few years ago the state of California opened up to drivers of hybrids. The effect of being able to use the lanes is the same; namely, to allow you to whip through rush hour traffic, leaving the drivers of the cars you are passing wondering why they aren't in an electric car too. Third, there are a lot of little financial perks: electric cars are exempt from highway tolls and even get to ride for free on the large network of ferries. They also can park for free in municipal parking lots. Finally – and this is very important – the owners of electric cars have made available to them a large and growing network of public charging stations. For the time being, they get to use these free of charge.[3]

None of the policy measures that Norway has put in place is particularly innovative or difficult to implement, and yet the combined effect has been substantial. To a large extent, it may simply be the wealth effect that matters most; the wealthier people are, the less the initial high price of an electric car really matters. Norwegians are wealthy people. Where I live in Switzerland, electric cars have also been relatively successful, and I think that this just has to do with the general wealth of the Swiss. Certainly, the government has not gone out of its way to promote electric cars.

One of the factors stimulating sales of Teslas in Switzerland has nothing to do with electric cars per se, but rather is a requirement that car importers limit their fleet's overall CO_2 emissions, similar in design to the corporate average fuel economy (CAFE) standards in the United States. This has two effects, both of which benefit electric cars. First, it means that selling an electric car, which has zero emissions, can bring down the overall fleet average and hence be of enormous value to the car company. Second, it means that manufacturers typically inflate the prices of the very large and luxury cars because the CAFE standards essentially limit the supply. A Tesla Model S can compete on price against a Jaguar or Mercedes or BMW of similar class because the latter cars' price is pushed up by the regulation operating in tandem with market forces.

There is one way in which the policy challenges of promoting electric cars is very different from and ultimately easier than changing the supply of power over from coal and gas to renewables. In the case of the power supply, there are two very big problems to overcome, and solving the first has just made the second one more acute: as PV and wind have been built en masse (problem 1), their low marginal costs and their intermittency have created disruptions to the power market more generally (problem 2). The power market can't solve the second problem on its own, and indeed, the more wind and PV grow, the harder the second problem becomes. The response of the players in that market has been to lobby to have the growth of PV and wind put to a stop. With electric cars, this just isn't the case. Electric cars face high initial barriers to market penetration, in terms of both high cost for batteries and the costs associated with building a network of charging stations. But the network problem, unlike the power market problem, becomes easier to solve the more prevalent electric cars become. It's analogous to mobile phones: the more people who own them, the greater is the market base from which to finance the construction of transmission masts. And the more masts, the greater the coverage and reliability and, in turn, the more useful a mobile phone becomes. It is a positive feedback process. Electric cars are the same.

So I am optimistic with electric cars. They aren't exploding in popularity like people hoped they would a few years ago, hopes that were themselves ill founded. But their growth is still impressive, holding steady in the double figures. The more people buy them, the lower the barriers will be for others to follow them. Already, more and more companies are bringing them to market and also bringing other kinds of electric vehicles, like electric scooters, into the mainstream. Soon, electric drive vehicles will be noticeably less expensive than those with internal combustion engines and noticeably better in other ways associated with handling, noise, and overall design. As this happens, the conditions for electric drive trucks will also start to fall into place as well. It may turn out to be the case that electric-drive trucks gain their power from a fuel cell, rather than a battery. But the same set of positive feedbacks dominates the spread of fuel cells and the network of hydrogen fueling stations as well.

There is another positive spillover, one that has to do with technological similarities between electric car batteries and stationary battery power storage. The more electric cars we build, the cheaper other kinds of batteries will become, and the easier the grid will become to manage. Producing hydrogen may become an effective way to deal with those times when the supply of power greatly exceeds the demand. In the meantime, we need more countries to do the kinds of things that Norway does. We know how to get people to buy electric cars – it isn't particularly difficult – and the more of them they buy, the better they will get. Soon we won't need to do anything at all.

Space heating

Heating is the single biggest source of CO_2 emissions associated with energy use. From a technology standpoint, it is worthwhile to distinguish between space heating and process heating. In the case of the former, which I cover in this section, it is useful to break things down into two different policy problems: (1) making sure that new construction employs the latest technologies that can allow buildings to be carbon neutral and (2) helping people to renovate existing buildings to become CO_2 neutral from now on. The good news is that governments have successfully implemented both sets of policies and found that both sets of actions pay for themselves.

When it comes to new construction, it is clear that making buildings energy efficient and able to utilize heat pumps rather than oil and gas burners makes simple economic sense. So why doesn't everyone do it? There are two reasons, and they are closely related.

The first stems from a basic irrationality that almost all of us share: we think that the pressures on our time and on our finances – making it seem like we don't have enough of either – are a temporary phenomenon. Behavioral economists describe this as being "present biased," in that we believe that our needs are most acute in the present period,[4] and they have even shown that different neural pathways get used to assess present and future needs.[5]

Let's say I go light bulb shopping. At the store is an energy-hungry and short-lived incandescent bulb for about 30 cents and an energy-saving and long-lived LED bulb for about $15. I know that the latter will save me money in the long run, but I also know that it is the end of the month, and my bank account is approaching a negative balance. So I say: "This time I will buy the cheap light bulb, but next time, when my bank account is less pressed from all sides, I will invest in the more expensive one." Except that next time, the story repeats itself. And the story also repeats itself when it comes time to buy a new house or office building. I know that an energy-efficient design with a heat pump heating system will save me money in the long run. But my problem isn't the long run: it is this year's finances. So I take the less expensive building option and plan on a retrofit in the future, or a better choice the next time I need to make a real estate investment. For decades, people have been making these kinds of smart long-term plans, and then ignoring them.

The second reason for the persistence of new construction that requires gas and oil for heating is the fact that everyone is aware of the first reason. Let us say that you are a real estate developer: you build buildings for residential and commercial customers. Your business model relies on your selling or renting the property as quickly as possible after it is finished. You know that smart consumers will think long and hard about heating and cooling costs and ultimately select a property that costs less in the long run. You also know that most people are not smart customers. To reach a broader customer base, enhancing the likelihood of a quick turnaround, you put your money into highly visible design elements and not the largely invisible aspects of energy efficiency and the heating system.

The combination of reasons one and two means that, even for those people who would like to purchase or rent an energy efficient building, they just aren't widely available. Indeed, for a long time, just about the only households and commercial building users who have occupied passive buildings have been those who have built them themselves. This in turn drives another problem: with so little demand for such buildings, not a lot of architects know how to design them.[6] Putting all the factors together, they have remained a niche element in the market.

The first step toward overcoming present-biased behavior is to recognize it as a genuine problem in the first case. Everyone would be better off if they could break out of it. Doing so typically requires making the present-biased choice less attractive or even impossible. One of the best examples of this is in the case of retirement savings. Most people, left to their own devices, manage to save virtually no money for retirement. In any given month, they may put some money aside but then they find something important to spend it on, digging into their savings to do so. One of the innovations that behavioral economists have touted to overcome the problem is the private retirement savings account: one decides that money (in the future) will automatically be deducted from one's salary and that there will be penalties for early withdrawal. The money ends up in retirement savings, and it stays there. But an even more fundamental innovation, and one developed far earlier, was simply for the state to mandate participation in social security or a similar pension system with a different name. Nobody has a choice but to have some of their monthly income placed into a state retirement income fund. Before social security, at least in the United States, most people dropped into acute poverty at the time they stopped working. Social security changed that, and people are better off for it.

What is needed for climate change is the building code equivalent of social security. The fact is, most building codes today have been designed to solve another problem that wouldn't exist if people were completely rational: fire safety.[7] Building codes typically do not address the issue of energy use, but they could, and in some places they do.[8] It is entirely reasonable to move toward building codes that mandate highly energy efficient buildings and heating systems that do not rely on fossil fuel combustion. Both would have the effect of saving their future occupants money. They would make everyone better off.

The good news here is that many governments have already started this process of rewriting their building codes, and a few of them are quite advanced in the process. California has already instituted building codes that mandate a clearly cost-effective set of energy efficiency options, such as insulation and improved windows, and has a target of updating its codes such that all new buildings constructed from 2021 onward will be energy neutral. Denmark has taken similar measures: starting in 2020 all new buildings will require "nearly zero" energy for their heating. The European Union (EU) has issued a directive requiring other countries to take steps similar to Denmark's. In 2012, the EU issued the Energy Performance of Buildings Directive, which requires that by 2019 all new public building will require nearly zero energy and that by 2021 this be extended to all new buildings. The effect of an EU directive is to require European member states to write their own national laws, and so far the various European countries are at different points in this process.

Economists like to point to one important flaw in this idea, which they call the *rebound effect*. Basically it is this: when people save money through energy efficiency, they end up spending that money elsewhere, and the things they spend money on use energy, too. Maybe they spend less money on home heating, which enables them to take a vacation to Thailand. The result is that the overall drop in energy use is somewhat smaller than originally anticipated.[9] In its worst incarnation, it results in *backfire*, namely an overall increase in the energy used as a result of a new efficiency.

As it turns out, a researcher who works with me, Nadya Komendantova, observed this precisely in the area of home heating. The Austrian government commissioned her to assess the effects of a program to help rural Austrians insulate their houses. What she observed was that people who lived in old farmhouses typically found the cost of heating to be prohibitive, and, as a result, they often occupied only a small fraction of the house during the winter and expanded into the additional space during the summer. After they had insulated, however, they discovered that for a much lower additional cost they could heat their whole farmhouse, thus using it year-round. As a result of doing so, their overall energy consumption actually went up a bit and not down. The effect of the government program was not to save energy at all. The effect was to give these people more living space.

The rebound effect is nothing to worry about as long as the policies to promote energy efficiency are part of a wider set of actions designed to move us beyond fossil fuels. Take the Austrian farmers from the last paragraph: if the government program had promoted not just home insulation but also the conversion of their heating systems to something that was carbon neutral and sustainable, such as wood pellets or district heating, then the increase in energy consumption brought about through the policy backfire would not have been a problem. There is nothing wrong with using energy. The problem is burning fossil fuels. If we are to achieve our policy goals with respect to climate change, nothing that anybody can spend their money on will lead to CO_2 emissions. So policies that promote energy efficiency, and hence give them more money to spend, are perfectly fine.

The case of policy backfire in rural Austria is also a good segue into the separate policy problem of retrofitting existing buildings. The Austrian program is an example of the kind of policies that many suggest is needed for the building sector: namely, public finance and grants to support building retrofit.[10] Indeed, this is a good policy intervention; yet, to be successful, it is important to consider the range of options that are appropriate in any given community, as well as the particular motivations and fears that building owners need to grapple with.

An example of this has come out in a project in Switzerland, in the community of Zernez. It is a beautiful town in the heart of the mountains, its village core full of historic buildings reflecting a distinctive architectural style. Community leaders there set the ambitious goal of making the entire town CO_2 neutral with respect to its heating needs by 2020. Working with researchers from my own university, they carefully assessed their options and came up with a set of insights. First, and most importantly, they realized that for the town's older building stock, the best option would be the development of a district heating system relying on waste (especially forest product waste) incineration to deliver steam to buildings in the core village. The entire village's charm (and tourism revenues) would be ruined by the application of thick insulation layers to the buildings' exteriors, and district heating would eliminate that need. Outside of the village center, the task of insulation would make more sense, and there would be value in locating new buildings as close as possible to existing ones in order to provide opportunities to share new heating systems, such as those utilizing heat pumps and solar collectors. Ultimately, the town leaders and researchers reached the conclusion that achieving CO_2 neutrality would be possible, and the town is moving forward in this direction. At the same time, community leaders have recognized that the entire process will take somewhat longer than the six years originally envisioned.[11] I am confident that they will succeed in their effort. I also believe that the success will depend in part from having made it a community project, rather than a set of isolated individual actions.

If the bureaucrats in Brussels have their way, the Zernez case could become increasingly more common across Europe. Their 2012 directive on building efficiency requires member states to develop national action plans for the renovation of existing building stock to improve energy efficiency. One of the major issues with respect to building retrofit is securing adequate finance. Clearly, the state can play a major role here, as it does when making grants to homeowners. In Germany, there exists a national financing agency, the Kreditanstalt für Wiederaufbau (KfW). Formed to finance reconstruction after World War II, KfW now is engaged in financing renewable energy and energy efficiency projects, making use of its ability to issue bonds to be able to provide finance at very attractive interest rates. There is also something to be learned from the success of the solar PV financing program, PACE, developed in Berkeley, California. That program made use of city bonds to raise the money to then loan to homeowners to install PV panels on their rooftops. The debt payments are then included in people's property taxes; moreover, the debt stays with the house, even over the course of a sale of the house to a new set of owners. In the case of PV panels, it is obvious to potential buyers that they are up there, and this can increase the value of the house, especially when the bank takes them into account when calculating the maximum mortgage. In the case of home insulation, it is important to mandate energy audits that present efficiency ratings for buildings, information that buyers have access to. A study in the EU found that, in the presence of such mandated audits, improvements in energy efficiency led to a slight increase in the building's rental value and a much larger rise in its overall resale value.[12]

Process heating for heavy industry

The final area of concern is industrial heat use and CO_2 emissions. Here, a wealth of studies has examined the potential for efficiency improvements. The good news is that there is a growing set of cost-effective options for using electricity, as well as solar thermal collectors, for low-temperature applications, as well as cost-effective options for efficiency improvements across both low- and high-temperature applications.[13] The bad news is that for the high-temperature applications, the options for complete decarbonization do not appear to be particularly cost effective. The alternative fuels certainly do exist in the form of waste materials, biomass, and biogas; the challenge is that these are, in most cases, more expensive than coal and likely to remain that way. Most of the scientific literature on the issue of process heat considers the policy options to improve energy efficiency, but there is much less that goes into the issue of fuel switching, and so the fact is that we simply do not know what policies will work.[14] What probably makes a lot of sense is a set of evolving regulatory standards for CO_2 emissions, very similar to those proposed for power plants in the United States, and potentially connected to subsidies to enable firms to invest in equipment retrofits or make new investments that are built around alternative fuels.

We are in a somewhat similar situation with respect to cement production and the emissions that come from its basic chemical reactions. The substitute technology exists – geopolymer cement – and yet there is a strong lock-in effect in favor of portland cement. That lock-in takes two forms. First, and most importantly, we lack the experience with geopolymer cement to be able to predict its structural properties under actual environmental conditions and how those properties change over time. We need this knowledge before engineers will feel comfortable specifying it for things like bridges and skyscrapers (and before people like me will feel comfortable using those bridges and skyscrapers). What policy makers can do is promote its use – through a combination of subsidies, mandates, and public procurement – in applications where safety is not a concern (such as sidewalks), thus allowing engineers to see how it ages. At the same time, policy makers need to provide the funds to collect the actual data, as well as to increase funding for basic research on its structural integrity. Second, the fact is that a switch from portland to geopolymer cement will make most of the world's cement factories obsolete: they would simply be located next to the wrong raw materials. Geoploymer cement comes from materials that are no less common than those in portland cement, but they are found under different mountains. The transition will be a long-term one. Policy makers need to develop long-term plans for the phase-out of portland cement, thus giving the main players in the cement industry the opportunity to plan their new investments accordingly.

Agriculture, deforestation, and biofuels

This book is primarily about an energy system transition, and yet the link between biofuels production and land use means that I cannot escape the latter issue. There is no getting around the fact that the problem of slowing down and eventually halting deforestation is a hard one. Making it even harder are the twin needs of feeding a growing global population at the same time that we are boosting biofuels production to the point where they can replace fossil fuels in several key applications – most importantly, in high-temperature process heat and aviation. The fact that this is a problem we face primarily in developing countries makes it particularly challenging for two additional reasons. First, developing countries have tended to lag behind in creating effective legal frameworks and the institutions to enforce them to regulate land use. Second, rural populations in developing countries are especially poor, and so whatever solutions are to be found need to enhance their economic and social prospects rather than detract from them. Many books have been written on all of these issues, and I cannot go into them with any degree of depth. What I can do, however, is to clearly identify the precise policy problems that we face in this area, if we think about this from a transitions framework, and identify some of the options that have been shown to solve each of these.

The world's net emissions from agriculture and forestry rose consistently from 1970 to 2000, but now appear to be falling.[15] Roughly half of the emissions, in CO_2 equivalent, stem from agricultural activities, with the methane produced in cows' bellies being the largest single contributor. The other half comes from the release of carbon stocks through deforestation and other forms of land-use change, as well as some from the draining and burning of peat.[16] Essentially all the reduction of emissions in recent years has come through a slowdown of deforestation and other adverse land-use changes: CO_2 emissions grew slowly from about 5 Gt (billion tons) per year during the 1970s to 5.5 Gt per year during the 1990s, but have fallen to below 4 Gt per year during the mid-2000s.[17] In 2010, the most recent year for which good data are available, they were about 3 Gt.[18] In other words, they have fallen by almost half since their peak in the 1990s. To meet a 2°C target, it will be important to bring this number down to zero over the next several decades.

Growing and maintaining forests in the manner needed to stop climate change does not mean simply ignoring forests and obtaining no economic value from them. To ameliorate climate change, we do not need wilderness, although having large patches of wilderness is important for many other reasons, including maintaining some kinds of biodiversity and providing a spiritual service to mankind. For climate change amelioration, we need forests that are managed in such a way as to maintain large stocks of biomass in the trees and in the soil. These can be working forests. They can supply us with timber, with firewood and pellets, with the feedstocks for biofuels production, and with countless other products and services of value to people. Since its birth more than a century ago in Germany and Japan,[19] the practice of sustainable forest management has developed methods for operating forests in ways that create clear synergies with climate protection and the creation of economic value and meaningful livelihoods.

Researchers have long observed a fairly consistent pattern of land-use change over time, what have been called *forest transitions*.[20] As people go through a process of economic development, one of the things they do is cut down their forests, converting the land to agriculture and other economically productive uses. Gradually, the rate of deforestation slows and eventually reverses itself altogether. Then the forests start to grow back, although typically not completely. A good example is Vermont, where I lived for many years. English settlers in the 1700s encountered a landscape almost completely covered with forests. They set to work chopping it down and burning it, converting the land primarily into extensive pasture. By the time they reached the turning point, in the late 1800s, only about 20 percent of the original forests remained. It was at about that point that the Industrial Revolution took hold, and the economic base of the state turned from agriculture to manufacturing. People began to abandon their farms and move to commercial centers, and, with this, the forests gradually returned to the former pastures. Today, the state is at about 75 percent forest cover, a number that continues to creep upward. In fact, every developed country on the planet has already passed its turnaround point. The problem is that a lot of developing countries, together accounting for a huge amount of the world's stored carbon (not to mention enormous reservoirs of biodiversity), have not.

The general pattern of the forest transition has repeated itself over and over, and yet the details have varied remarkably. Figure 10.1 illustrates two very different hypothetical forest transitions. Two countries can start out at a similar level of forest cover, but whereas Country A chops down most of this forest before reaching the turning point (the lowest point on the curve), Country B does not. One of the interesting trends that can be observed is a gradual shift over time from countries looking like Country A to looking more like Country B. Scotland, for example, reached its turning point in the early 1700s and, when it did so, less than 3 percent of the land area was covered in forest. Costa Rica reached its turning point very recently, with about 30 percent of its land covered in forest. Moreover, the variance in the degree of forest cover at the turning point has been increasing as well. Gambia reached its turning point near the end of the last century with about 10 percent

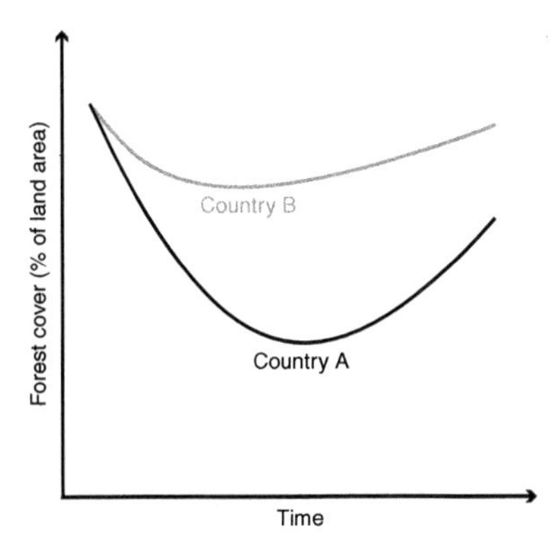

Figure 10.1. *Forest transitions.*

of the forests still standing, whereas South Korea did so with close to 60 percent.[21] The increase in variance may reflect the situation that factors associated with both socioeconomic development and governance institutions make a big difference and have become more heterogeneous over time.

Clearly, then, it is important to understand the factors that differentiate places looking like Country A from those resembling Country B. Researchers have identified three patterns that appear, today, to be causing countries to continue to deforest: poverty traps, war, and expanding markets. Countries like Ethiopia and Haiti fall into the first group: with very little financial or human capital available, and with high unemployment, the only way that rural people can survive is to try to farm more land. Farming, however, does not give them the income necessary to acquire capital, which they could use to move into more lucrative activities. Economic forces lead them to cut down forests in order to survive. Countries like Rwanda and El Salvador have fallen into the second group, those where war has made a difference. War has created an incentive to cut down valuable trees in order to raise immediate revenue and has also led to the disappearance of policing authority that would prevent this from happening. Finally, there are those countries that continue to cut forests down as part of increasing market access. Brazil, Indonesia, and Cameroon are all examples. In all these countries, money is to be made by converting forest into agriculture, whether it be for cattle grazing or palm oil production. The actors doing so are not poor at all. Governments, meanwhile, have often promoted the practice as a way of generating export revenues.[22]

The problem with biofuels production is that it contributes to this third pattern. Two chapters ago, I cited a number of reputable studies suggesting with high confidence that the potential to produce biofuel feedstocks exceeds the likely demand in the key sectors where they would be needed, as long as those sectors don't comprise the entire energy system. As a general rule, these studies take as a starting point that the production of biofuels would not take place on land currently standing as old-growth forests or needing to be devoted to agriculture, except to the extent that people use the waste products coming from this land. But there is no getting around the fact that biofuels production creates new market opportunities associated with growing stuff. Even if the biofuels themselves don't come from land converted from old-growth forest, they contribute to rising prices for agricultural commodities, making it more likely that people will chop down forests to grow these.

The same researchers have also identified two factors that lead to afforestation.[23] The first of these, seen in the Vermont example, is labor scarcity. The process of industrialization pulls people off the land into higher paying jobs, and the forests begin to grow back. The second factor is rising prices associated with forest products themselves, such as wood. One might think that this would lead to more rapid deforestation to get that wood, but in fact it does not. It generally leads to more sustainable forestry. This is especially the case where there are other ecosystem services at play that the government has an interest in protecting, such as drinking water provision, to which standing forests can contribute.

Thinking about forestry as a transition process begins to clarify several distinct issues that are important for the formulation of policy. First, we need to identify things that the international community can do to ameliorate the factors leading to deforestation while accelerating the factors leading to afforestation. Doing these things, whatever they are, would accelerate forest transitions, getting them to take place with a larger proportion of forests still standing. Second, we need to identify steps that national and subnational governments can take, perhaps with support from the international community, that would help to resolve conflicts of interest. It is easy to imagine national governments making the decision to halt deforestation and yet have local-level actors continuing to engage in the practice. Third, it is important to resolve the question of whether all these things together would be enough. Can one imagine the forces for afforestation outweighing those for deforestation soon enough, in the place where deforestation is currently most severe, to leave sufficient quantities of forest still standing to achieve our climate objectives? If the answer is yes, then the first two sets of measures – promoting a rapid turnaround and enhancing governance and enforcement – would be enough. If the answer is no, then we would need to augment these with something else, like a global treaty built around the principle of forests as contributing to a global public good.

Many of the policy options that have gained the most attention so far are associated with this final need. In other words, they look to solve the forest problem more or less from a static public goods perspective, echoing the basic ideas that I examined in Chapters 4 and 5 on market solutions and global treaties. Perhaps the best example of this is the idea of *payments for ecosystem services* (PES), embodied in a much broader governance framework known as *REDD* (Reduced Emissions from Deforestation and forest Degradation).[24] REDD has been agreed upon in principle within the United Nations Framework Convention on Climate Change (UNFCC), but negotiators are continuing to work out the final details of how it would operate. Many of the details turn out to be complicated and difficult to resolve.

The basic idea behind PES, and the original motivation for REDD, is that landowners have an economic incentive to convert their forests into agriculture, thus releasing huge stocks of carbon into the air. If you look at a particular piece of land and divide the actual amount of money to be earned from its conversion by the amount of carbon released, then you come up with some dollar value per ton of carbon. Researchers have assessed the distribution of this dollar value using different theory-driven models associated both with the economic incentives and with the carbon stocks released. Averaging across models, they have shown that, for most of the world's land facing the threat of deforestation, the economic incentive translates into less than about $20 per ton of CO_2, indeed perhaps less than $10 per ton of CO_2.[25] That implies that if you could give the owners of the land that amount of money *not* to deforest, they would gladly accept the deal. Meanwhile, this is less than the market price for CO_2 emissions that people see as necessary to stop climate changing using a carbon tax or cap-and-trade system, such as the European Emissions Trading System (ETS). This creates an apparent win-win opportunity for coming up with the money needed to pay the landowners not to cut their trees down; the money could come not from national governments, but rather from firms in industrialized countries that face a carbon cap and for whom it is expensive to reduce their own emissions. They could go on emitting, at least for a while, but pay the developing country landowners to compensate by not cutting down their trees. Model results have suggested that including PES in a carbon market could reduce the cost of achieving ambitious climate targets by up to 25 percent.[26]

The idea of PES seems great in theory, and a lot of people are enthusiastic about it. It also has a whole list of potential problems. How do you make sure that the payments really make a difference – namely, that the trees would have been cut down if it weren't for the payment? How do you make sure that by not cutting trees down in one place, you aren't influencing agricultural markets in such a way that it makes it more likely they will be cut down somewhere else? How do you create a system that is socially equitable and doesn't lead to poor people being pushed off the land so that others can collect money? Finally, how do you stop? If the only thing that is keeping people from deforesting is the payments, then once the payments stop, the trees will disappear. Ultimately, you are just buying time.

Buying time is how one should view the whole idea of PES. It is no substitute for policies that can eliminate the underlying desire to deforest in the first place by changing the drivers for deforestation and afforestation, as well as the ability of states to regulate land use with effective enforcement. Many other people have recognized this as well, which is why REDD now encompasses a lot more aspects than simply PES. To participate in the REDD program, for example, developing countries need to develop an overall forest management plan; doing so qualifies them for a whole set of international support elements. I believe that more research is needed into whether PES can actually be synergistic with this longer term perspective. As far as I can tell, nobody knows. If REDD does move forward, all indications are that the mechanism for raising finance for PES will not be carbon markets – since these are not able to maintain a high enough price – but rather direct government support through UNFCCC and bilateral financing mechanisms. That is a good thing. Using carbon markets as the means to raise finance for PES would mean delaying the energy system transition in developed countries. If that happens, then the harm could well outweigh any potential good.

So the idea of PES – if political leaders choose to pursue it – needs to be viewed as an add-on on top of a more fundamental set of policies, policies that can alter the underlying drivers of deforestation and afforestation in the first place. When you start to consider what these policies are, you discover that many of them have other underlying justifications. That is a good thing: they could potentially appeal to a wide variety of interests and be politically attractive. Many of the policies have come to be included in REDD as it has expanded from its initial focus purely on PES. Here is a partial list of what these policies are or could be:

- Continuing aggressive research to sustainably boost agricultural yields. Maintaining the increase in agricultural yields can have the net result of lowering the prices of agricultural commodities, thus reducing the financial incentive to convert forests to agricultural land. Unfortunately, in some locations, it can also increase the economic returns of converting forestland to highly productive agriculture. That is why it is not the only solution.
- Industrial growth and urban development. Historically, one of the main factors that have led to afforestation has been a contraction of the rural labor supply, so that people abandon farmland to take higher paying jobs in the cities. The process is ongoing in Europe and under way in many Asian countries.
- Improved access to secondary school education for girls, especially in rural areas. A great deal of research suggests that girls' education has a tremendous effect on rural population growth: the higher the proportion of girls who complete secondary school education, the lower the fertility rate.[27] The same evidence also suggests that this can create pathways out of rural poverty traps.
- Assessment of the locally and regionally derived ecosystem services from forestland. There is a tremendous amount of evidence, mainly from developed countries, that the local benefits of having forests are substantial, typically overshadowing the economic benefits from deforestation.[28] There is a relative dearth of assessment of such benefits in developing countries. Were this information widely available, it would likely accelerate the pace at which national governments decide that it is in their own interests to halt deforestation.
- Elimination of policies in other economic sectors that create an inefficient incentive to destroy forests. Examples include incentives to deforest land to create new mines or to expand hydropower production. The government ministries responsible for these policies need to consider the economic value that forests provide.
- Reliance on sustainably harvested woody biomass as a biofuel feedstock. There is clear evidence that as the value of forest

products such as wood increase, the area of land devoted to sustainable forest practices does as well, thus leading to net afforestation. A number of chemical processes can create biofuels out of biomass, and among the more effective rely on woody biomass as a feedstock. As second-generation biofuels develop to serve aviation and industry, there is room for standards requiring a minimum content from sustainably harvested wood.

- Support for land-use monitoring and enforcement programs in developing countries. In many countries, there exists a huge amount of illegal deforestation, something that may accelerate as states decide that it is in their interests to protect their forests, but private landowners continue to face an incentive to convert land to agriculture.

Will all of these policies be enough? We don't know. As I have already suggested, an important piece of good news is that the pace of deforestation has already started to slow down substantially.[29] If the reduction in the pace of deforestation observed over the past twenty years continues, it would be enough to solve this problem. But the truth is that nobody really knows whether it will continue, partly because nobody really knows why deforestation has slowed in the first place. It could be because the policies I have just suggested, most of which are already in place, are working. Much of the policy-oriented research on deforestation has focused on it from a more static point of view, looking at policy alternatives such as payments for ecosystem services. It may be that we need these to keep forests standing until developing countries and their citizens reach the decision that it is in their best interests to leave them there. But it would be a mistake to focus only this, ignoring those policies that could lead to a faster transition toward net afforestation. Fortunately, the REDD policy community has been moving steadily in this direction.

Staying the course with technology support

The past few years have seen a string of positive developments for climate change, if you look for them. The main positive storyline concerns the fact that two renewable energy technologies, wind and solar, continued a brisk rate of exponential growth, often exceeding 25 percent per year. They did this even as they emerged from tiny niche players to be noticeable features on the landscape. The growth survived a spike in commodity prices in 2007 and then a global economic crisis starting in 2008. Prices accordingly tumbled, and, for the first time, there emerged meaningful cases of investment into renewables even when not supported by subsidies. A string of scientific and engineering studies have demonstrated how they could form the heart of a reliable energy system. Other technologies began to take shape as well. It may not strike you as important, but in 2007 Apple released the first iPhone and with it the lithium ion battery revolution. This jumped to the electric car in 2008 with the release of the Tesla Roadster. Since then, battery performance has continued to improve, and battery prices have fallen. Pure electric and plug-in hybrid cars have hit the market, and although their sales may not have matched the early hype, they have been growing in popularity at an astonishing rate. Just this year, electricity has made the jump into the power train of large trucks. In the area of space heating, passive building design came of age. By 2010, more than 25,000 certified passive building have been built in Europe, and the concept is beginning to make the jump to North America.[30] Across both continents, there have been more and more zero energy buildings of all sizes built, with a growing number of architects and engineers capable of designing them.

If there is one core message to pull out of this chapter and the preceding one, it is that the policies we need to solve climate change are those that build on things we are doing already, things that have led to these success stories already and can do so at an increasingly faster pace in the future. With one exception – cement – we do not need a dramatic turnaround anywhere. True, our emissions have been rising. But several underlying trends, ones that can lead to rapid changes in investment behavior once particular thresholds are passed, are quite good. Most important are the falling technology costs in several different areas: renewable power production, electro-mobility, and building efficiency. The fact that new investment in wind and solar has exceeded new investment in fossil fuel electric generation, for example, is not an insignificant piece of trivia. It is a sign that things are changing in a way that can easily accelerate. The slowing pace of deforestation is also important, first because of the emissions that it causes directly, and second because it makes it possible to enhance our production of biofuels.

All these trends have resulted from a wide set of deliberate government actions and policy interventions in society. The last thing we want to do is abandon these, perhaps because we believe they aren't working or perhaps because we believe that carbon markets could work better. No. It may sound crazy, but what we need is to keep up the good work, continuing to improve the things that we are already doing.

We need, for now, to continue with feed-in tariffs (FITs) and other technology support for PV, wind, and CSP and augment this with policy attention to new grid connections and power storage options. We need continued R&D on batteries, both stationary and mobile. We need to continue a variety of measures to incentivize electro-mobility. Continue to revise our building codes to improve efficiency and require heat pumps and solar heating. Continue to expand the networks of district heating, and expand the incentives to retrofit existing buildings so that they don't rely on gas and oil for their heat. Continue the policies that make it the case in developing countries, as it has long been the case in industrialized ones, that people sustainably manage their forest resources. In none of these cases do we need radical new policies or radically expanded policies, as if there were a war going on. We do not need a war on climate change, and that is a good thing. Being a Buddhist, I would suggest that it is wrong even to pretend to be at war. What we need are incremental improvements to existing policy frameworks. Many of these incremental changes will be politically and technically challenging. All of them are possible.

Again, the only thing where we to do something radically new is with cement. We need to move geopolymer cement out of the laboratory and into practice. That will be a long process, and it is time to get started, most importantly with R&D. Ultimately, this is a story that I can see unfolding similar to the one for chlorofluorocarbons (CFCs) and ozone depletion, with an international treaty coordinating the phase-out of portland cement. I think it would be unwise to start the treaty process too soon. We are far more likely to get a meaningful treaty once it is clear that the technological substitute for portland cement exists and is feasible.

From industrialized to developing countries

Across all these different issue domains, with the exception of land use and biofuels, an important issue is the sequence of action across the range of national perspectives, from wealthy to developing countries. As I have already discussed in Chapters 4 and 5, the effectiveness and efficiency of carbon markets depend on their degree of universality, and yet that universality has been a major political stumbling block. Countries like the United States have balked at agreeing to a stringent treaty unless developing countries commit to doing the same, and yet these developing countries see themselves as climate victims and are unwilling to commit to costly actions. In the approach I am suggesting here, by contrast, the starting point is action to bring down the cost of decarbonized technologies, followed by institutional and infrastructure development to integrate large amounts of such energy into the overall system. As they become competitive, the local benefits of decarbonized technologies begin to exceed their costs, and developing countries have every reason to want to use them. But enabling that behavior may mean addressing another set of concerns, different from those I have mentioned so far.

In many developing countries, the central policy challenges that politicians face is creating opportunities for formal sector employment. People who are employed in the formal sector can lead more predictable lives, hold a bank account, borrow money to buy a house, and generally make positive investments in their and their children's future. They also pay taxes. There have been many promises that so-called green development can create new employment opportunities in the area of climate mitigation. But here, the devil is in the details. Nadejda Komendantova, whom I have mentioned before, has been looking into this for several years, in the case of solar development in North Africa. What she found is that the employment created from simply putting up CSP and PV generating stations is extremely modest, with most of the direct jobs being temporary ones in the construction sector.[31] This is a good thing because it suggests that renewable energy could be relatively inexpensive, something not possible if it required huge numbers of people to operate it. But it also means that expanding renewable energy would not do very much to solve the employment problem. However, the jobs created expand by a factor of five if the various components and parts of the solar energy systems are manufactured locally rather than imported. Some of these jobs required highly skilled labor, but most do not, and indeed fit the qualifications of local workers: things like fabricating metal frames for the solar panels to mount on. In the countries that we examined, there currently are a number of solar energy projects that are supported though bilateral support (such as from the German government) or multilateral development banks (such as the World Bank's Clean Technology Fund). What our results suggest is that these projects ought to include a requirement for local manufacture, even if it means decreasing the overall size of the solar project being financed. As local workers learn how to manufacture components, they can do so increasingly inexpensively and be in a competitive position once solar energy infrastructure really begins to gain momentum.

The second area where we have observed a strong need is in the mechanisms to manage the risks that project developers face. Typically, these risks are far higher in developing countries than in wealthy ones, and this has several pernicious effects. First, it scares off a lot of would-be investors. Second, it leads to inflated financing costs because project developers need to pay their investors a high rate of return. Third, it forces development agencies like the World Bank to bear a higher proportion of the financing, thus reducing the degree to which their finance can leverage capital investment from other sources. The source of risk that is the greatest in the countries we studied, in the Middle East and North Africa (MENA) region, was a poorly functioning and corrupt bureaucracy.[32] Because of this, project developers feared not being able to finish a project that they started or facing substantial delays in obtaining the needed permits simply because they hadn't paid the right attention (a euphemism for the right money) to the right people. The effect of this can be to multiply, by a factor of at least five, the public subsidies currently required to stimulate these projects.[33] Fortunately, there are sets of policies that can address these issues, well catalogued by organizations such as Transparency International, that are starting to be applied with apparent success.[34]

Closely related to this issue is the overall strength of the financial sector in developing countries. Being able to efficiently access sources of finance is incredibly important for renewable energy technologies precisely because these technologies require far more up-front investment, after which comes many years of low-cost operation with no fuel bills to pay. Most developing countries, however, have far less well-developed financial sectors than do wealthy countries, and this leads to far higher finance-associated costs, even after correcting for the perceived riskiness of projects. We looked into this issue for the case of PV. We found that the cost differential caused by the easy availability of finance was the single biggest factor determining the overall levelized cost of electricity (LCOE) of PV, far more important, for example, than the available sunshine.[35] Other researchers have reached similar conclusions.[36] Several years ago, I was consulting for DFID, the British Department for International Development, and the issue of financial service industry support came up as one of their strategies for helping countries reconstruct following natural disasters. It is an important strategy for climate mitigation as well.

A fourth area for policy development is with respect to environmental and health regulation. As I have already described, there are huge local benefits to be enjoyed from replacing fossil fuel energy systems with renewable ones and gasoline and diesel cars, trucks, and scooters with electrical ones. The most significant of these benefits come via reduced local air pollution,[37] although the protection of fresh water resources is also important. Local air pollution has real costs, both human and economic. People become sick and die on average earlier. Workers are less productive. The health care system becomes more expensive. Regulations to avoid these costs can promote human and economic development. They also accelerate the point at which a renewables-based power system and an electricity-based transportation system becomes economically attractive for private sector investors. Industrialized countries have been using the health consequences of pollution as a basis for regulating it for a few decades now, and there is evidence that developing countries are starting to do the same, with China being the most important.[38] This is a trend that it would be well to accelerate. Support for this through bilateral development assistance, as well as from relevant UN-associated bodies such as the World Health Organization, can play a critical role.

Key policy principles

Across all the policies I have described, two points deserve emphasis. First, the scale at which policies need to be implemented matches those where effective governance institutions already exist. Most of the policies are suited for countries to implement. Some are suited for cities. A few, especially concerning the power grid, may require action at the scale of the EU, within its already established areas of competency and authority. One case where something global may be required is in the setting of industrial standards (such as for electric car batteries); this is a task for the International Standards Organization, not necessarily the UN. Another is in promoting sustainable cement, and, indeed, this is the only policy area that would potentially require a binding obligation at the global level. Everything else that needs to happen to solve climate change is something that individual countries, regions, and cities already have an incentive to do, and, in fact, are already working on. Second, the policies are all politically feasible. The reason one can be confident of this is that they already exist in so many places. There is little I have proposed that is particularly new or radical; most of them represent incremental improvements to policies or planning processes already under way somewhere. For that reason as well, there is an evidence base from which to work. None of the policies is proposed simply because theory suggests that it would be effective. Rather, we can be confident that these policies would be effective because we have already observed that to be the case. Even for cement, perhaps the toughest nut to crack, it is possible to imagine a global protocol functioning similarly in scope to the Montreal Protocol.

Of course, theory doesn't fall out altogether. Three chapters ago, I suggested a set of theoretical principles that ought to guide our thinking about the policies we need. The basis of that theory was observation of how change occurs, and this took shape in several distinct ways: overcoming lock-in, overcoming behavioral biases toward sticking with the status quo, and instituting policies that allow broad-based support to build over time rather than requiring it as the precondition. Let us briefly look at how the set of policy options reflects these principles.

The issue of lock-in describes a feature of evolving technological systems. The more prevalent a technology is, the greater is the incentive for others to adopt it. If you want to get a new technology established, then it either has to be something that doesn't compete against technologies that already are common, or else you need to support that new technology very heavily at first, then back off on support as it gains market share. In fact, this is what technology-specific support, such as a subsidy or a FIT, precisely does, in stark contrast to carbon markets. For each such technology, the government's first task is to figure out how much money needs to be thrown at it in order to overcome its unique competitive disadvantage. As the support proceeds, the disadvantage shrinks, as can the level of support. Issues having to do with lock-in also form the basis behind an attention to removing the barriers to diffusion. The need to remove these barriers recognizes that a set of institutions and networks have grown up around the technologies we have been using that facilitate them working at large scale. An example is the network of gas stations or the design of the current electricity spot market. We need to grow a similar set of institutions and networks for the new technologies.

The issue of risk, and people's bias toward the status quo more generally, enters the picture in two respects. At a high level, it is in the overall strategy of establishing technological systems in wealthy countries first, before attempting such a change in developing countries. There is every reason to believe that the net economic impact of an energy system on a country undertaking it will be something close to nothing: perhaps a bit costly at first, but then growth-enhancing in the longer run. But this is not known for certain, and we can be confident that we will create some messes along the way. That is why it makes sense for countries able to take such risks be the first places for the transition to happen, as has been the case so far. At the level of particular policy instruments, the issue of risk and other behavioral biases enters as well. FITs have been so successful because they remove risk for the individual investor, thus facilitating the acquisition of finance. When FITs become obsolete because technologies like wind and PV can compete on the open market and no longer need a subsidy, there will still be a need to address the risk issue, such as through community financing schemes. In developing countries, dealing with investor risk is one of the prime challenges to overcome. This will require policies addressing how bureaucracies function, especially if corruption is a concern. It also requires growing the financial sector in general.

The final theoretical aspect has to do with diversity in worldviews and objectives, something described by cultural theory. This cuts across the policies suggested here, and perhaps most obviously in the area of land use, in the following way: every one of the policies for climate change is a policy that also addresses at least one other societal problem. Indeed, it is this feature of the policies that allows them to be implemented at (predominantly) the national level, without the need for a strong global governance regime. If climate change were solely a tragedy of the global commons, then such a global governance regime would be necessary. The nice fact is that addressing each of the separate challenges of eliminating CO_2 emissions happens to align with other societal objectives that operate at a lower spatial scale. The one exception to this is transforming the cement industry. In the area of land use, for example, a critical challenge is to reduce the pace of deforestation and ultimately eliminate the net loss of forest altogether. The fact is that forests provide a number of ecosystem services. Among these is carbon storage, and this offers global benefits. But other services provide local benefits, and it is clear that these are important to enough people to provide a sufficient reason to protect the forests, if those people are operating in a country with well-developed governance institutions. Protecting the forests means empowering those people through institution building. That institution building, at the same time, can appeal to a wider group of stakeholders for a wider set of reasons. It is happening already, but it has room to accelerate. Just about everything we are doing now has room to accelerate.

Carrots versus sticks

I believe that the most important debate occurring right now in climate policy is whether the best approach is one built around supporting new technologies or around penalizing the old ones. In this chapter and the previous one, I have argued for an approach built around supporting new technologies. This support needs to come in several ways: dealing with the separate problems of stimulating innovation and cost reduction, stimulating the establishment of the needed infrastructure and institutions, and finally stimulating the spread of renewables-based energy systems to developing countries. But, as I documented in Chapter 4, the far more prevalent belief is that our primary policy instruments need to take the form of a stick, rather than a carrot. Essentially, that belief boils down to three main arguments.

The first argument people make for using a stick instead of a carrot is based on theory, namely the body of economic theory that I summarized in Chapter 4. In theory, policies built around national caps (and carbon prices to achieve those caps) will get us the most efficient set of emissions reductions. In Chapter 4, I countered this argument in a number of ways. The dynamic nature of the energy transition problem, I have argued, means that solving it with carrots winds up being at least as efficient as with sticks. And even were it not as efficient, it wouldn't matter because the carrots are more likely to work. An inefficient policy that gets us decarbonized is better than an efficient one that fails to do so. Simply put: carrots get you creative innovation; sticks don't.

The second argument people make for using a stick instead of a carrot is a moral one and has to do with the "polluter pays" principle. The people who are destroying the world ought to pay the price of having done so. Market-based instruments begin to accomplish this immediately. But there are three good counterarguments. The first is that it doesn't matter who among the rich pays for the cost of change. The important thing is that change happens so that the burden to future generations is as little as possible. There is reason to believe that carrots can make this happen. The second counterargument is that, in fact, there is not enough blame to warrant punishment. The people who brought us fossil fuels did not do so to harm us, but rather to create a better quality of life and jobs. They may have become wealthy in the process, but one should not hold that against them, and they are not villains. The third counterargument is that, ultimately, the carrot approach will be the fastest way to make renewable energy sources competitive, which is what it will take to put fossil fuels out of business. That winds up being a pretty big punishment, even if it isn't coming from the tax collector.

The third argument people make for using a stick instead of a carrot is that even if supporting renewables leads these alternatives to become more affordable and more reliable, this will never be enough, on its own, to outcompete fossil fuels. As renewables-based energy systems become better, they argue, the fossil fuel industry will not sit idly by. It will find its own ways to bring down costs and to maintain its advantage. One can point to the shale gas revolution in the United States as evidence of this ability to do so. If more such revolutions were to happen, and fossil fuels were to maintain their advantage, then it would lead to two unfortunate outcomes. The first is that countries going down the path of supporting renewable energy through subsidies would have to keep doing so rather than being able to phase the subsidies out once cost parity is reached. This would be unfortunate, although to me it wouldn't be catastrophically so. After all, we continue to subsidize fossil fuels and many other things on a more or less permanent basis. It isn't ideal, but the cost of those subsidies isn't bringing the world to its knees. The second outcome, by contrast, could be catastrophic. It could mean that renewable energy systems never spread beyond those countries that are willing to support them with subsidies. If that happens, it means we will not solve climate change.

Of the three arguments, this is the one that by far worries me the most. The fact is, we don't really know the potential for cost reductions in the fossil fuel sector. Forecasts such as those coming from the International Energy Agency (IEA) or the U.S. Department of Energy generally suggest that costs for fossil fuels and the electricity derived from them will rise. Moreover, if you apply the idea of learning curves – that costs fall by a particular percentage every time total capacity doubles – then the potential for fossil fuel cost reductions is slight. Fossil fuels dominate the energy sector, and so doubling installed capacity would mean almost doubling our total energy use. That would take a while, and even so would only bring modest cost reductions. But forecasts like those of the IEA have been wrong in the past and could be wrong in the future. Moreover, the learning curve idea is very simplistic and may not even apply. Within fossil fuels as a whole, there may be particular new technologies that enable industry-wide cost reductions. That is what happened with shale gas. Ultimately, there is a very real possibility that the fossil fuels industry will be able to match the future cost reductions in renewables and maintain a slight cost advantage. That, the argument for carbon taxes rings out, is how everything could go wrong.

And yet it need not go wrong, for two reasons. The first reason has to do with the co-benefits of renewable energy. Renewable energy doesn't just protect the climate. Relative to fossil fuels, and coal combustion in particular, it also protects local air quality and, in some cases, local water resources. The more a developing country industrializes, the greater these effects become. The wealthier the people in these countries become as a result of that industrialization, the more they care about pollution. Already in China, policy makers are starting to scale back plans for the expansion of coal power stations for these two reasons.[39] If you add in the local and immediate costs associated with burning coal and oil, the likelihood of their maintaining a cost advantage over renewables is much less, perhaps even zero. For co-benefits to play a role requires countries to have strong policies in place protecting their domestic environmental quality. The number of such countries is growing. It would be a mistake to ignore that. It would also be a mistake to sit idly by and assume that. There is a need to promote the full accounting of local environmental effects of energy systems and to enact energy policies that take these into account.

The second reason why a continued cost advantage of fossil fuels may be less than fatal to climate mitigation lies in the potential for prohibition. Governments are in the business of prohibiting all sorts of behavior that harm the public good. We could put an end to climate change this very second if governments around the world were to prohibit the burning of fossil fuels and the emission of CO_2, just as they have prohibited the killing of whales and the trade in endangered species. But they are not prohibiting the use of fossil fuels because the economic and human consequences for society has been seen as catastrophic. Until very recently, the alternatives to fossil fuels were several times more expensive. We still do not have the infrastructure and the institutions in place to support a decarbonized energy system or even to know exactly what they would look like. It is important to recognize that even if fossil fuels do maintain an advantage over renewables, that advantage will grow smaller over time as both sets of technologies converge toward very low cost and very high reliability. At some point, it will be the case that the immediate consequences of phasing in a prohibition on fossil fuel combustion will be low enough to make this politically possible. That may be the time that an international treaty, built around prohibition, is both feasible and necessary. It may also be a time when a modest carbon tax or price instrument, perhaps as low as $10 or $20 per ton, could finally close the deal for an energy system transition. This would not be a bad outcome.

How can we most quickly get to the point where renewables beat fossil fuels outright, where they beat them once one takes co-benefits into account, or where they come close enough to beating them to make an absolute prohibition politically feasible or a politically feasible carbon tax absolutely effective? That is the relevant question. The answer lies in the policies I described in the previous two chapters.

11

Pulling it all together

Climate change is an incredibly important and urgent problem, one that could potentially destroy civilization as we know it unless we quickly curb our emissions of greenhouse gases (GHGs). For decades now, experts have been telling us that we need to make major changes to our lifestyles. At the same time, they have suggested that our governments need to agree to binding emissions cuts and, to achieve those cuts, need to implement market instruments like carbon taxes and tradable permits. These ideas are all good in the abstract. In the real world, however, they have gone nowhere and will continue to go nowhere. We should abandon them for now and approach the problem from a different angle. Instead of focusing on the immediate need to reduce emissions, we can view the core challenge as accelerating a set of societal and technical transitions that are already taking place. For that, we have a broad portfolio of policy instruments, most of which operate at the national scale and have already proved themselves to be effective. To solve climate change, we need to continue with these policy instruments, further enhancing their scope and geographic coverage.

That was it, the short synopsis of this book. The last ten chapters have provided the basis to support it. But those chapters have contained many different strands of ideas. In this final chapter, I pull them all together into a simple, clear storyline.

The big picture

The traditional story of climate change is that it is a natural and unintended consequence of modern industrial society operating in an unregulated state. To stop it, we need to enter into a more or less permanent state of intense regulation of one sort or another. This is what is required to keep our existing stocks of fossil fuels in the ground, rather than burning them to support additional consumption. The new story of climate change, the one I offer here, is that climate change is a natural and unintended consequence of one stage of society's becoming modern, but that stage is temporary. The next stage is one in which our energy comes from sources other than fossil fuels, and where we have stopped cutting down forests and started to let them grow back again. To stop climate change, we need to accelerate the transition so that it can happen as soon and as quickly as possible. Doing so does not require any fundamental restructuring of our economic, political, or industrial systems. Unlike a Marxist idea of a transition into some sort of utopia, moreover, it is important to recognize that the next stage, too, will be flawed and will be temporary; it will evolve over time into something else, impossible to foresee at this time.

Here is what the next stage looks like. We will generate our electricity from renewable resources, mainly Sun and wind, but also from falling water, biomass, and geothermal heat. To ensure that these are reliable, we will make use of a great deal of electricity storage capacity and have power markets operating at a more regional scale. All of this will save us money relative to our current system, while also reducing local air pollution and water consumption, although it may require us to devote somewhat more land to energy production than we do today. At the same time as we get our electricity from renewable sources, we will make use of electricity for as many applications as possible, including heating, industrial processes, and transportation. There are a few places where, for technical reasons, electricity will remain inappropriate, most importantly high-temperature heating and aviation. For these, we will produce the necessary biofuels. Biofuel production will not come at the expense of needed food production and forest enhancement because we will have improved the productivity and efficiency of the agricultural system and enhanced the societal incentives for people in developing countries to manage their forests sustainably, mirroring the incentives that already exist in all industrialized countries.

Part of the decarbonization puzzle lies outside the energy sector and will require long-term attention to cement production and its use within the construction sector and to the incentives for sustainable agriculture and forestry. Most of the puzzle, however, lies within the energy sector. There, we can break down the transition process into three basic steps that can come more or less in an overlapping sequence.

The first step is to bring a limited set of technologies to the point of being competitive, primarily in terms of bringing down their cost to be below those of fossil fuels. This results from R&D on immature technologies and from scaling up more mature technologies in the market, thus giving rise to learning-by-doing and economies of scale. The two most important sets of technologies are wind and solar because these can be scaled up enough to ultimately constitute the backbone of our energy system. Onshore wind and photovoltaics (PV) are already competitive in many places and applications, but there is still work to do with them and with offshore wind and concentrated solar power (CSP). The next set of technologies is that associated with electricity storage, both chemical and electrochemical, for both stationary applications to enhance grid reliability and mobile applications to provide power to vehicles. The final set of technologies is that associated with the production of biofuels, making it possible to generate these from increasingly more sustainable crops, again at costs that are below those of fossil fuels.

The second step is to enable these competitive technologies to scale up to the point where they can reliably, efficiently, and sustainably satisfy a country or region's entire energy demand. This is most difficult in the case of electricity, where the challenges become greater as the total share of renewables in the power system rises. At a technical level, it will require the construction of new power lines, the widespread adoption of smart grids, and the installation of storage capacity. At an institutional level, it will require the integration of national power markets, improvements to the merit order market clearing mechanism, the streamlining of permitting processes to ensure that they occur more quickly while responding to the needs of all stakeholders, and the enabling of public and private finance for new energy-sector investment.

Both of the first two steps involve substantial amounts of public investment and entail a good degree of risk taking on the part of states, firms, and private individuals. For these two reasons, it is reasonable to expect them to proceed largely in wealthy countries. Indeed, to a large extent, this is what we have observed so far, although there are some notable exceptions, such as China, India, South Africa, and Brazil. The third step, then, is to enable other developing countries, all the way to the very poorest, to benefit from these innovations. By the time that renewable energy systems are the least expensive and have been demonstrated to be sufficiently reliable, the goals of decarbonization and development will be aligned. Wealthy countries can help poorer countries to reap the benefits of this alignment, for everybody's good, in several distinct ways. These include contributing to bilateral and multilateral finance mechanisms, funding programs to educate workers and develop capacity in the production of energy systems, promoting transparency in government operations, and helping to establish robust financial sectors that are able to raise and distribute capital while managing risk.

The changes in actual emissions will not be linear throughout this process. During the first step, we will not notice any reduction in emissions; indeed, they will continue to rise. That trend will reverse during the second step and accelerate into the third. By the time the third step is over, we will be at zero net emissions. It is hard to say how long this will all take, but my best guess is that the first step should take about thirty to forty years to complete, based on what we have observed with respect to the technological learning rates. There is reason to believe that the second step would take another twenty to thirty years to complete, assuming that many of the changes had already started during the first step and are limited in their pace by the turnover of infrastructure.[1] The third step would take another ten to twenty years to complete, based on the same two assumptions.

There may need to be a fourth step, which is to lock us into the new, decarbonized technologies by passing regulations that gradually prohibit the combustion of fossil fuels. If we need such a step, it will occur when the decarbonized energy system is functioning smoothly in a given economic sector. Whether we need it all will depend on how large the competitive advantages of the decarbonized technologies turn out to be, compared to fossil fuels.

Fortunately, we aren't starting now, but did so in the late 1980s, when Denmark was getting busy with wind and Germany just starting to look at renewables. That means we are already twenty-five years into the process. We have a lot to do, but are on track to complete decarbonization in industrialized countries by mid-century and globally a couple decades after. That is roughly what it would take to achieve a 2°C temperature target with a high degree of probability. We could vastly improve our chances by taking active steps to remove CO_2 from the air. At a policy level, this would probably require governments to directly fund the process,[2] but it is really too early to say anything specific.

A summary of immediate policy needs

So, we have a broad storyline covering about seventy-five years, of which we are about a third of the way through. Most of the work so far has been on the first step, with a little on the second, and the very beginnings of the third. Given where we are, what are the high-priority items for right now and the next few years? Here is a list of fifteen, culled from the text of Chapters 9 and 10:

1. We need to promote a limited set of renewable energy technologies, most importantly wind and solar, in a manner that reduces risk and can drive a further round of cost reductions. The most effective and efficient way to do so is with feed-in tariffs (FITs), modeled after those in Germany and perhaps incorporating some new design elements. Other policies also work, including renewable portfolio standards (RPSs) and tradable green certificates, as well as direct government subsidies, often taking the form of tax rebates.
2. We need to promote the construction of smart grids, super grids, and electrical storage. For this, we need a strong government hand creating the financial mechanism to get stuff built, most likely passing the financing costs along to consumers, as in the Texas model. These costs will be more than outweighed by the cost savings associated with a diminished need for excess capacity. With respect to storage technologies, we also need to continue aggressive public funding for R&D.
3. We need to streamline the regulatory practices for siting new power infrastructure – primarily wind farms and power lines – to empower local residents and reduce a reactionary response against project proposals while also accelerating the pace at which permitting processes function. For the latter, we simply need to put dispute processes on a fast track, as has been done in many jurisdictions for corporate law disputes. For the former, we need to revise regulations to ensure that all stakeholders have the opportunity to participate in the process of infrastructure planning from the very beginning, at the stage where the needs for new infrastructure are identified. In that planning process, we need to make use of the most recent generation of energy systems models, which can predict the full range of local environmental costs and benefits associated with a given piece of infrastructure, including the portfolio of generation options that new power lines enable. Finally, we need to promote local ownership for power generation and distribution.
4. In Europe and its neighboring regions, we need greater international cooperation on infrastructure permitting processes,

power market integration, and policy support for key renewable technologies like CSP. This will require continuing the process of shifting formal competence in the area of power system planning from European member states to the European Commission in Brussels, at the same time as Brussels engages in a set of proactive bilateral negotiations with third countries.

5. We need to reform the structure of wholesale power markets so that they can guarantee adequate profits to power producers in the context of an energy system where high investment costs and low operating costs are universal. One option that could succeed is to move toward a more regulated market as we had in the past, where government regulators set the prices that different power providers received and imposed obligations in terms of the needs for reserve capacity. There may also be successful options that preserve the key benefits of market liberalization, and we need economic research to develop these and test their effectiveness and efficiency.
6. We need to give electric vehicles, light-duty and heavy-duty, a solid but temporary push. With respect to light-duty vehicles, we can use a portfolio of incentives, including tax breaks and subsidies, corporate efficiency or emissions standards, and access to high-speed commuter lanes and free charging stations. These have proved successful in the past to promote hybrid vehicles and are proving successful in Norway for electric and plug-in hybrid vehicles. For heavy-duty vehicles, we have no experience, but it is likely that a similar set of financial and road incentives could also work. Especially in the area of heavy-duty vehicles, we need a similar set of incentives to promote hydrogen fuel cells.
7. We need building codes that mandate energy efficiency and electricity-driven heating and cooling systems in new construction and aggressive programs to promote and finance the retrofit of the existing building stock. For the former, several jurisdictions are well into the planning phase, with a launch date of 2020. For the latter, the European Union has issued a directive requiring the development of national action plans. We also have examples of community-led programs to decarbonize the existing building stock in a cost-effective and integrated manner.
8. We need to accelerate the deployment of drop-in biofuels for airplanes while developing standards to mandate their use. A

number of airlines have already engaged in testing efforts in a voluntary manner, and the results so far have been encouraging, but there is room for much stronger government support, especially in testing the use of 100 percent biofuels in commercial jetliners. As the issues of safety and reliability become clear, there is also room to institute mandatory standards for biofuels. These have been effective to promote biofuels for light-duty vehicles, but did not make sense given the preferable path of relying on other energy carriers. In the case of aviation, there is no other immediately feasible path, and mandatory requirements would make sense.

9. We need to continue R&D on advanced biofuels production to improve the cost and efficiency of such processes and to maximize the amount that will be available for aviation, high-temperature industrial processes, and, potentially, shipping. Most of the biofuels currently being employed are first-generation, meaning that they rely on oils, starches, and sugars from a number of crops. A high priority is to accelerate the deployment of second-generation biofuels, especially those that can be obtained from woody plant material. For the more distant future, we also need to find ways of producing inexpensive fuels from abiotic processes, such as artificial photosynthesis or power-to-liquid technologies.
10. We need to accelerate the pace with which developing country governments and the people who live there decide that it is in their own best interests to reverse the processes of deforestation and land degradation. We have policies that theory and a limited amount of evidence suggest should work. These are built primarily around accelerating the pace of social and economic development on the one hand and increasing the economic value of sustainable forest management on the other. All of these are win-win options. At the same time, we need to continue the process of improving agricultural yields, including reducing the yield gap between industrialized and developing countries.
11. We need research to figure out whether there is an additional and temporary need for payments for ecosystem services. If we find out that there is, we need to implement this based on finance from bilateral and multilateral development assistance. For this, we will need research, policy negotiation, and, ultimately, financial participation in the Reduced Emissions from Deforestation

and forest Degradation (REDD+) framework of the United Nations Framework Convention on Climate Change (UNFCC).

12. We need to address the regulatory and bureaucratic risks that renewable energy project developers face in developing countries. The single largest risk that project developers face is born out of inefficient and corrupt bureaucratic processes. A variety of policy options, promoted by donor countries as well as by non-governmental organizations (NGOs) such as Transparency International, are proving effective at changing these practices. A gradual trend toward more democratic and accountable government would help in this regard while lessening the risks associated with state expropriation. The latter, of course, is a long-term project largely driven by local desires.
13. We need to make finance available for renewable energy projects in developing countries at the lowest possible interest rates. For this, bilateral and multilateral financing mechanisms, including those within the UNFCCC, can play an important role. We also need, however, to develop the financial services sector within developing countries into one that is competitive and robust.
14. We need developing countries to raise the level of their human capital so that they can create meaningful domestic employment from the construction of renewable energy infrastructure. Bilateral and multilateral development support can help in several ways. It can promote primary and secondary schooling, for both boys and girls, thus enhancing overall rates of functional literacy. It can support specific job training programs in the renewable energy supply chain industries. Finally, it can support the enhancement of public health to enhance worker productivity, as well as welfare more generally, including the development and enforcement of legal frameworks to regulate air, ground, and water pollution.
15. We need to gradually eliminate the use of portland cement in concrete. Almost certainly, this will involve the substitution of geopolymer cement. The highest immediate priority is to enhance government support for R&D into the structural properties of geopolymer cement and to push for its immediate use in low-risk applications where we can observe its structural integrity over time. In the long run, and once our knowledge base is greater, we may need an international legal framework to mandate the phase-out of portland cement across the entire construction industry.

Those are the immediate needs. Each of them is urgent, some more urgent than others. It is a long and diverse list. In most cases, we have examples of exactly how to respond effectively to the policy need. In the remainder, we have a solid body of theory to point us forward, but we will have to do so in a manner that enhances learning, making sure that the policies we implement actually bring about the desired result and making adjustments as necessary. None of the items on the list requires that governments raise new and substantial sources of revenue. Each and every one of them will bring some set of immediate and visible rewards to at least one or two different sets of politically important stakeholders. In short, every item on the list is economically and politically possible.

Frequently asked questions

I am confident that we can and will do the fifteen things. The fact that every item on the list is feasible drives my optimism about the chances of success. But I have to admit that, among experts on climate change and climate policy, I am one of the few whom I know who is so optimistic. Is such optimism is a pipe dream? As this book draws to a close, I want to address the questions that people frequently ask, questions that can easily kill any sense of hope.

"Climate change is a global problem. Doesn't it require a global solution?": Yes, climate change requires a global solution. Everybody on the planet needs to stop doing things that result in the emissions of CO_2 and other GHGs. Starting with our grandchildren's generation, they must never do these things again. But the fundamental issue is not about whether it will take a global solution, but whether it will take a *coordinated* global solution. Here, the answer is more nuanced. A relatively uncoordinated approach will work for climate change, as it has worked for all sorts of other problems.

Consider HIV-AIDS. In the past decade, the disease has gone from being a global crisis, hitting Africa hardest of all, to a tragedy that we are gradually bringing under control. We are bringing it under control through a combination of new drugs that reduce both its symptoms and its rate of transmission and through public health programs to get people to wear condoms, avoid unprotected sex, and become aware if they actually have the disease and could transmit it to their loved ones.

All of this has happened through the actions of a diversity of actors, including private foundations and other NGOs, multinational agencies such as the World Health Organization, and governments at the local, national, and regional scale.

Consider extreme poverty. In the past decade, the number of people living on less than $1 or $2 per day has fallen dramatically. It has been largely a result of economic growth in Asia but has also included large parts of Africa. There have been policy goals set at the global level, such as the Millennium Development Goals set forth around the turn of the century. These have been largely achieved. But they have not been achieved by a single set of policies enacted at the global level through a single institutional framework like the UN. Rather, they have been achieved as a result of a lot of different things taking place in different countries.

Of course, there are also global problems that we have tried to solve in an uncoordinated manner and have failed. Protecting fisheries and the oceans more generally is one of these.

A critical requirement for an uncoordinated approach to work is that the incentives at the local or national scale align with those at the global scale. The core feature of a tragedy of the commons – which people argue fits the climate problem – is that they do not. Framed as a global commons problem, all of us in the world want climate change solved, but all of us want the solution to primarily come from someone else. Second best for each of us is that we all work together to solve the problem. So we need to sign a global treaty to ensure that the second-best solution happens instead of the worst thing of all – namely, none of us doing anything.

When you think about climate change in a manner that does not include technological or societal change, then it appears very clearly to be a tragedy of the commons. When you include technological and social changes in your frame of reference, then it looks like something very different. Recognizing that solving climate change involves eliminating CO_2 emissions altogether forces the need for technological and social change. We are moving toward having a set of energy technologies and a set of practices for forest management that are better than our current ones in several ways, only one of which is the elimination of GHG emissions. The other ways in which these are better are felt at the local and national levels. Hence, individual countries have an incentive to help their own people reach this future faster. They all need to work on this, but they don't necessarily need to work together.

"Haven't people proved that market instruments work?": Journalists around the world write articles on a daily basis suggesting that the evidence is clear that market instruments work: they reduce emissions, they do so at the least possible costs, and they stimulate innovation. As with the need for global cooperation, it is a core mantra of climate policy. And if market instruments are the things that work, then our failure to enact them at a stringent enough level can be very depressing.

Economic theory offers some very strong arguments for why market instruments ought to work. And yet that theory rests on a number of counterfactual assumptions. These include the assumption that marginal costs are rising for most goods, that people treat risk and time delay in a clear and consistent manner, and that firms across multiple economic sectors have access to better information about energy technology costs than does the government. The presence of such assumptions doesn't necessarily make the predictions based on that theory wrong, but it raises the possibility. That, in turn, implies that it is important to actually look at the evidence. When you actually look, the evidence turns out to be very thin.

What we do know is that a market instrument was used in the United States in the effort to reduce acid rain and that the costs of achieving that objective turned out to be much less than people had predicted. Those predictions were in turn based on cost estimates for end-of-pipe solutions and failed to consider how the costs for rail transportation might drop sharply in the future, making it possible to burn a less polluting type of coal. In fact, those costs did decline and, in the presence of the market instrument, became the cost-effective solution to the problem of reducing emissions. There is no doubt that if policy makers had mandated the end-of-pipe solutions and continued to mandate them even as it became clear that rail transport of low-sulfur coal was becoming much less expensive, then solving acid rain with instruments other than the market instrument would have cost more. But why would policy makers have stuck with the end-of-pipe solutions? As far as I can tell, policy makers are not generally stupid.

We also know that market instruments have been implemented in several jurisdictions and that emissions in these jurisdictions have fallen in a manner that has not brought economic ruin. Examples include British Columbia, Sweden, and Europe more generally. The Regional Greenhouse Gas Initiative (RGGI) states don't really count because their market instrument has been so weak as to make essentially no difference. But even for the positive examples, we simply don't know whether the market instruments have been much of a cause either for the decline in emissions or for the lack of economic troubles. Indeed, the weak body of evidence that does exist suggests that the economic instruments have played an extremely minor role in each case.

Even if we did have evidence that market instruments had worked so far to reduce emissions at the least possible cost, such evidence would not necessarily be much of a guide for the future. There are two reasons. First, one of the main benefits of market instruments is that they take advantage of heterogeneous abatement costs, relying on market forces to allocate emissions reductions to those countries, sectors, and firms where the costs are lowest. But that possibility to reduce costs through an efficient abatement allocation mechanism disappears once one realizes that the abatement needs to be 100 percent. Second, the primary way that market instruments have so far led to any change in emissions has been through improvements in energy efficiency. All the evidence suggests that they have done relatively little to stimulate innovation in new technologies, at least compared to other kinds of policy instruments. To eliminate our emissions, however, we will need to rely on these new technologies.

Market instruments are a great idea in theory and a good idea in some circumstances. They are an especially good idea for pollution problems where all the relevant technologies are more or less mature and where the desirable level of pollution is just a little bit less than what we have today. But they are not a good idea for promoting the kind of technological transition that will eliminate our CO_2 emissions in the next few decades.

"Can the societal and technological transitions happen fast enough?": A few pages ago, I suggested that it could happen fast enough based on assumptions concerning turnover in infrastructure. Supporting this as well is the whole set of energy system modeling results that I alluded to in Chapters 2, 4, and 6. These models have been used to construct different internally consistent scenarios of the future, and they allow the user to specify a level of climate policy. The models show that with aggressive policy action, we can turn around our energy system in time to achieve a level of warming, 2°C or less, that is probably safe. They show that it is possible to do so at a net economic cost that is so close to zero that it will not be noticeable. And they include the things that make this difficult, like the recent decline in natural gas prices in the United States.

It is true that the policy instruments that the models simulate are nearly exclusively a carbon price or a carbon cap, rather than instruments like subsidies. But that doesn't really matter. The way these models work, the simulated policy intervention makes two things happen. First, it leads the agents in the model to make use of decarbonized sources of energy rather than fossil fuels, even though there is a short-term cost penalty attached to doing so. Second, it leads the agents to innovate, making the decarbonized technologies eventually become cheaper than fossil fuels. All of the modelers I know use carbon prices and caps in part because they believe these to be the most effective and the most efficient means of making these two things happen.

But they aren't the most effective and most efficient means. The most effective and efficient means are technology-specific subsidies, such as the FIT developed in Germany and copied elsewhere. The main reason is that such subsidies do a much better job than market instruments at reducing investors' risks. In fact, in countries that have instituted FITs, the rate of new investment that they have stimulated, and the pace of innovation that has resulted, has exceeded anybody's expectations. In the real world, we can use these instruments to get exactly the same level of mitigation at costs that are similar to or lower than those that modelers have been able to create in their models.

At the same time, the question of whether societal and technological transitions can happen fast enough is a bit of a misleading question in the first place. It would be great if we achieve a 2°C target. It would be even better if we were to achieve a 1.5°C target. But if we achieve neither, then limiting climate change to 2.5°C is still a lot better than limiting it to 3°C, which is in turn better than 3.5°C. To my mind, the importance of the 2°C target is not in the magic of that particular temperature change, but rather in what achieving it implies. It implies that we eliminate all net GHG emissions, everywhere, about as fast as anyone believes is politically, economically, and technologically feasible. The 2°C target leaves us no room for delay, and it leaves no room for anybody to avoid eventually taking part.

The real question, then, is what set of policies can get us to eliminate GHG emissions in the fastest way politically, economically, and technologically possible. It's not with market instruments. Even if they are OK economically, they fare pretty poorly politically and don't do very much for technology. All of the policies that I have suggested here in this book have already shown themselves to be politically acceptable at roughly the scale needed, have shown themselves to be economically affordable, and have acquired an excellent track record of stimulating exactly the kind of technological progress that we need. Building on this success is our best chance to eliminate emissions in the shortest possible time.

"Won't politics get in the way?": Politics can always get in the way and, especially in the American political system it seems, lead to inaction. Often, when political action does replace inaction, it is in response to an immediate political or social crisis. A policy-making window of opportunity opens up, politicians make a new law to solve the immediate crisis, and then within a few weeks the public's attention moves on to other things and everything is forgotten. Except that a new law is now on the books.

Concerning climate change, one hears all the time that the one thing we need right now, more than anything else, is political will. The will to sign a stringent treaty. The will to enact a tough carbon tax. So, we the people need to lobby our elected leaders. We need to march on the streets, showing them how much we care, so that they will enact strong climate policies. Moreover, we the policy analysts and policy entrepreneurs need to be ready with a demonstrably good policy solution, so that when a crisis does occur – whether it be a war in the Middle East or a typhoon in the Pacific – policy makers can grab it and make a meaningful change while the opportunity is there.

I actually believe both of these conclusions. I do think that we need to increase the overall level of political motivation and that we need to be ready with solutions to problems for those times when the moment is right. But I also believe two other things. First, the political will to solve climate change is already quite large, but it has to coexist with the political will to solve lots of other problems as well. Second, over the coming years and decades, the precise challenges associated with climate policy will change. To eliminate emissions altogether, we need one set of laws now, but in all likelihood a slightly different set of laws in ten or twenty years' time.

Given these two additional beliefs, it then becomes clear that we need laws that can serve multiple policy objectives at the same time. Serving multiple policy objectives makes it more likely that the laws will get enacted in the first place. FITs became the law in Germany because they served different agendas in two different political parties. Serving multiple objectives also makes it more likely that policy makers will build on the legal frameworks on into the future, improving them over time rather than eliminating them when conditions change. Again, the FIT is a good example: when it became clear that it wasn't serving one of its implicit objectives – building an export industry in PV panels – it became vulnerable to attack. The more different interests a policy instrument or legal instrument can serve, the better.

One of the problems with market instruments and binding national targets is that just about the only policy objective they serve is climate change. They don't particularly gain the support of any industrial lobby and, indeed, have more frequently engendered opposition. Especially with market instruments, their very flexibility makes it unclear which technological pathway firms will actually take to comply with the cap or reduce their carbon tax burden, and this means again that it is harder to point to any clear winners. Against this, they create clear and obvious losers in the form of industries supplying and relying on fossil fuels.

Almost all of the policies I have described in this book aim to promote something new, rather than discourage something old. Mathematically, the two may be equivalent, but it's not mathematical proofs that drive politics. It is feelings and emotions like fear, hope, and trust. More importantly, almost all of the policies I have suggested are ones that already exist in one country or another, at roughly the level of ambition that we need. This serves two political purposes. First, it can resolve uncertainty as to the costs to society or unintended consequences of the law. A lot of countries instituted FITs, for example, after they observed their success in Germany and the fact that their cost was not inordinately high. Second, it is simply an indicator that the policy can make it through a contested political environment; if it is able to do so in one country, chances are that it can do so in another. By contrast, there is only one example of a successful carbon tax at the level of ambition that would be required, and this is in Sweden, which is a bit of a special case when it comes to taxation. There are no examples of global caps on anything with the level of ambition or economic consequences that an effective climate treaty would require. In short, politics matters. And because it matters, the policies I have proposed make sense.

In truth, the place where politics are most likely to get in the way, as I have already flagged, is at the local level. Here, the decisions are not so much about things like FITs or support for biofuels, but rather about the desirability of dedicating land to new uses, most importantly wind and solar generation facilities and power lines. People often use the term NIMBY (Not in My Back Yard) to describe the source of such objections, but the truth is that NIMBY is a poor descriptor of what is going on. People object to infrastructure projects that they perceive as unnecessarily harmful to a landscape that they cherish or that are unfair and unrepresentative in their planning and execution. As I have already described, we need policies enacted at the national level that lessen the chances of disputes arising at the local level.

"If we count on technologies to solve climate change, won't these in turn lead to other problems?": Yes, of course, they will. It is impossible to do anything without having unintended consequences. Some of these can be predicted and some cannot. We will never get to a place where all the problems in the world have been solved and there is nothing left to fix. The best that we can hope for is to avoid solving today's problems in ways that create even bigger problems in the future. Ideally, the new problems that we create should be substantially smaller and more manageable than the ones we solve.

When it comes to market instruments or binding national emissions targets, it is difficult to identify particular new problems that will emerge precisely because there are potentially many different ways that firms or countries can respond to the tax or cap. The only thing we do know is that if these instruments are successful at stopping climate change, then it is because they motivate people to do something. We don't know exactly what that something is, but it is almost certain to involve the same technologies that I am suggesting we support directly. By contrast, when it comes to supporting particular technologies directly, as I have suggested here, it is far easier to put one's finger on particular negative outcomes. PV panels, for example, contain a lot of material that is in limited supply and that can lead to toxic pollution. The same is true for lithium ion batteries. CSP can lead to water scarcity, create land-use conflicts, and, in some configurations, kill birds. Windmills don't fry birds, but they do whack them to death. They also are a visual disturbance to many people. The list of negative consequences goes on.

However, none of these negative consequences is of anywhere near the seriousness of climate change, and all of them are things that we can, in one way or another, solve. Far more birds will become threatened from climate change than we could ever kill with renewable energy technologies. As a better example, the dominant environmental issue of the late 1970s and early 1980s in the United States was toxic chemicals, both their use and their improper disposal. At the time, it was seen as a national crisis. I think that it is unlikely that any of the technologies we will rely on to solve climate change will cause problems related to toxic chemicals half as severe as what we faced two decades ago. At the same time, the problems that we faced two decades ago have also faded in importance in comparison to climate change. One reason they have faded is because the consequences are simply much less important and catastrophic, at least at the global scale. The other reason they have faded is that we have figured out how to solve them.

An important fact to recognize is that the technologies that we will rely on to solve climate change will probably not be the same ones we are using 100 years from now. We will need lithium ion batteries to power the first generation of electric cars, the generation that takes over the roadways from gas and diesel powered vehicles. But it is difficult to imagine some other technology not taking over from lithium ion in another few decades. If the environmental consequences of lithium ion batteries are particularly severe, then in all likelihood it will be sooner rather than later.

Bringing it into our own lives

As the closing credits to *An Inconvenient Truth* were rolling, a list of ideas flashed across the screen, ideas about what the people watching the film could do to stop climate change. Being clear about how individual people can make a difference is incredibly important. All of us derive pleasure and satisfaction from doing good. By getting involved, we generate satisfaction and improve our lives. More importantly, all of us feel more committed and excited about the things that we feel we are involved in. We are more likely to support the policies to stop climate change, prioritizing these over other objectives or even simple political apathy, if we feel personally engaged in the effort. But I am convinced that the list of things in that film wasn't quite right. Most of them involved small personal sacrifices or else ways of being careful about energy use, with the idea that the results of that sacrifice and care would make a difference for the climate. I have argued in this book that this just isn't the case. None of this will make much of a difference for stopping climate change, and none of it is really necessary.

I need to be clear: I am not arguing against saving energy, against reducing consumption, against being mindful in how we live. All of these things can improve the quality of our lives, and the lives of the people around us. I very much hope it happens across society, just as I am trying to make it happen in my own life. But engaging in a greater degree of mindfulness and everything that goes with it is a deeply spiritual practice. As such, it is highly personal, guided by our own sense of morality and our own quest for life satisfaction and meaning. It needs to be born out of free choice, not a lack thereof. The evidence shows that it does not need to be tied to solving climate change. It would be best to keep it separate.

So, is there anything that an individual person can do that can make a difference? I believe that there are three things.

The first is to consume new things. More precisely: be an early adopter of new, low-carbon technologies, the kind that we need to become successful. If you spend the money to put a PV panel on your roof, for example, the immediate effect on CO_2 emissions will be small. It will, however, contribute to the market development of PV, thus leading to lower costs in the future. That, in turn, will make it more likely that your neighbor will do the same thing. The more people who are visibly using a new technology, and doing so without problems, the lower the perceived risks for the rest of us. Of course, if you do use a new technology and encounter problems, then please make use of that knowledge to help the technology improve. Make the knowledge public. You can make a difference in making good technologies successful.

The second thing to do is to be more mindful about your existing consumption, although not in the ways that most people would suggest, which is to pay attention to the total amount of energy you use. Instead, pay attention to your consumption choices in other respects. Are you living well within your means, even below your means? The fact is that being an early adopter of new technologies is typically expensive. If you are going to be able to do it, then you will need the resources to be available. So join a carpool. Take your vacations closer to home. Cut down on beef consumption. All of these save substantial amounts of money. Use the money you save to replace your current car with an electric one, put PV panels on your roof, or buy the more expensive food that has been grown without excessive nitrogen fertilizer. Next, pay attention not to the quantity of your consumption, but to the quality: to the particular types of energy and the particular materials that you are using. We want to encourage investment in the electric power sector, so, as much as possible shift to using electricity instead of other energy carriers in areas like cooking, heating, and getting about. We want to discourage new investments into oil and gas exploration and into new factories that produce portland cement. That investment responds to consumer demand. Try to use less of these three things. Try to use more wood, at least when you have some confidence that it has been sustainably harvested. In other words, pay attention to what sort of long-term societal trends your consumption contributes to.

The third thing is not about consumption at all, but about your state of mind and how you express that state of mind to others. To solve climate change, there is still so much for all of us to learn, often in unexpected places. Germans from Texans, for example, and vice versa. Each of us has the ability to foster this kind of learning in our own communities, whether they be geographically based or online, keeping our communication channels open. That requires us to practice a certain degree of compassion and nonjudgment.

In one sense, all of us share the blame for climate change, at least those of us living in conditions of wealth. Almost all of us live longer, safer, and probably more enjoyable and fulfilling lives than we could have hoped for before the Industrial Revolution took place. We benefit from the energy systems and the land-use practices that are causing climate change, and almost none us will voluntarily do without these things, at least as long as there are no alternatives. In another sense, none of us shares that blame. We are all individual people who were born into our lives long after the Industrial Revolution had launched us down a pathway of using fossil fuels, a pathway that has proved difficult so far to move away from. But what matters most is how we move forward. How we stop climate change, in a manner that is quick, socially just, and empowering of the human spirit.

It is hardest of all to learn and to work together when it seems that everything is going wrong and when one feels personally powerless to change that. That is when we start blaming other people for the problems we are facing and feel like the world would simply be better if others would get out of our way. It is easiest to be open-minded, to listen, to accept when you are sitting in a position of strength, power, and confidence, when you are optimistic about the future. We are far more likely to solve climate change if we believe that success is possible. Our motivation will be greatest if we believe that success will be the inevitable result of the actions we take now and that success will still happen even if we make some mistakes along the way.

We have made some mistakes. I am willing to be shown to be wrong, but I believe that the Kyoto Protocol has turned out to be a big mistake. The Emissions Trading System, the Clean Development Mechanism, the Regional Greenhouse Gas Initiative: mistakes. None of them has proved harmful in their own right, but all of them, I think, have distracted us from the types of policies that could have been far more effective. We will certainly make more mistakes in the future.

And yet, if we learn from each of our mistakes and correct them, the chances of stopping climate change are good. Our chances are a lot better now than they looked to be ten, twenty, or even thirty years ago, even though the emissions numbers suggest otherwise. Optimism is justified. For the first time since humanity went down the road of burning fossil fuels, we have a set of technologies that can replace them, ones that can make life better in the present, even as they are saving the planet for the future. Countries have experimented with a variety of policy instruments to move these technologies into the mainstream, and we have started to learn which ones work best. We recognize the immediate challenges we face, and we have the means, the desire, and the courage to overcome them.

Notes

1 From optimism to pessimism and back again

1. A. Lovins, *World Energy Strategies: Facts, Issues, and Opinions* (London: Friends of the Earth, 1973).
2. A. Lovins, *Soft Energy Paths: Towards a Durable Peace* (San Francisco: Friends of the Earth, 1977).
3. D. Yergin, *The Quest: Energy, Security, and the Remaking of the Modern World* (New York: Penguin Press, 2011).
4. C. Pope, "Amory Lovins," *Time*, April 30, 2009.
5. A. Lovins, M. Odum, and J. Rowe, *Reinventing Fire: Bold Business Solutions for the New Energy Era* (White River Junction, VT: Chelsea Green, 2011).
6. D. Adam, "World Will Not Meet 2C Warming Target, Climate Change Experts Agree," *The Guardian*, April 14, 2009, sec. Environment.
7. J. Rogelj et al., "Copenhagen Accord Pledges Are Paltry," *Nature* 464 (2010): 1126–28; Jasper van Vliet et al., "Copenhagen Accord Pledges Imply Higher Costs for Staying below 2°C Warming," *Climatic Change* 113, no. 2 (July 1, 2012): 551–61, doi:10.1007/s10584-012-0458-9; Malte Meinshausen et al., "Greenhouse-Gas Emission Targets for Limiting Global Warming to 2°C," *Nature* 458, no. 7242 (April 2009): 1158–62, doi:10.1038/nature08017.
8. F. Harvey, "Worst Ever Carbon Emissions Leave Climate on the Brink," *The Guardian*, May 29, 2011, sec. Environment.
9. Ibid.
10. M. Wald, "On Not Reaching Carbon Goals," *New York Times*, June 11, 2012, sec. Green: A Blog About Energy and the Environment.
11. Ibid.
12. C. Berg, "We Can't Stop Climate Change - It's Time to Adapt," *The Drum Opinion*, accessed June 22, 2012, www.abc.net.au/unleashed/3997798.html.

13. J. Eilperin and P. Craighill, "Global Warming No Longer Americans' Top Environmental Concern, Poll Finds," *Washington Post*, July 3, 2012, www.washingtonpost.com/national/health-science/global-warming-no-longer-americans-top-environmental-concern-poll-finds/2012/07/02/gJQAs9IHJW_story.html.
14. Ibid.
15. Beate M. W. Ratter, Katharina H. I. Philipp, and Hans von Storch, "Between Hype and Decline: Recent Trends in Public Perception of Climate Change," *Environmental Science & Policy* 18, no. 0 (April 2012): 3–8, doi:10.1016/j.envsci.2011.12.007; Eilperin and Craighill, "Global Warming No Longer Americans' Top Environmental Concern, Poll Finds."
16. F. Newport, "Americans' Global Warming Concerns Continue to Drop: Multiple Indicators Show Less Concern, More Feelings That Global Warming Is Exaggerated" (Gallup, 2010), www.gallup.com+Americans-Global-Warming-Concerns-Continue-Drop.aspx.
17. Ratter, Philipp, and von Storch, "Between Hype and Decline: Recent Trends in Public Perception of Climate Change."
18. Ibid.
19. Sara Peach, "2012 GOP Candidates Demonstrate Dramatic Political Shift on Climate," *Yale Forum on Climate Change and the Media*, November 8, 2011, www.yaleclimatemediaforum.org/2011/11/2012-gop-candidates-demonstrate-dramatic-political-shift-on-climate/.
20. Ibid.
21. State of North Carolina, "Climate Change Imitative at the Department of Environment and Natural Resources," accessed July 16, 2012, www.climatechange.nc.gov/pages/ClimateChange/CC_DENR_Initiative.htm.
22. K. Sheppard, "North Carolina Tries to Wish Away Sea-Level Rise," *The Guardian*, June 1, 2012, sec. Guardian Environment Network, www.guardian.co.uk/environment/2012/jun/01/north-carolina-sea-level-rises.
23. "About Climategate.com," *Climategate: Anthropogenic Global Warming, History's Biggest Scam*, 2010, www.climategate.com/about.
24. Paul Krugman, "Climate Change Economics, What Economists Agree On," *New York Times Magazine*, June 7, 2010, www.nytimes.com/2010/04/11/magazine/11Economy-t.html.
25. Thomas Friedman, "The Power of Green," *New York Times Magazine*, April 15, 2007, www.nytimes.com/2007/04/15/magazine/15green.t.html?src=tp&pagewanted=all.

26. Editorial staff, "New Day on Climate Change," *New York Times*, January 26, 2009, sec. Editorial, www.nytimes.com/2009/01/27/opinion/27tue1.html?ref=selecteditorialsonclimatechange.
27. Editorial staff, "Climate Change," *New York Times*, February 21, 2010, www.nytimes.com/2010/02/22/opinion/22mon1.html.
28. Editorial staff, "A Deal in Durban," *The Economist*, December 17, 2011, www.economist.com/node/21541806.
29. Editorial staff, "Climate Change: Ambition Gap," *The Guardian*, December 12, 2011, www.guardian.co.uk/commentisfree/2011/dec/12/durban-climate-change-conference-2011-climate-change.
30. Editorial staff, "Cut the Sprawl, Cut the Warming," *New York Times*, October 6, 2008, www.nytimes.com/2008/10/07/opinion/07tue2.html.
31. Editorial staff, "Moment of Truth on Congestion Pricing," *New York Times*, March 26, 2008, www.nytimes.com/2008/03/26/opinion/26wed4.html.
32. Ramez Naam, *The Infinite Resource: The Power of Ideas on a Finite Planet* (Lebanon NH: University Press of New England, 2013).
33. Maggie Koerth-Baker, *Before the Lights Go out: Conquering the Energy Crisis before It Conquers Us* (Hoboken NJ: John Wiley & Sons Inc., 2012).

2 The natural and social science of climate change

1. P. Forster et al., "Changes in Atmospheric Constituents and in Radiative Forcing," in *Climate Change 2007: Impacts, Adaptation and Vulnerability. Contribution of Working Group II to the Fourth Assessment Report of the Intergovernmental Panel on Climate Change* (Cambridge UK: Cambridge University Press, 2007).
2. Gerard H. Roe and Marcia B. Baker, "Why Is Climate Sensitivity So Unpredictable?," *Science* 318, no. 5850 (2007): 629–32.
3. T. P. Guilderson and D. Schrag, "Abrupt Shift in Subsurface Temperatures in the Tropical Pacific Associated with Chances in El Niño," *Science* 281, no. 5374 (1998): 240–43.
4. This is, at least, what Mark Erdmann told me. You may have heard of him, by the way. He is perhaps best known in the scientific community for his discovery of a coelocanth fish off the waters of Manado. See M. Erdmann, R. Caldwell, and M. Moosa, "Indonesian 'King of the Sea' Discovered." *Nature*, 395, no. 6700 (1998): 335.

5. One major difference between research funding from the U.S. government and governments in Europe is that the Americans will under no circumstances pay for alcoholic beverages, whereas the Europeans take the attitude that hard work requires hearty meals, and people of culture drink a glass of wine with their meal, substitutable with beer or hard alcohol. To be clear, each of us paid for our own beer and scuba diving fees, out of our salaries and not out of the U.S. National Science Foundation research grant that was funding this trip.
6. K. Hughen et al., "El Niño during the Last Interglacial Period Recorded by a Fossil Coral from Indonesia," *Geophysical Research Letters* 26, no. 20 (1999): 3129–32.
7. Myles R. Allen and David J. Frame, "Atmosphere: Call Off the Quest," *Science* 318, no. 5850 (2007): 582–83.
8. Gerald A. Meehl et al., "Global Climate Projections," in *Climate Change 2007: The Physical Science Basis. Contribution of Working Group I to the Fourth Assessment Report of the Intergovernmental Panel on Climate Change*, ed. S. Solomon et al. (Cambridge/New York: Cambridge University Press, 2007), 747–846.
9. N Nakicenovic and Rob Swart, *Special Report on Emission Scenarios* (Geneva: Intergovernmental Panel on Climate Change, 2000).
10. Richard Moss et al., "The Next Generation of Scenarios for Climate Change Research and Assessment," *Nature* 463, no. 7282 (2010): 747–56, doi:10.1038/nature08823.
11. S. Solomon et al., *Climate Change 2007: The Physical Science Basis. Contribution of Working Group I to the Fourth Assessment Report of the Intergovernmental Panel on Climate Change* (Cambridge/New York: Cambridge University Press, 2007).
12. William Nordhaus, *Managing the Global Commons* (Cambridge, MA: MIT Press, 1994).
13. Simon Dietz and Nicholas Stern, "Why Economic Analysis Supports Strong Action on Climate Change: A Response to the Stern Review's Critics," *Review of Environmental Economics and Policy* 2, no. 1 (January 1, 2008): 94–113.
14. Richard S. J. Tol, "Estimates of the Damage Costs of Climate Change: Part II. Dynamic Estimates," *Environmental and Resource Economics* 21 (2002): 135–60; Chris Hope, John Anderson, and Paul Wenman, "Policy Analysis of the Greenhouse Effect: An Application of the PAGE Model," *Energy Policy* 21, no. 3 (1993): 327.

15. Samuel Randalls, "History of the 2°C Climate Target," *Wiley Interdisciplinary Reviews: Climate Change* 1, no. 4 (2010): 598–605, doi:10.1002/wcc.62.
16. Detlef P. Van Vuuren et al., "Low Stabilization Scenarios and Implications for Major World Regions from an Integrated Assessment Perspective," *Energy Journal* 31 (2010): 165–92.
17. O. Edenhofer, N. Bauer, and E. Kriegler, "The Impact of Technological Change on Climate Protection and Welfare: Insights from the Model MIND," *Ecological Economics* 54, no. 2–3 (2005): 277–92; H. Held et al., "Efficient Climate Policies under Technology and Climate Uncertainty," *Energy Economics* 31, no. Suppl. 1 (2009): S50–61; Brigitte Knopf and O. Edenhofer, "The Economics of Low Stabilization: Implications for Technological Change and Policy," in *Making Climate Change Work for Us: European Perspectives on Adaptation and Mitigation Strategies*, ed. M. Hulme and H. Neufeldt (Cambridge, UK: Cambridge University Press, 2010), 291–318.
18. Brigitte Knopf et al., "Report on the First Assessment of Low Stabilization Scenarios" (Potsdam: Potsdam Institute for Climate Impact Research, 2008); Detlef P. van Vuuren et al., "Preliminary ADAM Scenarios" (Eindhoven: Netherlands Environmental Assessment Agency, 2007).
19. N. Stern, *The Economics of Climate Change* (Cambridge, UK: Cambridge University Press, 2007).
20. Knopf and Edenhofer, "The Economics of Low Stabilization: Implications for Technological Change and Policy."
21. M. L. Parry et al., *Climate Change 2007: Impacts, Adaptation and Vulnerability. Contribution of Working Group II to the Fourth Assessment Report of the Intergovernmental Panel on Climate Change* (Cambridge, UK: Cambridge University Press, 2007).
22. T. M Lenton et al., "Tipping Elements in the Earth's Climate System," *Proceedings of the National Academy of Sciences* 105, no. 6 (2008): 1786.

3 The solution space and its distractions

1. A. G. Patt et al., "Adaptation in Integrated Assessment Modeling: Where Do We Stand?," *Climatic Change* 99 (2010): 383–402, doi: 10.1007/s10584-009-9687-y.
2. M. Fischetti, "Protecting New Orleans," *Scientific American*, 2006.

3. Mark Burton and Michael Hicks, "Hurricane Katrina: Preliminary Estimates of Commercial and Public Sector Damages" (Huntington, WV: Marshall University, Center for Business and Economic Research, 2005).
4. J. Reed, *Potential Economic Impacts of Hurricane Katrina* (Washington, DC: Congressional Printing Office; Democrats, United States Congress Joint Economic Committee, 2005).
5. M. Hulme et al., "Relative Impacts of Human-Induced Climate Change and Natural Climate Variability," *Nature* 397 (1999): 688–91; Mike Hulme, Saffron J. O'Neill, and Suraje Dessai, "Is Weather Event Attribution Necessary for Adaptation Funding?," *Science* 334, no. 6057 (November 11, 2011): 764–65, doi:10.1126/science.1211740.
6. Roger Pielke Jr., "Misdefining 'Climate Change': Consequences for Science and Action," *Environmental Science & Policy* 8, no. 6 (décembre 2005): 548–61, doi:16/j.envsci.2005.06.013; Roger Pielke Jr., "Mistreatment of the Economic Impacts of Extreme Events in the Stern Review Report on the Economics of Climate Change," *Global Environmental Change* 17, no. 3–4 (2007): 302.
7. Kerry Emanuel, "Increasing Destructiveness of Tropical Cyclones over the Past 30 Years," *Nature* 436 (2005): 686–88; Kerry Emanuel, R. Sundararajan, and J. Williams, "Hurricanes and Global Warming: Results from Downscaling IPCC AR4 Simulations," *Bulletin of the American Meteorological Society* 89, no. 347–67 (2008).
8. Knopf and Edenhofer, "The Economics of Low Stabilization: Implications for Technological Change and Policy"; Brigitte Knopf et al., "Managing the Low-Carbon Transition: From Model Results to Policies," *Energy Journal* 31 (2010): 223–45.
9. World Bank, "The Economics of Adaptation to Climate Change: Synthesis Report" (Washington, DC: The International Bank for Reconstruction and Development, 2010).
10. UNFCCC, "Investment and Financial Flows to Address Climate Change" (Paris: United Nations Framework Convention on Climate Change Secretariat, 2007); S. Agrawala et al., "Economic Aspects of Adaptation to Climate Change: Costs, Benefits and Policy Instruments" (Paris: Organization for Economic Cooperation and Development (OECD), 2008).
11. World Bank, "DAC Members' Net Official Development Assistance in 2010," 2012.
12. Richard J. T. Klein et al., "Resilience or Vulnerability: Coastal Dynamics or Dutch Dikes?," *The Geographical Journal* 164, no. 3 (1998): 259–68.

13. Joyeeta Gupta et al., "Mainstreaming Climate Change in Development Cooperation Policy: Conditions for Success," in *Making Climate Change Work for Us*, ed. M. Hulme and H. Neufeldt (Cambridge, UK: Cambridge University Press, 2010), 319–39.
14. Global Subsidies International, *Relative Subsidies to Energy Sources: GSI Estimates* (Vancouver: International Institute for Sustainable Development, 2011), www.iisd.org/gsi/sites/default/files/relative_energy_subsidies.pdf.
15. Mark Cane, G. Eshel, and R. Buckland, "Forecasting Zimbabwean Maize Yield Using Eastern Equatorial Pacific Sea Surface Temperatures," *Nature* 370 (1994): 204–5.
16. Michael Glantz, *Currents of Change: Impacts of El Niño and La Niña on Climate and Society*, vol. II (Cambridge, UK: Cambridge University Press, 2001).
17. M. A. Cane, S. Zebiak, and S. Dolan, "Experimental Forecasts of El Niño," *Nature* 321 (1986): 827–32.
18. National Oceanic and Atmospheric Administration, "An Experiment in the Application of Climate Forecasts: NOAA-OGP Activities Related to the 1997–98 El Niño Event" (NOAA Office of Global Programs, U.S. Department of Commerce, 1999); NOAA, "NOAA's Contribution to Furthering Our Understanding of Climate Variability Studied," 1996, www.publicaffairs.noaa.gov/pr96/dec96/noaa96-080.html.
19. Anthony G. Patt, "Trust, Respect, Patience, and Sea Surface Temperatures: Useful Climate Forecasting in Zimbabwe," in *Global Environmental Assessments: Information and Influence.*, ed. Ronald Mitchell et al. (Cambridge, MA: MIT Press, 2006), 241–69.
20. Ibid.
21. Michael Glantz, *Once Burned, Twice Shy? Lessons Learned from the 1997–98 El Niño* (Tokyo: UNEP/NCAR/UNU/WMO/ISDR, 2000).
22. Patt, "Trust, Respect, Patience, and Sea Surface Temperatures."
23. A. G. Patt, "Understanding Uncertainty: Forecasting Seasonal Climate for Farmers in Zimbabwe," *Risk Decision and Policy* 6 (2001): 105–19.
24. A. G. Patt, Angie Dazé, and Pablo Suarez, "Gender and Climate Change Vulnerability: What's the Problem, What's the Solution?," in *The Distributional Effects of Climate Change: Social and Economic Implications*, ed. M. Ruth and M. Ibarrarán (Cheltenham, UK: Edward Elgar, 2009), 82–102.

25. A. G. Patt and Chiedza Gwata, "Effective Seasonal Climate Forecast Applications: Examining Constraints for Subsistence Farmers in Zimbabwe," *Global Environmental Change* 12, no. 3 (2002): 185–95.
26. A. G. Patt, Pablo Suarez, and Chiedza Gwata, "Effects of Seasonal Climate Forecasts and Participatory Workshops among Subsistence Farmers in Zimbabwe," *Proceedings of the National Academy of Sciences of the United States of America* 102 (2005): 12623–28.
27. Ibid.
28. A. G. Patt, Laban Ogallo, and Molly Hellmuth, "Learning from 10 Years of Climate Outlook Forums in Africa," *Science* 318 (2007): 49–50.
29. M. Dreyfus and A. Patt, "The European Commission White Paper on Adaptation: Appraising Its Strategic Success as an Instrument of Soft Law," *Mitigation and Adaptation Strategies for Global Change*, in press, doi:10.1007/s11027-011-9348-0.
30. S. Pfenninger et al., *Report on Perceived Policy Needs and Decision Contexts* (Mediation Project Deliverable Report, 2011).
31. Anita Wreford, M. Hulme, and W. N. Adger, "Strategic Assessment of the Impacts, Damage Costs, and Adaptation Costs of Climate Change in Europe" (Norwich, UK: Tyndall Centre for Climate Change, 2007).
32. Patt, Suarez, and Gwata, "Effects of Seasonal Climate Forecasts and Participatory Workshops among Subsistence Farmers in Zimbabwe."
33. M. Boko et al., "Africa," in *Climate Change 2007: Impacts, Adaptation and Vulnerability*. Contribution of Working Group II to the Fourth Assessment Report of the Intergovernmental Panel on Climate Change, ed. M. L. Parry et al. (Cambridge UK: Cambridge University Press, 2007), 433–67.
34. A. G. Patt et al., "Estimating Least-Developed Countries' Vulnerability to Climate-Related Extreme Events over the Next 50 Years," *Proceedings of the National Academy of Sciences* 107 (2010): 1333–37.
35. Erich Striessnig, Wolfgang Lutz, and Anthony G. Patt, "Effects of Educational Attainment on Climate Risk Vulnerability," *Ecology and Society* 18, no. 1 (2013), doi:10.5751/ES-05252-180116.
36. Hallie C. Eakin and Anthony Patt, "Are Adaptation Studies Effective, and What Can Enhance Their Practical Impact?," *Wiley Interdisciplinary Reviews: Climate Change* 2, no. 2 (2011): 141–53, doi:10.1002/wcc.100.
37. National Oceanic and Atmospheric Administration, "An Experiment in the Application of Climate Forecasts: NOAA-OGP Activities Related to the 1997–98 El Niño Event."

38. H. Alderman and T. Haque, "Insurance against Covariate Shocks: The Role of Index-Based Insurance Protection in Low-Income Countries" (Washington, DC: The World Bank, 2007).
39. J. Linnerooth-Bayer, R. Mechler, and G. Pflug, "Refocusing Disaster Aid," *Science* 309 (2005): 1044–46.
40. R. Kunzig and Wallace Broeker, Fixing Climate: The Story of Climate Change – and How to Stop Global Warming (London: Profile Books, 2008).
41. Oliver Morton, "Net Benefits: The Idea of Pulling Carbon Dioxide out of the Atmosphere Is a Beguiling One. Could It Ever Become Real?," *The Economist*, March 17, 2012.
42. Rachel Nuwer, "A Way to Trap Carbon Deep in the Ocean," *New York Times*, July 19, 2012, http://green.blogs.nytimes.com/2012/07/19/a-way-to-trap-carbon-deep-in-the-ocean/.
43. James Kanter, "Debating the Climate Benefits of 'Biochar,'" *New York Times*, April 15, 2009, http://green.blogs.nytimes.com/2009/04/15/debating-the-climate-benefits-of-biochar/.
44. Michael Specter, "The Climate Fixers: Is There a Technological Solution to Global Warming?," *The New Yorker*, May 14, 2012, www.newyorker.com/reporting/2012/05/14/120514fa_fact_specter?currentPage=all.
45. "Mt. Pinatubo's Cloud Shades Global Climate," *Science News*, July 18, 1992, 7www.thefreelibrary.com/Mt.+Pinatubo%92s+cloud+shades+global+climate.-a01246705.
46. J. Blackstock et al., *Climate Engineering Responses to Climate Emergencies* (Santa Barbara, CA: Novim, 2009), http://arxiv.org/pdf/0907.5140.
47. Ibid.
48. Ibid.
49. J. Blackstock and J. C. S. Long, "The Politics of Geoengineering," *Science* 327 (2010): 527–28.
50. Roger Bradbury, "A World without Coral Reefs," *New York Times*, July 14, 2012.
51. Blackstock et al., *Climate Engineering Responses to Climate Emergencies*; Blackstock and Long, "The Politics of Geoengineering."

4 Getting the prices right

1. www.who.int/tobacco/global_data/regional_databases/en/. Accessed January 22, 2015.

2. Andreas and Sara, "International Cigarette Prices," 2012, www.cigaretteprices.net.
3. I have not been able to find data or analyses to show whether rates of smoking in Austria have changed since the introduction of the new law.
4. Eric Frey, "Pro Nichtraucherschutz: Kunst Des Kompromisses," *Der Standard*, May 30, 2012, http://derStandard.at/1336698316046/Pro-Nichtraucherschutz-Kunst-des-Kompromisses.
5. Krugman, "Climate Change Economics, What Economists Agree On."
6. In theory, this may not be the outcome. The legal professor Ronald Coase, in a famous article, demonstrated that private parties ought to be able to negotiate their way to the optimal solution, as long as property rights were clarified and the transaction costs are low. In reality this has seldom, if ever, been observed.
7. As the economist Nico Bauer pointed out to me, it is important to note that this increase in profits comes about only if the regulatory action is to restrict supply without requiring firms to purchase their share of that supply. The latter, as I will describe soon, would be an auctioned cap-and-trade system, which would not have the effect of boosting firm profits.
8. It is worth noting that this equality is a pure accident, resulting from the particular numbers I made up for this example. There is nothing guaranteeing that there would be such an equality in markets more generally.
9. This equality emerges when the government has full information about the market and so is able to determine that 800 permits is the correct number. If the government had incomplete information about the market, it might set the number of permits differently, and then the outcomes would diverge.
10. Lawrence H. Goulder, "Environmental Taxation and the Double Dividend: A Reader's Guide," *International Tax and Public Finance* 2, no. 2 (August 1, 1995): 157–83, doi:10.1007/BF00877495.
11. Among consumers, however, there might be some transfer of wealth when the auctioned permit system comes into place. Compared to the unregulated market, those people who are consuming chairs will have to pay more for them as the price rises from $100 to $110 and will suffer a corresponding loss of wealth. The beneficiaries of the reduction in other taxes may or may not be these same people. If, for example, the other tax that is reduced is the income tax, and the reduction does so in a manner that

preserves the same degree of progressivity in the income tax (whereby the wealthy pay a larger share of their income than the poor), then the wealthy may well benefit more than the poor from the reduction in the income taxes.

12. One indicator that I do not show is the price of the chairs, which aligns with the total quantity. In the first three columns, fifty chairs are sold, and the price is $100. In the latter three columns, forty chairs are sold, and the price is $110. So all of the market instruments do have an effect of raising prices.
13. John Dryzek, *The Politics of the Earth: Environmental Discourses* (Oxford, UK: Oxford University Press, 1997).
14. This simplicity with respect to the tax – that it can simply stay at $1 – is also an artifact of my simple example. If the cost of each dead fish were to change as a function of the total number of dead fish – something economists often assume – then the tax would have to change somewhat with technological change.
15. The source of the information in this paragraph is a keynote lecture that Jane Lubchenco gave at the Sustainability Science Conference, October 2014, in Copenhagen, Denmark. Dr. Lubchenco is the former director of the U.S. National Oceanic and Atmospheric Administration.
16. Aquaculture might appear to be such a technology. Unfortunately, it often makes the problem worse. Fished raised on farms have to eat something, and that something is typically fish meal, produced from wild caught fish of a different species.
17. I base this review of the legislative history on Erlandson, D., "The BTU Tax Experience: What Happened and Why It Happened." *Pace Environmental Law Review* 12, no. 1 (1994): 173–84.
18. Anonymous, "Carbon Tax" (Wikipedia, 2012), http://en.wikipedia.org/wiki/Carbon_tax#Europe.
19. Potomac Economics, *Annual Report on the Market for RGGI CO2 Allowances: 2009*, 2010, www.rggi.org/docs/MM_2009_Annual_Report.pdf.
20. Ibid.
21. Potomac Economics, *Annual Report on the Market for RGGI CO2 Allowances: 2010* (Washington, DC: Potomac Economics, 2011), www.rggi.org/docs/MM_2010_Annual_Report.pdf.
22. Potomac Economics, *Annual Report on the Market for RGGI CO2 Allowances: 2011* (Washington, DC: Potomac Economics, 2012), www.rggi.org/docs/MM_2011_Annual_Report.pdf.

23. Carbon Trust, *Conversion Factors* (London: Carbon Trust, 2011), www.carbontrust.com/media/18259/ctl153_conversion_factors.pdf.
24. NYSERDA, "Monthly Average Retail Price of Electricity – Residential" (State of New York, 2012), www.nyserda.ny.gov/en/Page-Sections/Energy-Prices-Supplies-and-Weather-Data/Electricity/Monthly-Avg-Electricity-Residential.aspx.
25. The average New Yorker directly consumes about 2,225 kWh hours of electricity per year for residential use and another 1,825 kWh indirectly, through the use of electricity in the commercial and industrial sectors.

    That means that the average New Yorker spends about $400 per year on electricity for the home and another $325 indirectly. The RGGI carbon price adds another $2.25 and $1.82 to each of these respective sums, about $4 in total.
26. Potomac Economics, *Annual Report on the Market for RGGI CO2 Allowances: 2011*.
27. Paul J. Hibbard and Susan F. Tierney, "Carbon Control and the Economy: Economic Impacts of RGGI's First Three Years," *The Electricity Journal* 24, no. 10 (December 2011): 30–40, doi:10.1016/j.tej.2011.10.020.
28. Potomac Economics, *Annual Report on the Market for RGGI CO2 Allowances: 2011*.
29. Hibbard and Tierney, "Carbon Control and the Economy: Economic Impacts of RGGI's First Three Years."
30. M. Pahle, "Germany's Dash for Coal: Exploring Drivers and Factors," *Energy Policy* 38 no. 7 (2010): 3431–42; published online March 2010.
31. European Commission, "Emissions Trading 2007: 2007 Verified Emissions from EU ETS Businesses" (European Commission, 2006), http://europa.eu/rapid/pressReleasesAction.do?reference=IP/08/787&format=HTML&aged=0&language=EN&guiLanguage=en.
32. Pahle, "Germany's Dash for Coal: Exploring Drivers and Factors."
33. BUND, "Geplante Und Im Bau Befindliche Kohlekraftwerke" (Berlin: Friends of the Earth Germany, 2009).
34. Since 2009, many of the planned plants have been canceled because they have failed to win local regulatory and permitting approval.
35. I will describe CCS in detail in Chapter 8.
36. Pahle, "Germany's Dash for Coal: Exploring Drivers and Factors."

37. Giulio Cainelli, Massimiliano Mazzanti, and Simone Borghesi, *The European Emission Trading Scheme and Environmental Innovation Diffusion: Empirical Analyses Using Italian CIS Data* (University of Ferrara, Department of Economics, January 2012), http://ideas.repec.org/p/udf/wpaper/201201.html.
38. Gwladys Fouche, "Sweden's Carbon Tax Solution to Climate Change Puts It Top of the Green List," *The Guardian*, April 29, 2008, www.guardian.co.uk/environment/2008/apr/29/climatechange.carbonemissions.
39. Ibid.
40. J. Aldy and R. Stavins, "The Promise and Problems of Pricing Carbon: Theory and Experience," HKS Faculty Research Working Paper Series RWP11-041 (John F. Kennedy School of Government, 2011), http://dash.harvard.edu/bitstream/handle/1/5347069/RWP11-041_Aldy_Stavins.pdf?sequence=1.
41. Boqiang Lin and Xuehui Li, "The Effect of Carbon Tax on per Capita CO_2 Emissions," *Energy Policy* 39, no. 9 (September 2011): 5137–46, doi:10.1016/j.enpol.2011.05.050.
42. The control group was Austria, Belgium, the Czech Republic, France, Greece, Hungary, Iceland, Ireland, Luxembourg, Poland, Portugal, Slovakia, and Spain.
43. Lin and Li, "The Effect of Carbon Tax on per Capita CO_2 Emissions."
44. It is also worth noting that the most recent IPCC assessment report, in its chapter on national policy instruments, cited the Lin and Li (2011) paper as one of several papers in support of the proposition that there is some evidence that Scandinavian carbon taxes have a beneficial impact. They did not distinguish the Lin and Li results from other results, and, indeed, of the other papers they did cite, none provides empirical evidence to contradict the Lin and Li findings.
45. Aldy and Stavins, "The Promise and Problems of Pricing Carbon: Theory and Experience."
46. Aldy and Stavins, "The Promise and Problems of Pricing Carbon: Theory and Experience."
47. Ministry of Environment, "B.C. GHG Inventory Report" (Government of British Columbia, 2011), www.env.gov.bc.ca/cas/mitigation/ghg_inventory/index.html.
48. Gasbuddy.com, "Historical Price Charts," 2012, http://gasbuddy.com/gb_retail_price_chart.aspx. Accessed August 2012.
49. Ibid.

50. Roger Pielke, *The Climate Fix* (New York: Basic Books, 2010).
51. Gwladys Fouche, "Where Tax Goes up to 60 per Cent, and Everybody's Happy Paying It," *The Guardian*, November 16, 2008, sec. The Observer, www.guardian.co.uk/money/2008/nov/16/sweden-tax-burden-welfare.
52. Lin and Li, "The Effect of Carbon Tax on per Capita CO_2 Emissions."
53. Here, I begin discussing the social cost of CO_2, matching the fact that most market instruments in existence place a tax or a price on tons of CO_2. The literature on the social cost of carbon typically speaks of the social cost of carbon in tons of C. Given that oxygen has a typical molecular weight of 16 and carbon a molecular weight of 12, one ton of C translates into about 3.7 tons of CO_2.
54. R. Mendelsohn et al., "Country-Specific Market Impacts of Climate Change," *Climatic Change* 45, no. 3–4 (2000): 553–69; R. Mendelsohn, "The Social Cost of Carbon: An Unfolding Value" (presented at the Social Cost of Carbon Conference, London, UK, 2003).
55. Richard S. J. Tol, "The Social Cost of Carbon: Trends, Outliers and Catastrophes," *Economics: The Open-Access, Open-Assessment E-Journal*, 2008, doi:10.5018/economics-ejournal.ja.2008-25.
56. Martin L. Weitzman, "Fat-Tailed Uncertainty in the Economics of Catastrophic Climate Change," *Review of Environmental Economics and Policy* 5, no. 2 (July 1, 2011): 275–92, doi:10.1093/reep/rer006.
57. N. Goodwin et al., *Microeconomics in Context* (Armonk, NY: Sharpe, 2014).
58. The mathematics of elasticities is a bit complicated because elasticity refers to effects of tiny changes in price. If the price of the Ford were to rise by 10 percent, the demand would not fall by 28 percent, but rather by more like 17 percent. The effect is similar to that of compounding interest.
59. Tomas Havranek, Zuzana Irsova, and Karel Janda, "Demand for Gasoline Is More Price-Inelastic than Commonly Thought," *Energy Economics* 34, no. 1 (January 2012): 201–7, doi:10.1016/j.eneco.2011.09.003.
60. There have been other meta-analyses of the elasticity data, and, by and large, these have suggested the price elasticity of gasoline demand to be somewhat greater than the study by Havranek, Irsova, and Janda. Havranek et al., however, suggested that these previous studies had systematically discounted some estimates of price elasticity, resulting in a bias in the direction of seeing greater

elasticity. Even if this were not the case and demand were more elastic, as per the previous studies, this would not change the qualitative arguments that I make in this chapter.

61. Mark G. Lijesen, "The Real-Time Price Elasticity of Electricity," *Energy Economics* 29, no. 2 (March 2007): 249–58, doi:10.1016/j.eneco.2006.08.008.
62. Energy in Sweden, as in most countries, has long been taxed at a much higher rate than in the United States. The Swedish carbon tax, then, comes on top of a higher price, meaning that the proportionate change in prices would be less. Adding a $3.50 per gallon tax in a country such as the United States would double the price of gasoline; in Europe, this would only raise it by 50 percent. The change in demand as a result of the tax would hence be lower.
63. Lovins, Odum, and Rowe, *Reinventing Fire: Bold Business Solutions for the New Energy Era.*
64. S. Lüthi, "Effective Deployment of Photovoltaics in the Mediterranean Countries: Balancing Policy Risk and Return," *Solar Energy* 84, no. 6 (June 2010): 1059–71, doi:10.1016/j.solener.2010.03.014.
65. The values for hydropower, geothermal electricity, biomass electricity, and ocean electricity are copied directly from the IPCC report. I have altered the values for wind and solar from those presented in the IPCC report based on more recent cost estimates, the details of which I cover in Chapter 8. I have added my own estimates for nuclear and CCS based on published values. The translation between the added cost of CCS and the equivalent carbon price is contingent on the percentage of CO_2 emissions actually captured. CCS can never remove 100 percent of the CO_2, and most studies assume that, in practice, it will capture 85–90 percent. The full references are: IPCC, *Special Report on Renewable Energy* (Cambridge UK: Cambridge University Press, 2011); Peter Bradford, "Energy Policy: The Nuclear Landscape," *Nature* 483, no. 7388 (March 8, 2012): 151–52, doi:10.1038/483151a; Arnulf Grubler, "The Costs of the French Nuclear Scale-up: A Case of Negative Learning by Doing," *Energy Policy* 38, no. 9 (September 2010): 5174–88, doi:10.1016/j.enpol.2010.05.003; Edward S. Rubin, Chao Chen, and Anand B. Rao, "Cost and Performance of Fossil Fuel Power Plants with CO_2 Capture and Storage," *Energy Policy* 35, no. 9 (September 2007): 4444–54, doi:10.1016/j.enpol.2007.03.009.

66. K. Galbraith, “Gulf Coast Wind Farms Spring Up, as Do Worries,” *New York Times*, February 10, 2011, www.nytimes.com/2011/02/11/us/11ttwind.html?_r=1.
67. Antonella Battaglini et al., “Development of SuperSmart Grids for a More Efficient Utilization of Electricity from Renewable Resources,” *Journal of Cleaner Production* 17 (2009): 911–18.
68. K. Galbraith, “Lines Go up to Ferry Wind Energy to Major Cities,” *New York Times*, October 21, 2011, www.nytimes.com/2011/10/21/us/lines-go-up-to-ferry-wind-energy-to-major-cities.html.
69. G. Schellekens et al., “Moving towards 100% Renewable Electricity in Europe and North Africa by 2050: Evaluating Progress in 2010” (London: PricewaterwaterhouseCoopers, 2011).
70. C. Carraro and A. Favero, “The Economic and Financial Determinants of Carbon Price,” *Czech Journal of Economics and Finance* 59, no. 5 (2009): 396–409; Julien Chevallier, “Carbon Futures and Macroeconomic Risk Factors: A View from the EU ETS,” *Energy Economics* 31, no. 4 (July 2009): 614–25, doi: 10.1016/j.eneco.2009.02.008; Zhen-Hua Feng, Le-Le Zou, and Yi-Ming Wei, “Carbon Price Volatility: Evidence from EU ETS,” *Applied Energy* 88, no. 3 (March 2011): 590–98, doi:10.1016/j.apenergy.2010.06.017. J. Zhao, “Irreversible Abatement Investment under Cost Uncertainties: Tradable Emission Permits and Emissions Charges,” *Journal of Public Economics* 87, no. 12 (December 2003): 2765–89, doi: 10.1016/S0047-2727(02)00135-4.
71. William Blyth et al., “Investment Risks under Uncertain Climate Change Policy,” *Energy Policy* 35, no. 11 (November 2007): 5766–73, doi: 10.1016/j.enpol.2007.05.030; Ming Yang et al., “Evaluating the Power Investment Options with Uncertainty in Climate Policy,” *Energy Economics* 30, no. 4 (July 2008): 1933–50, doi: 10.1016/j.eneco.2007.06.004.
72. William Blyth et al., “Policy Interactions, Risk and Price Formation in Carbon Markets,” *Energy Policy* 37, no. 12 (December 2009): 5192–207, doi: 10.1016/j.enpol.2009.07.042; Lin Fan, Benjamin F. Hobbs, and Catherine S. Norman, “Risk Aversion and CO_2 Regulatory Uncertainty in Power Generation Investment: Policy and Modeling Implications,” *Journal of Environmental Economics and Management* 60, no. 3 (November 2010): 193–208, doi: 10.1016/j.jeem.2010.08.001; Lin Fan, Catherine S. Norman, and Anthony G. Patt, “Electricity Capacity Investment under Risk Aversion: A Case Study of Coal, Gas, and Concentrated Solar Power,” *Energy*

Economics 34, no. 1 (January 2012): 54–61, doi:10.1016/j.eneco.2011.10.010.

73. Blyth et al., "Investment Risks under Uncertain Climate Change Policy"; P. Reinelt and D. Keith, "Carbon Capture Retrofits and the Cost of Regulatory Uncertainty," *The Energy Journal* 28 (2007): 101–27.
74. Henry D. Jacoby and A. Denny Ellerman, "The Safety Valve and Climate Policy," *Energy Policy* 32, no. 4 (March 2004): 481–91, doi: 10.1016/S0301-4215(03)00150-2; Peter John Wood and Frank Jotzo, "Price Floors for Emissions Trading," *Energy Policy* 39, no. 3 (March 2011): 1746–53, doi: 10.1016/j.enpol.2011.01.004; Jana Szolgayova, Sabine Fuss, and Michael Obersteiner, "Assessing the Effects of CO_2 Price Caps on Electricity Investments-A Real Options Analysis," *Energy Policy* 36, no. 10 (October 2008): 3974–81, doi: 10.1016/j.enpol.2008.07.006.
75. G. Eskeland et al., "Transforming the European Energy System," in *Making Climate Change Work for Us: European Perspectives on Adaptation and Mitigation Strategies*, ed. M. Hulme and H. Neufeldt (Cambridge, UK: Cambridge University Press, 2010), 165–99.
76. Yongfu Huang and Terry Barker, "The Clean Development Mechanism and Sustainable Development: A Panel Data Analysis," *Energy Economics* 34: (2012) 1033–40.
77. A.G. Patt et al., "What Can Social Science Tell Us about Meeting the Challenges of Climate Change? Five Insights from Five Years That Might Make a Difference," in *Making Climate Change Work for Us: European Perspectives on Adaptation and Mitigation Strategies*, ed. M. Hulme and H. Neufeldt (Cambridge, UK: Cambridge University Press, 2010), 369–88.
78. Gregory F. Nemet and Daniel M. Kammen, "U.S. Energy Research and Development: Declining Investment, Increasing Need, and the Feasibility of Expansion," *Energy Policy* 35, no. 1 (January 2007): 746–55, doi:10.1016/j.enpol.2005.12.012.
79. Mary Jean Bürer and Rolf Wüstenhagen, "Which Renewable Energy Policy Is a Venture Capitalist's Best Friend? Empirical Evidence from a Survey of International Cleantech Investors," *Energy Policy* 37, no. 12 (December 2009): 4997–5006, doi: 10.1016/j.enpol.2009.06.071.
80. Anne Held, Mario Ragwitz, and Reinhard Haas, "On the Success of Policy Strategies for the Promotion of Electricity from Renewable Energy Sources in the EU," *Energy & Environment* 17, no. 6 (November 1, 2006): 849–68, doi:10.1260/095830506779398849.

81. E. Knight, “The Economic Geography of Clean Tech Venture Capital” (Oxford: University of Oxford, 2010).
82. N. Johnstone, I. Hascic, and D. Popp, “Renewable Energy Policies and Technological Innovation: Evidence Based on Patent Counts,” *Environmental and Resource Economics* 45, no. 1 (2010): 133–55.
83. G. Nemet, “Beyond the Learning Curve: Factors Influencing Cost Reductions in Photovoltaics,” *Energy Policy* 34 (2006): 3218–32.
84. J. Perloff, *Microeconomics: Theory and Applications with Calculus* (New York: Prentice Hall, 2008).
85. I can’t speak for the whole country, but I can speak for the outdoor public pools in the town where I lived, Perchtoldsdorf, and the neighboring city, Mödling. Both have ashtrays available for people sunbathing next to the poolside, despite no smoking signs at the entrance. I think people interpret the signs to only cover the inside spaces, like the locker rooms. Outside, a thin cloud of smoke coats the surface of the water, right where the swimmers need to breathe with each stroke.
86. T. Friedman, “Get It Right on Gas,” *New York Times*, August 4, 2012, www.nytimes.com/2012/08/05/opinion/sunday/friedman-get-it-right-on-gas.html?src=rechp.
87. Pielke, *The Climate Fix*; Gwyn Prins et al., “The Hartwell Paper” (London School of Economics and Oxford University, 2010).
88. D. Fullerton, A. Leicester, and S. Smith, *Environmental Taxes*, Dimensions in Tax Design (Oxford, UK: Oxford University Press, 2010); Goulder, “Environmental Taxation and the Double Dividend: A Reader’s Guide.”
89. Aldy and Stavins, “The Promise and Problems of Pricing Carbon: Theory and Experience.”
90. I. Parry and R. Williams, “What Are the Costs of Meeting Distributional Objectives for Climate Policy?” (NBER Working Paper No. 16486, 2010), www.nber.org/papers/w16486.

5 Striking a global bargain

1. M. Grosjean, “Der Taleinungsbrief” (University of Bern, date unknown), www.giub.unibe.ch/cde/projects/griwa/viel/la_talein.htm.
2. Ibid.
3. Dryzek, *The Politics of the Earth: Environmental Discourses*.

4. J. Nash, "Equilibrium Points in N-Person Games," *Proceedings of the National Academy of Sciences of the United States of America* 36 (1950): 48–49.
5. G. Hardin, "The Tragedy of the Commons," *Science* 162, no. 3859 (1968): 1243–48.
6. E. Ostrom, *Governing the Commons: The Evolution of Institutions for Collective Action* (New York: Cambridge University Press, 1990).
7. Prins et al., "The Hartwell Paper."
8. One very large and very important Annex 1 country, the United States, is not a party to the Kyoto Protocol.
9. Axel Michaelowa and Frank Jotzo, "Transaction Costs, Institutional Rigidities and the Size of the Clean Development Mechanism," *Energy Policy* 33, no. 4 (2005): 511; Huang and Barker, "The Clean Development Mechanism and Sustainable Development: A Panel Data Analysis."
10. Michael Jacobs, "The Doha Climate Talks Were a Start, but 2015 Will Be the Moment of Truth," *The Guardian*, December 10, 2012, sec. Comment, www.guardian.co.uk/commentisfree/2012/dec/10/doha-climate-talks-global-warming.
11. M. den Elzen et al., *The Emissions Gap Report: Are the Copenhagen Accord Pledges Sufficient to Limit Global Warming to 2 Deg. C or 1.5 Deg. C? A Preliminary Assessment* (Nairobi: United Nations Environment Programme, 2010).
12. Matthew Hoffmann, *Climate Governance at the Crossroads: Experimenting with a Global Response after Kyoto* (Oxford, UK: Oxford University Press, 2011).
13. Gwyn Prins and Steve Rayner, "Time to Ditch Kyoto," *Nature* 449, no. 973–75 (2007).
14. Richard Benedick, *Ozone Diplomacy: New Directions in Safeguarding the Planet* (Cambridge, MA: Harvard University Press, 1991).
15. Paul Krugman, "Expectations and the Confidence Fairy," *New York Times*, September 23, 2012, sec. Opinion, http://krugman.blogs.nytimes.com/2012/09/23/expectations-and-the-confidence-fairy/.
16. D. Victor, *Global Warming Gridlock* (New York: Cambridge University Press, 2011).
17. John Baylis, Steve Smith, and Patricia Owens, *The Globalization of World Politics: An Introduction to International Relations*, Fifth Edition (Oxford, UK: Oxford University Press, 2011); Kurt Taylor Gaubatz, "Democratic States and Commitment in International Relations," *International Organization* 50, no. 01 (1996): 109–39, doi:10.1017/S0020818300001685; Kal Raustiala and Anne-Marie Slaughter,

"International Law, International Relations and Compliance," Working Paper (Princeton, NJ: Princeton University Press, 2007), http://papers.ssrn.com/sol3/papers.cfm?abstract_id=347260.

18. R. Vincent, *Human Rights and International Relations* (Cambridge, UK: Cambridge University Press, 1986).
19. Baylis, Smith, and Owens, *The Globalization of World Politics: An Introduction to International Relations*.
20. Raustiala and Slaughter, "International Law, International Relations and Compliance."
21. L. Nass, P. Pereira, and D. Ellis, "Biofuels in Brazil: An Overview," *Crop Science* 47 (2007): 2228–37.
22. E. Broughton, A. Brent, and L. Haywood, "Application of a Multi-Criteria Analysis Approach for Decision-Making in the Energy Sector: The Case of Concentrating Solar Power in South Africa," *Energy & Environment* 23, no. 8 (December 1, 2012): 1221–32, doi:10.1260/0958-305X.23.8.1221; M. Carolin Mabel and E. Fernandez, "Growth and Future Trends of Wind Energy in India," *Renewable and Sustainable Energy Reviews* 12, no. 6 (August 2008): 1745–57, doi:10.1016/j.rser.2007.01.016.
23. ZhongXiang Zhang, "China in the Transition to a Low-Carbon Economy," *Energy Policy* 38, no. 11 (November 2010): 6638–53, doi:10.1016/j.enpol.2010.06.034.
24. Robert Wright, *Nonzero: The Logic of Human Destiny* (New York: Random House, 2000), 209.
25. Knopf et al., "Managing the Low-Carbon Transition: From Model Results to Policies"; Van Vuuren et al., "Low Stabilization Scenarios and Implications for Major World Regions from an Integrated Assessment Perspective."
26. M. Hulme, *Why We Disagree about Climate Change* (Cambridge, UK: Cambridge University Press, 2010).
27. Sheila M. Olmstead and Robert N. Stavins, "Three Key Elements of a Post-2012 International Climate Policy Architecture," *Review of Environmental Economics and Policy* 6, no. 1 (Winter 2012): 65–85, doi:10.1093/reep/rer018.
28. Mustafa H. Babiker, "Climate Change Policy, Market Structure, and Carbon Leakage," *Journal of International Economics* 65, no. 2 (March 2005): 421–45, doi:10.1016/j.jinteco.2004.01.003.
29. Richard N. Cooper, "Financing for Climate Change," *Green Perspectives* 34, Supplement 1, no. 0 (November 2012): S29–33, doi:10.1016/j.eneco.2012.08.040.

30. Knopf and Edenhofer, "The Economics of Low Stabilization: Implications for Technological Change and Policy."
31. Arthur P. J. Mol, "Carbon Flows, Financial Markets and Climate Change Mitigation," *Environmental Development* 1, no. 1 (January 2012): 10–24, doi:10.1016/j.envdev.2011.12.003.
32. Steffen Böhm, Maria Ceci Misoczky, and Sandra Moog, "Greening Capitalism? A Marxist Critique of Carbon Markets," *Organization Studies* 33, no. 11 (November 1, 2012): 1617–38, doi:10.1177/0170840612463326.
33. Gert Spaargaren and Arthur P. J. Mol, "Carbon Flows, Carbon Markets, and Low-Carbon Lifestyles: Reflecting on the Role of Markets in Climate Governance," *Environmental Politics* 22, no. 1 (February 1, 2013): 174–93, doi:10.1080/09644016.2013.755840.
34. R. M. Martin, *State of the World's Forests* (Rome: Food and Agriculture Organization of the United Nations, 2012), www.fao.org/docrep/016/i3010e/i3010e.pdf.
35. Lera Miles and Valerie Kapos, "Reducing Greenhouse Gas Emissions from Deforestation and Forest Degradation: Global Land-Use Implications," *Science* 320, no. 5882 (June 13, 2008): 1454–55, doi:10.1126/science.1155358.
36. Constance L. McDermott, Kelly Levin, and Benjamin Cashore, "Building the Forest-Climate Bandwagon: REDD+ and the Logic of Problem Amelioration," *Global Environmental Politics* 11, no. 3 (August 5, 2011): 85–103, doi: 10.1162/GLEP_a_00070.
37. Patt et al., "Estimating Least-Developed Countries' Vulnerability to Climate-Related Extreme Events over the next 50 Years."
38. Striessnig, Lutz, and Patt, "Effects of Educational Attainment on Climate Risk Vulnerability."
39. World Bank, "The Economics of Adaptation to Climate Change: Synthesis Report."
40. It is worth pointing out that David Victor's critique of the UNFCCC and Kyoto architecture goes back to the years when those documents were being negotiated. In 1991, he published a commentary in the journal *Nature* and an Op-Ed in the *Washington Post* raising many of the issues that later appear in his book. According to him, these articles earned him a stiff rebuke from the U.S. negotiators going to Rio. For his ideas at the time, see D. Victor, "How to Slow Global Warming," *Nature* 349 (February 7, 1991): 6309.

6 Changing the way we live

1. Will Steffen et al., "The Anthropocene: Conceptual and Historical Perspectives," *Philosophical Transactions of the Royal Society A: Mathematical, Physical and Engineering Sciences* 369, no. 1938 (March 13, 2011): 842–67, doi:10.1098/rsta.2010.0327.
2. Lester Brown, "Gross World Product 1950–2011" (Washington, DC: Earth Policy Institute, 2012), www.earth-policy.org/datacenter/xls/indicator2_2012_01.xls.
3. O. Edenhofer, R. Pichs-Madruga, and Y. Sokona, *IPCC, 214: Climate Change 2014: Mitigation of Climate Change* (Cambridge, UK/New York: Cambridge University Press, 2014).
4. J. Rockström et al., "A Safe Operating Space for Humanity," *Nature* 461, no. 7263 (2009): 472–75.
5. Stewart Cohen et al., "Climate Change and Sustainable Development: Towards Dialogue," *Global Environmental Change* 8, no. 4 (November 1998): 341–71, doi:10.1016/S0959-3780(98)00017-X; Anke Fischer et al., "Energy Use, Climate Change and Folk Psychology: Does Sustainability Have a Chance? Results from a Qualitative Study in Five European Countries," *Symposium on Social Theory and the Environment in the New World (dis)Order* 21, no. 3 (August 2011): 1025–34, doi:10.1016/j.gloenvcha.2011.04.008; Fritz Reusswig and Wiebke Lass, "Post-Carbon Ambivalences – The New Climate Change Discourse and the Risks of Climate Science," *Science, Technology & Innovation Studies* 6, no. 2 (2010): *Ambiguous Progress*, 2011, www.sti-studies.de/ojs/index.php/sti/article/view/42.
6. Jay Forrester, "Lessons from System Dynamics Modeling," *System Dynamics Review* 3, no. 2 (1987).
7. Jay Forrester, *Urban Dynamics* (Cambridge, MA: MIT Press, 1969).
8. Donella Meadows et al., *The Limits to Growth* (Universe Books, 1972).
9. Jared Diamond, *Collapse: How Societies Choose to Fail or Succeed* (New York: Viking, 2004).
10. For more information on the ICLEI program, I highly recommend the Wikipedia site: http://en.wikipedia.org/wiki/Cities_for_Climate_Protection_program
11. www.ci.berkeley.ca.us/Mayor/Home/Berkeley_FIRST_(Financing_Initiative_for_Renewable_and_Solar_Technology).aspx
12. Michele Betsill and Harriet Bulkeley, "Looking Back and Thinking Ahead: A Decade of Cities and Climate Change Research," *Local Environment* 12, no. 5 (October 2007): 447–56.

13. I calculated this myself, more or less on the back of an envelope, using data on emissions released by the U.S. Energy Information Administration.
14. Sustainable Development Commission, *Prosperity without Growth* (London: British Sustainable Development Commission, 2009); Sustainable Development Commission, *Redefining Prosperity: Resource Productivity, Economic Growth and Sustainable Development* (London: British Sustainable Development Commission, 2003), www.sd-commission.org.uk/data/files/publications/030627percent20Redefining%20percent20prosperity,percent20resource percent20productivity.pdf.
15. Leaf Van Boven, "Experientialism, Materialism, and the Pursuit of Happiness," *Review of General Psychology* 9, no. 2 (2005): 132–42; Bruno S. Frey and Alois Stutzer, "Happiness, Economy and Institutions," *The Economic Journal* 110, no. 466 (2000): 918; Alois Stutzer and Bruno Frey, "Recent Developments in the Economics of Happiness: A Selective Overview," IZA Discussion Paper No. 7078 (Basel: Institute for the Study of Labor, University of Basel 2012) http://papers.ssrn.com/sol3/papers.cfm?abstract_id=2192854.
16. BBC, "Well-Being Survey: Three-Quarters Satisfied with Lives," *BBC News UK*, December 2, 2011, www.bbc.co.uk/news/uk-15989841.
17. Ibid.
18. Robert Putnam, *Bowling Alone: The Collapse and Revival of the American Community* (New York: Simon and Schuster, 2000).
19. Robert Skidelsky and Edward Skidelsky, *How Much Is Enough: Money and the Good Life* (New York: Other Press, 2012), 167.
20. Skidelsky and Skidelsky, *How Much Is Enough: Money and the Good Life*.
21. Sustainable Development Commission, *Prosperity without Growth*.
22. Ibid.
23. Edenhofer, Pichs-Madruga, and Sokona, *IPCC, 214: Climate Change 2014: Mitigation of Climate Change*.
24. IPCC WG3, AR5, chapter 5, page 30 of the post-plenary prepublication draft.
25. Arnulf Grübler, N. Nakicenovic, and D. Victor, "Dynamics of Energy Technologies and Global Change," *Energy Policy* 27 (1999): 247–80.
26. Kerstin Damerau et al., „Direct Impacts of Alternative Energy Scenarios on Water Demand in the Middle East and North Africa." Published online February 2015 at *Climatic Change*.

27. Stefan Pfenninger et al., "Potential for Concentrating Solar Power to Provide Baseload and Dispatchable Power," *Nature Climate Change* advance online publication (June 22, 2014), http://dx.doi.org/10.1038/nclimate2276.
28. K. Damerau et al., "Costs of Reducing Water Use of Concentrating Solar Power to Sustainable Levels: Scenarios for North Africa," *Energy Policy* 39 (2011): 4391–98.
29. Fabian Wagner et al., "Sectoral Marginal Abatement Cost Curves: Implications for Mitigation Pledges and Air Pollution Co-Benefits for Annex I Countries," *Sustainability Science* 7, no. 2 (July 1, 2012): 169–84, doi:10.1007/s11625-012-0167-3.

7 Theories of transitions

1. For example: Prins and Rayner, "Time to Ditch Kyoto"; Prins et al., "The Hartwell Paper"; Naam, *The Infinite Resource: The Power of Ideas on a Finite Planet*; Pielke, *The Climate Fix*. Also, Andrew Revkin of the *New York Times* has started to frame climate change this way in his Dot Earth blog.
2. Frank Geels, *Technological Transition and System Innovations: A Co-Evolutionary and Socio-Technical Analysis* (Cheltenham, UK: Edward Elgar, 2005).
3. C. Darwin, *On the Origin of Species by Natural Selection* (London: John Murray, 1859).
4. Richard Dawkins, *The Selfish Gene* (Oxford, UK: Oxford University Press, 1976).
5. K. Marx and F. Engels, *Manifest Der Kommunisticschen Partei* (London: J. E. Burghard, 1848).
6. T. Veblen, *The Theory of the Leisure Class* (New York: Macmillan, 1899).
7. J. Schumpeter, *Theorie Der Wirtschaftlichen Entwicklung. Eine Untersuchung Über Unternehmergewinn, Kapital, Kredit, Zins, Und Den Konjunkturzyklus* (Berlin, 1912).
8. R. Nelson and S. Winter, *An Evolutionary Theory of Economic Change* (Cambridge, MA: Harvard University Press, 1982).
9. W. Brian Arthur, "Competing Technologies, Increasing Returns, and Lock-in by Historical Events," *The Economic Journal* 99 (1989): 116–31; W. B. Arthur, *Increasing Returns and Path Dependence in the Economy* (Anne Arbor: University of Michigan Press, 1994).

10. Chevallier, "Carbon Futures and Macroeconomic Risk Factors: A View from the EU ETS."
11. Richard Thaler and E. J. Johnson, "Gambling with the House Money and Trying to Break Even: The Effects of Prior Outcomes on Risky Choice," *Management Science* 36 (1990): 643–60.
12. Daniel Kahneman and Amos Tversky, "Prospect Theory: An Analysis of Decision under Risk," *Econometrica* 47 (1979): 263–91.
13. As of June 27, 2014, Google Scholar told me that the paper had been cited more than 29,000 times.
14. Kahneman and Tversky, "Prospect Theory: An Analysis of Decision under Risk."
15. Jonathan Baron and Ilana Ritov, "Reference Points and Omission Bias," *Organizational Behavior and Human Decision Processes* 59 (1994): 475–98.
16. A. G. Patt and Richard Zeckhauser, "Action Bias and Environmental Decisions," *Journal of Risk and Uncertainty* 21, no. 1 (2000): 45–72.
17. This is illustrated by the story a colleague of mine told me. He was teaching a statistics class and asked his class what the probability was of rolling a six-sided die while playing Monopoly and having it come up as a six.

    "About fifty percent," a student replied.

    "Explain that one," my colleague said, amazed at what he was hearing.

    "Well," said the student," either it comes up a six, or it doesn't. So to me it's fifty-fifty."
18. William Samuelson and Richard Zeckhauser, "Status Quo Bias in Decision Making," *Journal of Risk and Uncertainty* 1 (1988): 7–59.
19. Narayanan Kandasamy et al., "Cortisol Shifts Financial Risk Preferences," *Proceedings of the National Academy of Sciences*, February 18, 2014, doi:10.1073/pnas.1317908111.
20. M. Thompson, "The Quest for 'Clumsy Solutions' in Nepal's Mountains," *IIASA Options Magazine*, Winter 2011.
21. Michael Thompson, Richard Ellis, and Aaron B. Wildavsky, *Cultural Theory* (New York: Westview Press, 1990).
22. Marco Verweij and Michael Thompson, *Clumsy Solutions for a Complex World: Governance, Politics, and Plural Perceptions* (New York: Palgrave Macmillan, 2006).
23. E. Kolbert, "The Island in the Wind: A Danish Community's Victory over Carbon Emissions," *The New Yorker*, July 7, 2008.

24. ETH stands for Eidgenössische Technische Hochschule, which translates into English as Swiss Federal Institute of Technology.
25. A. O'Connor, "A Call for a Low-Carb Diet that Embraces Fat," *New York Times*, September 1, 2014, sec. Health.
26. Lydia A. Bazzano et al., "Effects of Low-Carbohydrate and Low-Fat Diets: A Randomized Trial," *Annals of Internal Medicine* 161, no. 5 (September 2, 2014): 309–18, doi:10.7326/M14-0180.
27. C. L. Ogden et al., "Prevalence of Childhood and Adult Obesity in the United States, 2011–2012," *JAMA* 311, no. 8 (February 26, 2014): 806–14, doi:10.1001/jama.2014.732.
28. CDC Newsroom, "New CDC Data Show Encouraging Development in Obesity Rates among 2 to 5 Year Olds," February 25, 2014, www.cdc.gov/media/releases/2014/p0225-child-obesity.html.
29. C. McDougall, *Born to Run: A Hidden Tribe, Superathletes, and the Greatest Race the World Has Never Seen* (New York: Knopf, 2009).
30. Two examples come immediately to mind. One also centers on the diet issue, in terms of salt intake. For years, we have been told that reducing the intake of salt leads to lower blood pressure and a reduced risk of heart disease. This belief was built on the simple fact that when people do eat a lot of salt, their bodies maintain the salinity of the blood by increasing the water in the blood as well, which leads to a temporary increase in blood pressure. Recent evidence suggests strongly that limiting salt intake to levels recommended in the United States may actually lead to health complications, and although excessive consumption has negative outcomes, salt levels in a traditional diet appear to be healthy. A second example concerns highway safety. For years, it was believed that widening and straightening roads would lead to greater safety based on the simple intuition that drivers would have to make fewer inputs to avoid hitting obstacles. It turns out that this has had the opposite effect. On wide, straight roads, drivers stop paying attention and are much more likely to fall asleep at the wheel. The modern thinking in road safety is to design roads and village centers in such a way that drivers need to pay close attention to what is in front of them in order to avoid hitting things and, in doing so, tend to slow down. T. Vanderbilt, *Traffic* (New York: Knopf, 2008); M. Bittman, "Can You Eat Too Little Salt?," *New York Times*, May 28, 2013.
31. Sustainable Development Commission, *Prosperity without Growth.*

8 Strategic technologies

1. To keep it both simple and relevant while matching data availability, this figure mixes and matches both the geographic scale being considered and the sources of data. The overall picture for the energy system is for the OECD countries in the aggregate, in 2009, with the source of the data being the International Energy Agency (IEA). For electricity, it portrays the United States in 2013, based on data from the Energy Information Administration (EIA). For the transportation sector, the data are for the United States for 2012 and come from Oak Ridge National Laboratory. Note that the data related to heat do not include heat generated by electricity; these are covered under the electricity data.
2. That is why one often sees very low LCOE values for old coal and gas power plants and why these power plants are often the most profitable to operate. The company has depreciated them over a period of time that is consistent with the accounting rules, calculating an LCOE during those years based on this expected plant lifetime. Once the plant is fully depreciated, the capital cost of the plant falls to zero, and it is primarily the operating costs that dictate its LCOE from that point forward. Of course, the later LCOE can only be so low because the earlier LCOE was too high. The true LCOE would be the average over its whole operating life.
3. U.S. EIA, *Annual Energy Outlook 2014* (Washington, DC: US DoE, 2014).
4. K. Branker, M. J. M. Pathak, and J. M. Pearce, "A Review of Solar Photovoltaic Levelized Cost of Electricity," *Renewable and Sustainable Energy Reviews* 15, no. 9 (December 2011): 4470–82, doi:10.1016/j.rser.2011.07.104.
5. EIA, *International Energy Statistics, Electricity Generation* (Energy Information Administration, 2009), http://tonto.eia.doe.gov/cfapps/ipdbproject/IEDIndex3.cfm?tid=2&pid=2&aid=12.
6. See www.tepco.co.jp/en/challenge/energy/nuclear/plants-e.html. Accessed July 31, 2014.
7. By way of comparison, France has fifty-eight nuclear power reactors at twenty-nine sites with a combined capacity of 66 GW. Arnulf Grubler, "The Costs of the French Nuclear Scale-up: A Case of Negative Learning by Doing," *Special Section on Carbon Emissions and Carbon Management in Cities with Regular Papers* 38, no. 9 (September 2010): 5174–88, doi:10.1016/j.enpol.2010.05.003.

8. At least this is what Wikipedia tells us: http://en.wikipedia.org/wiki/List_of_Volkswagen_Group_factories. Accessed July 31, 2014.
9. See www.volkswagengroupamerica.com/facts.html. Accessed July 31, 2014.
10. See www.tepco.co.jp/en/challenge/energy/nuclear/plants-e.html
11. The person who said this was Lewis Strauss, chairman of the U.S. Atomic Energy Commission, in 1954. He was referring to the potential of hydrogen fusion and not of uranium fission, the latter being the process used in today's nuclear power plants. Nevertheless, his opinion took hold in the popular consciousness to describe all of nuclear power. Richard Pfau, *No Sacrifice Too Great: The Life of Lewis L. Strauss* (Charlottsville: University Press of Virginia, 1984).
12. Grubler, "The Costs of the French Nuclear Scale-up: A Case of Negative Learning by Doing," September 2010.
13. Y. Du and J. Parsons, *Update on the Cost of Nuclear Power* (Cambridge, MA: MIT Press, 2009).
14. R. Black, "Nuclear Power Gets Little Support Worldwide," *BBC News*, November 25, 2011.
15. M. Cooper, "Nuclear Power Loses Support in New Poll," *New York Times*, March 22, 2011.
16. Naam, *The Infinite Resource: The Power of Ideas on a Finite Planet*; Prins et al., "The Hartwell Paper."
17. J. Lilliestam, J. Bielicki, and A. G. Patt, "Comparing Carbon Capture and Storage (CCS) with Concentrating Solar Power (CSP): Potentials, Costs, Risks, and Barriers," *Energy Policy* 47 (2012): 447–55.
18. It is worth pointing out that the total amount of CO_2 that would need to be stored, at least by weight, is much more than the amount of fuel that we would burn in the first place. Coal is made up of mostly carbon atoms, each having an atomic mass of twelve. Natural gas is CH_4 molecules, with an atomic mass of sixteen. CO_2, by contrast, has an atomic mass of forty-four.
19. J. Dooley et al., *A First-Order Global Geological CO_2-Storage Potential Supply Curve and Its Application in a Global Integrated Assessment Model* (College Park, MD: Battelle/Pacific Northwest National Laboratory, 2003).
20. Lilliestam, Bielicki, and Patt, "Comparing Carbon Capture and Storage (CCS) with Concentrating Solar Power (CSP)."
21. Ibid.
22. IPCC, *Special Report on Renewable Energy*.

23. T. Dhansay, M. De Wit, and A. Patt, "An Evaluation for Harnessing Low-Enthalpy Geothermal Energy in the Limpopo Province, South Africa," *South African Journal of Science* 110, no. 3/4 (2014): 1–10.
24. Adrian Cho, "Energy's Tricky Tradeoffs," *Science* 329, no. 5993 (August 13, 2010): 786–87, doi:10.1126/science.329.5993.786.
25. Ibid.
26. Observ'ER, *Worldwide Electricity Production from Renewable Electricity Sources* (France, 2013), www.energies-renouvelables.org/observ-er/html/inventaire/pdf/15e-inventaire-Chap02.pdf.
27. C. Vittrup, *2013 Was a Record-Setting Year for Danish Wind Power* (Risø, Denmark: Energinet, 2014), http://energinet.dk/EN/El/Nyheder/Sider/2013-var-et-rekordaar-for-dansk-vindkraft.aspx.
28. U.S. DoE, *2013 Wind Report* (Washington, DC: United States Department of Energy, 2014), http://energy.gov/2013-wind-report.
29. U.S. EIA, *Annual Energy Outlook 2014*.
30. Geraint Ellis, John Barry, and Clive Robinson, "Many Ways to Say No, Different Ways to Say Yes: Applying Q-Methodology to Understand Public Acceptance of Wind Farm Proposals," *Journal of Environmental Planning and Management* 50, no. 4 (2007): 517–51; Jan Zoellner, Petra Schweizer-Ries, and Christin Wemheuer, "Public Acceptance of Renewable Energies: Results from Case Studies in Germany," *Energy Policy* 36, no. 11 (November 2008): 4136–41, doi:10.1016/j.enpol.2008.06.026.
31. K. Williges, Johan Lilliestam, and A. G. Patt, "Making Concentrated Solar Power Competitive with Coal: The Costs of a European Feed-in Tariff," *Energy Policy* 38 (2010): 3089–97, doi:10.1016/j.enpol.2010.01.049.
32. N. Komendantova, A. G. Patt, and K. Williges, "Solar Power Investment in North Africa: Benefits of Reducing Risk Perceptions," *Renewable and Sustainable Energy Reviews* 15, no. 9 (2011): 4829–35.
33. Nadejda Komendantova, Stefan Pfenninger, and Anthony Patt, "Governance Barriers to Renewable Energy in North Africa," *The International Spectator* 49, no. 2 (April 3, 2014): 50–65, doi:10.1080/03932729.2014.907627.
34. Stefan Pfenninger et al., "Potential for Concentrating Solar Power to Provide Baseload and Dispatchable Power," *Nature Climate Change* 4, no. 8 (August 2014): 689–92.
35. P. Claudy, M. Gerdes, and J. Ondraczek, *Germany's Photovoltaic Industry at the Crossroads – Challenges and Opportunities for German*

PV Companies along the Value Chain (Frankfurt, Germany: PricewaterhouseCoopers, 2011).

36. Branker, Pathak, and Pearce, "A Review of Solar Photovoltaic Levelized Cost of Electricity."
37. There are all sorts of possible sources for this value. I got it by using a solar cost calculator developed by the National Renewable Energy Laboratory (NREL) available online at http://pvwatts.nrel.gov.
38. Data from Bloomberg New Energy Finance. August 2014.
39. http://greenzone.co/2014/01/29/solar-panels-cost-1-per-watt-pay-5-per-watt-get-solar-system/. Accessed August 20, 2014.
40. H. Wirth, *Aktuelle Fakten Zur Photovoltaik in Deutschland* (Freiburg, Germany: Fraunhofer ISE, July 28, 2014).
41. Again, I used the PV cost calculator from NREL to come up with these numbers. http://pvwatts.nrel.gov.
42. An important exception to this is in much of the United States during the summer months, when air conditioning accounts for the largest share of power demand.
43. www.swissolar.ch/de/fuer-bauherren/solardachrechner/. Accessed August 20, 2014.
44. The remaining 70 percent of the costs are also tax deductible. If I were not a U.S. citizen, this would represent an additional cost savings. As a U.S. citizen, however, I need to pay U.S. income taxes on my total income, claiming my Swiss income taxes (which are somewhat lower than my U.S. taxes) as a credit toward my U.S. income taxes. The effect of taking the Swiss tax deduction would be to reduce my Swiss tax liability, but also the size of the credit, thereby raising my net U.S. tax liability and leaving my overall tax liability unchanged. The United States also has tax credits and deductions available for those who install PV systems, but these are limited to PV systems actually installed within the United States and hence inapplicable to me.
45. David MacKay, *Sustainable Energy – without the Hot Air* (Cambridge, UK: UIT, 2009).
46. Damerau et al., "Costs of Reducing Water Use of Concentrating Solar Power to Sustainable Levels: Scenarios for North Africa."
47. Damerau et al., "Direct Impacts of Alternative Energy Scenarios on Water Demand in the Middle East and North Africa," unpublished.
48. Zoellner, Schweizer-Ries, and Wemheuer, "Public Acceptance of Renewable Energies: Results from Case Studies in Germany."

49. We haven't published the results yet because we first want to complete the Jordan survey. In Morocco, we asked people for their level of support for the CSP facility being built there, with the answers on a five-point scale (strongly favor, moderately favor, neither favor nor oppose, moderately oppose, strongly oppose). More than 90 percent of the respondents indicated that they favored the project either strongly or moderately.
50. S. Inage, *Modelling Load Shifting Using Electric Vehicles* (Paris: International Energy Agency, 2010), www.iea.org/publications/freepublications/publication/load_shifting.pdf.
51. T. Pregger et al., "Resources, Capacities and Corridors for Energy Imports to Europe," *International Journal of Energy Sector Management* 5, no. 1 (2011): 125–56; Peter Viebahn, Yolanda Lechon, and Franz Trieb, "The Potential Role of Concentrated Solar Power (CSP) in Africa and Europe – A Dynamic Assessment of Technology Development, Cost Development and Life Cycle Inventories until 2050," *Energy Policy* 39, no. 8 (August 2011): 4420–30, doi:10.1016/j.enpol.2010.09.026.
52. W. Kempton et al., "Electric Power from Offshore Wind via Synoptic-Scale Interconnection," *Proceedings of the National Academy of Sciences* 107, no. 16 (2010): 7240–45.
53. Gregor Czisch, "Szenarien Zur Zukünftigen Stromversorgung: Kostenoptimierte Variationen Zur Versorgung Europas Und Seiner Nachbarn Mit Strom Aus Erneuerbaren Energien," ed. Dietmar Hein (Kassel: Universität Kassel, Elektroteknik, 2005).
54. Pfenninger et al., "Potential for Concentrating Solar Power to Provide Baseload and Dispatchable Power," August 2014.
55. European Climate Forum, "Roadmap 2050: A Practical Guide to a Prosperous, Low-Carbon Europe" (The Hague: European Climate Forum, 2010).
56. Sebastiaan Deetman et al., "Deep Greenhouse Gas Emission Reductions in Europe: Exploring Different Options," *Special Section: Long Run Transitions to Sustainable Economic Structures in the European Union and Beyond* 55, no. 0 (April 2013): 152–64, doi:10.1016/j.enpol.2012.11.047.
57. NREL, *Exploration of High-Penetration Renewable Electricity Futures* (Golden, CO: National Renewable Energy Laboratory, 2012).
58. To be fair, the Swiss are not the only ones making money. According to the U.S. DoE Energy Information Administration, the French themselves have more than twice the pumped hydro capacity as the Swiss, 4.3 GW peak capacity compared to 1.8,

whereas the Austrians have 4.4 GW and the Germans 6.7. Both the Swiss and the Austrian cases are particularly interesting because their capacities are quite large relative to the overall size of the power system.

59. The first utility-scale CAES facility entered service in 1978; to date, there are only a handful of such plants. Several more are in the planning stages.
60. http://energystorage.org/energy-storage/technologies/hydrogen-energy-storage
61. Benedikt Battke et al., "A Review and Probabilistic Model of Lifecycle Costs of Stationary Batteries in Multiple Applications," *Renewable and Sustainable Energy Reviews* 25, no. 0 (September 2013): 240–50, doi:10.1016/j.rser.2013.04.023.
62. NREL, *Exploration of High-Penetration Renewable Electricity Futures*; Deetman et al., "Deep Greenhouse Gas Emission Reductions in Europe: Exploring Different Options"; European Climate Forum, "Roadmap 2050: A Practical Guide to a Prosperous, Low-Carbon Europe"; M. Delucchi and M. Jacobson, "Providing All Global Energy with Wind, Water, and Solar Power, Part II: Reliability, System and Transmission Costs, and Policies," *Energy Policy*, in press.
63. This was the cost listed on the Internet at the time of writing. I am now writing this footnote two months later, in the process of copyediting, and in the intervening time, partly as a result of researching this chapter, I have actually bought a Renault Zoe for my own use. I was able to negotiate a discount of CHF 1,000 on the price. I suspect that I could have negotiated about the same on the price of a Clio, leaving the core result in this chapter unchanged.
64. http://cleantechnica.com/2013/07/08/40-drop-in-ev-battery-prices-from-2010-to-2012/. Accessed August 24, 2014.
65. www.mckinsey.com/insights/energy_resources_materials/battery_technology_charges_ahead. Accessed August 24, 2014.
66. www.technologyreview.com/news/516876/forget-battery-swapping-tesla-aims-to-charge-electric-cars-in-five-minutes/. Accessed August 25, 2014.
67. Duncan Kushnir and Björn A. Sandén, "The Time Dimension and Lithium Resource Constraints for Electric Vehicles," *Resources Policy* 37, no. 1 (March 2012): 93–103, doi:10.1016/j.resourpol.2011.11.003; Paul W. Gruber et al., "Global Lithium

Availability," *Journal of Industrial Ecology* 15, no. 5 (October 1, 2011): 760–75, doi:10.1111/j.1530-9290.2011.00359.x.

68. P. Wriggelsworth, "The Car of the Perpetual Future," *The Economist*, September 4, 2008.
69. Ibid.
70. IPCC, *Special Report on Renewable Energy*.
71. The fact is that solar panels convert a much higher share of sunlight into usable energy than do plants.
72. In fact, the top estimate was slightly more than 1,500 EJ. I am told by an expert on biofuels that the 1,500 EJ estimate should be disregarded: it relies on assumptions including the housing of farm animals in multistory stables in order to save more land for biocrops, and its author privately regrets its publication.
73. A. Eisentraut, *Sustainable Production of Second-Generation Biofuels* (Paris: International Energy Agency, 2010).
74. G. Fischer and L. Schrattenholzer, "Global Bioenergy Potentials Through 2050," *Biomass and Bioenergy* 20 (2001): 151–59.
75. IPCC, *Special Report on Renewable Energy*.
76. F. Cherubini et al., "Energy- and Greenhouse Gas-Based LCA of Biofuel and Bioenergy Systems," *Resources, Conservation and Recycling* 53 (2009): 434–47.
77. Data from the U.S. Energy Information Administration, August 2014.
78. Ibid.
79. It is worth noting that the energy content of different fuels is not the same. A liter of today's jet fuel has an energy content slightly higher than a liter of diesel fuel or gasoline, which is in turn slightly higher than that of biofuels currently available. The differences are not enough to influence the qualitative conclusions from the calculations I do here.
80. Technology Quarterly, "Electric Ships: Making Waves," *The Economist*, December 6, 2007.
81. www.aero-news.net/index.cfm?do=main.textpost&id=a3309cef-59ee-4742-8f24-bb7d39a86cf8. Accessed August 29, 2014.
82. George Marsh, "Biofuels: Aviation Alternative?," *Renewable Energy Focus* 9, no. 4 (July 2008): 48–51, doi:10.1016/S1471-0846(08)70138-0; S. Gössling and P. Upham, *Climate Change and Aviation: Issues, Challenges and Solutions* (London: Earthscan, 2009).

83. www.balqon.com/electric-vehicles/nautilus-xe30/. Accessed September 2, 2014.
84. L. von Meiss, "Der Geräuschlose Truck," *Tages Anzeiger*, July 11, 2013.
85. From what I can tell, there hasn't been a lot written on the potential for eliminating emissions from long-distance trucking. The most recent IPCC assessment report was similarly silent on the topic, reporting only on the potential for improving efficiency within the trucking sector while sticking with the internal combustion engine.
86. www.technologyreview.com/news/415773/next-stop-ultracapacitor-buses/?a=f. Accessed September 2, 2014.
87. Technology Quarterly, "Electric Ships: Making Waves."
88. Wärtsilä Corporation, "Pipeline Pumping," 2011. www.wartsila.com/file/russia/1278526926772a1267106724867-pipeline-pumping.pdf
89. M. Beerepoot and A. Marmion, *Policies for Renewable Heat* (Paris: International Energy Agency, 2012).
90. E. Taibi, D. Gielen, and M. Bazilian, *Renewable Energy in Industrial Applications: An Assessment of the 2050 Potential* (Vienna: United Nations Industrial Development Organization, 2011).
91. Ibid.
92. Ibid.; Beerepoot and Marmion, *Policies for Renewable Heat.*
93. The name "portland" cement dates back to its original invention, close to 200 years ago. It refers to the Isle of Portland in the English Channel. Portland cement was seen to create a material with similar properties to the limestone that forms the island's bedrock.
94. I. Amato, "Green Cement: Concrete Solutions," *Nature* 494, no. 7437 (2013): 300–301.
95. One of the reasons that the Netherlands stands out in this respect is that the country has very little land available for waste disposal. As a consequence, policy makers have sought out productive ways to burn it, such as in cement production and also in district heating plants.
96. Thomas Van Dam, *Geopolymer Concrete* (Washington, DC: Federal Highway Administration, U.S. Department of Transportation, 2010).

9 *Energiewende* in the German power sector

1. J. Gillis, "Sun and Wind Alter Global Landscape, Leaving Utilities Behind," *New York Times*, September 13, 2014.
2. A. Berchem, "Das Unterschätzte Gesetz," *Die Zeit*, September 25, 2006.
3. Ibid.
4. Ryan H. Wiser, K. Porter, and Robert Grace, "Evaluating Experience with Renewable Portfolio Standards in the United States," *Mitigation and Adaptation Strategies for Global Change* 10 (2005): 237–63.
5. The source of the data on wind power, seen in the left-hand graph, is the U.S. Energy Information Administration. The right-hand graph combines these data with data on total electricity consumption, which come from the International Energy Agency.
6. There are other countries that deserve to be in this graph, but that I have omitted in the interests of keeping things simple. The United States was the early leader in the 1970s and early 1980s, and then forfeited that role to Denmark. Denmark still accounts for the highest proportion of wind power, although its total capacity is well below that of all three countries shown in the graph. In recent years, other countries such as the United Kingdom and Ireland have also made major investments in wind.
7. At the Berlin airport, in the heart of wind power development in northern Germany, the average wind speed is about 8 knots. Search for any U.S. city between the Appalachian and Rocky Mountains, and you are likely to find a value that is higher. Fargo, North Dakota, for example, averages about 12 knots, Chicago 11, Dallas 11; www.windfinder.com. Given that the power derived from a windmill rises with the cube of the wind speed, this implies that potential power to be derived from the wind is at least twice as high around these American cities as around Berlin.
8. The source of these data is a series of annual reports, from 2003 to 2013, issued by the International Energy Agency's Photovoltaic Power Systems Program, in Paris.
9. pace.org. Accessed 4 September 2014.
10. B. Burger, *Electricity Production from Solar and Wind in Germany 2014* (Freiburg, Germany: Fraunhofer ISE, 2014).
11. Ibid.

12. Miguel Mendonça, *Feed-in Tariffs: Accelerating the Deployment of Renewable Energy* (London: Earthscan, 2007); Lucy Butler and Karsten Neuhoff, "Comparison of Feed-in Tariff, Quota and Auction Mechanisms to Support Wind Power Development," *Renewable Energy* 33, no. 8 (August 2008): 1854–67, doi: 10.1016/j.renene.2007.10.008.
13. In other countries, the FIT system has not entirely eliminated risk. In Spain, for example, the FIT was structured so that the government bore its cost rather than electricity rate payers directly. In the wake of the financial crisis, the Spanish government changed its law to reduce the government payments, including to renewable energy projects that were already collecting FIT revenues. This demonstrates that there is always such a risk, namely, of retroactive policy changes. In Germany, these have not materialized, and there is no evidence that anybody believes they will.
14. S. Lacey, "U.S. Solar Projects Boom as PV Prices Keep Falling," *Grist*, September 13, 2011, http://grist.org/solar-power/2011-09-12-u-s-solar-projects-boom-as-pv-prices-keep-falling/.
15. Ibid.
16. Bürer and Wüstenhagen, "Which Renewable Energy Policy Is a Venture Capitalist's Best Friend? Empirical Evidence from a Survey of International Cleantech Investors"; Sonja Lüthi and Rolf Wüstenhagen, "The Price of Policy Risk – Empirical Insights from Choice Experiments with European Photovoltaic Project Developers," *Energy Economics*, no. 0 (in press), doi:10.1016/j.eneco.2011.08.007; Held, Ragwitz, and Haas, "On the Success of Policy Strategies for the Promotion of Electricity from Renewable Energy Sources in the EU."
17. Pablo del Rio and Miguel A. Gual, "An Integrated Assessment of the Feed-in Tariff System in Spain," *Energy Policy* 35, no. 2 (2007): 994; Greenwire, "A Cautionary Tale about Feed-in Tariffs: For a Brief, Shining Moment, Spain Was the Best Solar Market in the World," *Greenwire Journal*, 2009. Paul Voosen, "Spain's Solar Market Crash Offers a Cautionary Tale about Feed-In Tariffs," *New York Times*, Energy and Environment section, 2009 August 18.]]
18. Earth Policy Institute, *Annual Solar Photovoltaics Production by Country, 1995–2012* (Washington, DC: Author, 2013).
19. www.bmwi.de/DE/Themen/Energie/Erneuerbare-Energien/eeg-reform.html. Accessed September 26, 2014.

20. Wirth, *Aktuelle Fakten Zur Photovoltaik in Deutschland*; H. Wirth, *Recent Facts about Photovoltaics in Germany* (Freiburg, Germany: Fraunhofer ISE, 2014).
21. Wirth, *Aktuelle Fakten Zur Photovoltaik in Deutschland*.
22. Ibid.
23. I arrived at this number by multiplying 1.4 cents, which I reported in the last paragraph, by 20 percent; that yields 0.28 cents, which I rounded up to 0.3 cents. One could also get it by multiplying the amount by which a kWh of PV exceeds the market price, 8 cents, by the fraction of total demand that the additional PV would supply, about 4 percent.
24. The Economist, "European Utilities: How to Lose Half a Trillion Euros," *The Economist*, October 12, 2013.
25. Eskeland et al., "Transforming the European Energy System"; G. Eskeland et al., "Policy Appraisal for the Electricity Sector" (Oslo: CICERO, 2008).
26. As an executive of a large Swiss power company made clear to me, there is one class of power company that is affected very differently: namely, the provider of carbon-free conventional power (i.e., hydropower or nuclear, which is exactly what the Swiss power companies provide). If a stringent ETS cap were causing wind and solar to be competitive, it would result in a wholesale market price at least as high as the carbon price. Producers of fossil-based power would have to subtract out the carbon price, meaning that their net revenue might well be close to zero. Providers of nuclear and hydro, however, would not. This would clearly benefit them. In the case of a FIT, the effect is for the wholesale market price to be low. This hurts coal and gas providers just as does the FIT, but it also hurts owners of nuclear and hydro.
27. P. Joskow, "Lessons Learned from Electricity Market Liberalization," *The Energy Journal* 29, no. 2 (2008): 9–42.
28. M. Wald, "Texas Is Wired for Wind Power, and More Farms Plug in," *New York Times*, July 23, 2014.
29. Zoellner, Schweizer-Ries, and Wemheuer, "Public Acceptance of Renewable Energies: Results from Case Studies in Germany."
30. Charles R. Warren and Malcolm McFadyen, "Does Community Ownership Affect Public Attitudes to Wind Energy? A Case Study from South-West Scotland," *Forest Transitions Wind Power Planning, Landscapes and Publics* 27, no. 2 (April 2010): 204–13, doi:10.1016/j.landusepol.2008.12.010.

31. http://rael.berkeley.edu/switch. Accessed October 27, 2014.
32. Antonella Battaglini et al., "Perception of Barriers for Expansion of Electricity Grids in the European Union," *Energy Policy* 47, no. 0 (August 2012): 254–59, doi:10.1016/j.enpol.2012.04.065.
33. See, again, the PhD thesis of Bendikt Battke, not yet published.
34. E. Goosens, "Germany to Support Solar Backup Batteries with Subsidy," *Bloomberg*, April 17, 2013.
35. Ibid.
36. S. Gupta et al., "Cross-Cutting Investment and Finance Issues," in *Climate Change 2014* (New York: Cambridge University Press, 2014).
37. The authors also report on a separate set of results concerning energy efficiency. Here, the models suggest the potential need for huge amounts of new investment, ranging up to more than $600 billion per year, with a median estimate of roughly $300 billion per year. These investments would take place outside of the energy sector, rather in the building, transportation, and manufacturing sectors. They do not report on what the baseline level of investment is in these sectors.
38. Trend:research, *Definition und Marktanalyse von Bürgerenergie in Deutschland* (Lüneburg, Germany: Trend:research and Leuphana Universität Lüneburg, 2013).
39. Frontier economics/Consentec, *Folgenabschätzung Kapazitätsmechanismnen* (Berlin: Bundesministerium für Wirtschaft und Energie, 2014).

10 Policies beyond power

1. http://newclimateeconomy.report. Accessed October 8, 2014.
2. http://thornews.com/2014/03/21/battery-powered-nissan-leaf-norways-most-sold-car/. Accessed September 5, 2014.
3. Ibid.
4. David Laibson, "Golden Eggs and Hyperbolic Discounting," *Quarterly Journal of Economics* 112, no. 2 (1997): 443–78; David Laibson, Andrea Repetto, and Jeremy Tobacman, "Estimating Discount Functions from Lifecycle Consumption Choices," 2004. David Laibson, Andrea Repetto, and Jeremy Tobacman, "Estimating discount functions with consumption choices over the lifecycle," NBER Working Paper No. 13314, Cambridge, MA, 2007.

5. Samuel McClure et al., "Separate Neural Systems Value Immediate and Delayed Monetary Rewards," *Science* 306 (2004): 503–7.
6. I need to mention a friend of mine, my stepfather, Peter Papesch, who is an architect and an architecture educator. He noticed the fact that, in the United States, training in energy-efficient design was simply not an obligatory part of a student's education in architecture schools. He has been lobbying for several years for the adoption of standards that would require this but encountering a tepid response. The fact is that the market for architects does not yet place much value on these skills.
7. Karen Ehrhardt-Martinez and John A. Laitner, "Rebound, Technology and People: Mitigating the Rebound Effect with Energy-Resource Management and People-Centered Initiatives," *Proceedings of the 2010 ACEEE Summer Study on Energy Efficiency in Buildings*, 2010, 76–91.
8. O. Lucon et al., "Buildings," in *Climate Change 2014* (New York: Cambridge University Press, 2014).
9. Karen Ehrhardt-Martinez and John A. Laitner, "Rebound, Technology and People."
10. Lucon et al., "Buildings."
11. See https://www.ethz.ch/de/news-und-veranstaltungen/eth-news/news/2014/07/ein-dorf-mit-co2-neutralen-gebaeuden-ist-moeglich.html. Accessed October 6, 2014.
12. DG Energy, *Energy Performance Certificates in Buildings and Their Impact on Transaction Prices and Rents in Selected EU Countries* (Brussels: European Commission, 2013).
13. M. Fischedick et al., "Industry," in *Climate Change 2014* (New York: Cambridge University Press, 2014).
14. Ibid.
15. It is important to note that all the estimates of emissions from agriculture and forestry are imprecise. These emissions are not something that one can directly measure, but instead must be estimated based on observations of agricultural practices and changes in land use combined with models of how these practices lead to fluxes in greenhouse gases. There is a considerable amount of uncertainty with respect to the latter.
16. P. Smith et al., "Agriculture, Forestry and Other Land Use (AFOLU)," in *Climate Change 2014* (New York: Cambridge University Press, 2014).
17. Ibid.

18. https://climatedataguide.ucar.edu/climate-data/carbon-emissions-historical-land-use-and-land-use-change. Accessed 9 October 2014.
19. Diamond, *Collapse: How Societies Choose to Fail or Succeed.*
20. Thomas K. Rudel, Laura Schneider, and Maria Uriarte, "Forest Transitions: An Introduction," *Land Use Policy* 27, no. 2 (April 2010): 95–97, doi:10.1016/j.landusepol.2009.09.021; Thomas K. Rudel et al., "Forest Transitions: Towards a Global Understanding of Land Use Change," *Global Environmental Change* 15, no. 1 (April 2005): 23–31, doi:10.1016/j.gloenvcha.2004.11.001.
21. Rudel et al., "Forest Transitions: Towards a Global Understanding of Land Use Change."
22. Ibid.
23. Policy makers and other experts actually use a variety of words describe the act of increasing the size of forests, as well as increasing the amount of ecosystem services (including carbon storage) those forests provide. Various terms include *forest restoration, reforestation,* and *afforestation.* To a large extent, the appropriateness of a given term depends on what the historical land use in that precise place was. Since I am not going into that level of detail, I feel comfortable using a single term to refer to all these processes. I use "afforestation" to refer to the act of creating new and valuable forests where such forests do not currently exist.
24. Actually, the term REDD took on a specific set of meanings a few years ago, and, in UNFCCC negotiations since then, people have wanted to add some additional elements. Having done so, the term that people now primarily use is REDD+, where the + refers to a growing list of more recent elements. I use REDD to mean REDD+ as well.
25. Georg Kindermann et al., "Global Cost Estimates of Reducing Carbon Emissions through Avoided Deforestation," *Proceedings of the National Academy of Sciences* 105, no. 30 (July 29, 2008): 10302–7, doi:10.1073/pnas.0710616105.
26. Valentina Bosetti et al., "Linking Reduced Deforestation and a Global Carbon Market: Implications for Clean Energy Technology and Policy Flexibility," *Environment and Development Economics* 16, no. Special Issue 04 (2011): 479–505, doi:10.1017/S1355770X10000549.
27. Wolfgang Lutz, J. Crespo Cuaresma, and Warren Sanderson, "The Demography of Educational Attainment and Economic Growth," *Science* 319 (2008): 1047–48.

28. D. Waltner-Toews et al., “Perspective Changes Everything: Managing Ecosystems from the Inside Out,” *Frontiers in Ecology and the Environment* 1, no. 1 (2003): 23–30; Gretchen C. Daily, “Management Objectives for the Protection of Ecosystem Services,” *Environmental Science & Policy* 3, no. 6 (December 2000): 333–39, doi:10.1016/S1462-9011(00)00102-7.
29. Smith et al., “Agriculture, Forestry and Other Land Use (AFOLU).”
30. T. Zeller, “Can We Build in a Brighter Shade of Green?,” *New York Times*, September 25, 2010.
31. Nadejda Komendantova and Anthony Patt, “Employment under Vertical and Horizontal Transfer of Concentrated Solar Power Technology to North African Countries,” *Renewable and Sustainable Energy Reviews* 40, no. 0 (December 2014): 1192–1201, doi:10.1016/j.rser.2014.07.072.
32. N. Komendantova et al., “Perception of Risks in Renewable Energy Projects: Power in North Africa,” *Energy Policy* 40 (2012): 103–9; N. Komendantova and A. G. Patt, “Could Corruption Prove to Be a Barrier to Roll-out of Renewable Energy in North Africa?,” in *Transparency International Global Corruption Report: Climate Change* (London: Earthscan, 2011).
33. Komendantova, Patt, and Williges, “Solar Power Investment in North Africa: Benefits of Reducing Risk Perceptions”; Tobias S. Schmidt, “Low-Carbon Investment Risks and de-Risking,” *Nature Climate Change* 4, no. 4 (April 2014): 237–39.
34. Komendantova, Pfenninger, and Patt, “Governance Barriers to Renewable Energy in North Africa”; Anthony Patt et al., “Regional Integration to Support Full Renewable Power Deployment for Europe by 2050,” *Environmental Politics* 20, no. 5 (2011): 727–42, doi:10.1080/09644016.2011.608537.
35. Janosch Ondraczek, Nadejda Komendantova, and Anthony Patt (2015). WACC the dog: the effect of financing costs on the levelized cost of solar PV power. *Renewable Energy* 75: 888-898.
36. Tobias S. Schmidt, Robin Born, and Malte Schneider, “Assessing the Costs of Photovoltaic and Wind Power in Six Developing Countries,” *Nature Climate Change* 2, no. 7 (July 2012): 548–53, doi:10.1038/nclimate1490.
37. Wagner et al., “Sectoral Marginal Abatement Cost Curves: Implications for Mitigation Pledges and Air Pollution Co-Benefits for Annex I Countries.”

38. D. Gardner, “China’s Environmental Awakening,” *New York Times*, September 14, 2014.
39. Ibid.

11 Pulling it all together

1. Several scientific results support these two assumptions. Ben Haley, “Pathways to 2050: Deep Carbonization Strategies in the United States” (presented at the IARU Sustainability Science Conference, Copenhagen, Denmark, 2014); G. Schellekens et al., “100% Renewable Electricity: A Roadmap to 2050 for Europe and North Africa” (PriceWaterhouse Coopers, 2010); Schellekens et al., “Moving towards 100% Renewable Electricity in Europe and North Africa by 2050: Evaluating Progress in 2010”; Johan Lilliestam et al., “An Alternative to a Global Climate Deal May Be Unfolding before Our Eyes,” *Climate and Development* 4, no. 1 (2012): 1–4, doi:10.1080/17565529.2012.658273.
2. One could argue that the funding for carbon sequestration could come from carbon markets. This would work as long as there are actors within those markets legally entitled to buy the right to emit CO_2. If we are to move along a trajectory of falling emissions, then, a market instrument to support sequestration would be most effective during the early stages, when emissions would be high enough to support substantial market transactions. That would, in turn, mean that we begin carbon sequestration efforts now, phasing them out by about mid-century, when emissions themselves would be scheduled to cease. This is unrealistic from a technological point of view, for all the reasons I have discussed concerning CCS. Additionally, it contradicts the timing suggested by virtually all the cost optimization studies examining deep emissions reduction scenarios, such as those reported in the IPCC AR5.

Index